Moh'd A. Radaideh

Software Project Management

Also of interest

Loss Data Analysis.
The Maximum Entropy Approach
2nd extended edition
Gzyl, Mayoral, Gomes-Gonçalves, 2023
ISBN 978-3-11-104738-6, e-ISBN (PDF) 978-3-11-104818-5,
e-ISBN (EPUB) 978-3-11-104818-5

Cloud Analytics for Industry 4.0
Potluri, Nandan Mohanty, Mohammad, Shitharth (Eds.), 2022
ISBN 978-3-11-077149-7, e-ISBN (PDF) 978-3-11-077157-2,
e-ISBN (EPUB) 978-3-11-077166-4

Advances in Industry 4.0.
Concepts and Applications
Niranjanamurthy, Peng, Naresh, Jayasimha, Jayasimha, Balas (Eds.),
2022ISBN 978-3-11-072536-0, e-ISBN (PDF) 978-3-11-072549-0,
e-ISBN (EPUB) 978-3-11-072553-7

Noise Filtering for Big Data Analytics
Bhattacharyya, Ghosh (Eds.), 2022
ISBN 978-3-11-069709-4, e-ISBN (PDF) 978-3-11-069721-6,
e-ISBN (EPUB) 978-3-11-069726-1

Computational Intelligence in Software Modeling
Jain V., Chatterjee, Bansal, Kose, Jain A. (Eds.), 2022
ISBN 978-3-11-070543-0, e-ISBN (PDF) 978-3-11-070924-7,
e-ISBN (EPUB) 978-3-11-070934-6

Knowledge Management and Web 3.0.
Next Generation Business Models
Kautish, Singh, Polkowski, Mayura, Jeyanthi (Eds.), 2021
ISBN 978-3-11-072264-2, e-ISBN (PDF) 978-3-11-072278-9,
e-ISBN (EPUB) 978-3-11-072293-2

Moh'd A. Radaideh

Software Project Management

With PMI, IEEE-CS, and Agile-SCRUM

DE GRUYTER

Author
Moh'd A. Radaideh, Ph.D., MCPM
Department of Software Engineering
Jordan University of Science and Technology
P.O. Box 3030
22110 Irbid
Jordan
maradaideh@just.edu.jo

ISBN 978-3-11-120646-2
e-ISBN (PDF) 978-3-11-120686-8
e-ISBN (EPUB) 978-3-11-120761-2

Library of Congress Control Number: 2023943022

Bibliographic information published by the Deutsche Nationalbibliothek
The Deutsche Nationalbibliothek lists this publication in the Deutsche Nationalbibliografie; detailed bibliographic data are available on the Internet at http://dnb.dnb.de.

Cover image: Khanchit Khirisutchalual/iStock/Getty Images Plus
Typesetting: Integra Software Services Pvt. Ltd.
Printing and binding: CPI books GmbH, Leck

www.degruyter.com

Acknowledgments

I would like to thank **De Gruyter** for giving me the opportunity to publish this textbook through them. In particular, I would like to thank **Steven Elliot** (senior editor, De Gruyter, Boston) for introducing me to De Gruyter, along with his colleague **Daniel Tiemann**. Special thanks go to **Dr. Damiano Sacco** (editor engineering/computer science, De Gruyter, Berlin) for facilitating the publishing agreement with De Gruyter. A big thanks goes to **Melanie Götz** (content editor – Books STEM, De Gruyter, Berlin) for her support throughout the period of preparing and developing this textbook.

I would like to acknowledge and thank my university, **Jordan University of Science and Technology (JUST)**, for their ongoing support throughout the past 6 years since I joined their Software Engineering Department.

I would like to dedicate this textbook to **my beloved wife Rawiah; sons Saeb, late Ahmad, Saleh, Khaled, and Abdul-Kareem; and daughters Haneen and Ro'yah** for their love and support. **My mother Hajjah Samiha Radaideh** deserves my appreciation for her ongoing blessing prayers that God saves and blesses me and my family, and **my late father Sheikh Dr. Ahmad Radaideh** deserves to have this textbook dedicated in his memory.

Last, but not least, my former students **Nadil Mohammad** and **Maya Muqbil** deserve my sincere thanks and appreciations for their help in creating the drawings and figures included throughout this textbook. Also, my current student **Fadi Fahed Ayoub** for proof-reading the banks of questions associated with this text book.

Kind regards,
Moh'd A. Radaideh, Ph.D., MCPM
Department of Software Engineering
Faculty of Computer and Information Technology
Jordan University of Science and Technology
Irbid, Jordan
Email: maradaideh@just.edu.jo / radaideh03@gmail.com
LinkedIn: https://www.linkedin.com/in/radaideh03
YouTube: https://www.youtube.com/radaideh03

https://doi.org/10.1515/9783111206868-202

Preface

Course initiation

As I was wondering back in September 2017 about the topics that should be included in my software project management course at Jordan University of Science and Technology (JUST), I decided to include a generic Project Management Institute (PMI)-flavored project management module; a second module on the IEEE Computer Society (IEEE-CS) software engineering knowledge areas, particularly the software engineering management knowledge area; and a third module on project management function of planning.

Since then, I have revised the course context for several times, by restructuring its content, adding new topics and/or merging some of these topics. I continued that scenario for about 4 years, until the second semester of 2020–2021 when I decided to restructure the course and inject another module, this time on the Agile and SCRUM project management methods.

Course most recent status

By the second semester of 2020–2021, I transformed the course into a composition of the project management and software project management perspectives of PMI, IEEE-CS, and the SCRUM.org society. These three perspectives are presented in this textbook along with my own view of software project management.

ABET and IET accreditations

Although the course went through lot of changes throughout the past 6 years, its structure and context remained compliant with the accreditations' requirements of the Institute of Engineering and Technology (IET) as well as with the accreditations' requirements of the Accreditation Board of Engineering and Technology (ABET).

Course delivery method

I have been delivering the course in a blended format starting from the second semester of 2020–2021. With such format, I typically deliver about 65% of the course lectures synchronously online, while I record the remaining 35% of them offline before making them available asynchronously to my students.

https://doi.org/10.1515/9783111206868-203

Students' satisfaction

Students' performance and satisfaction continued to rise during the past 5 academic years, and the numbers of enrolled students have been increasing, particularly during the past few semesters.

Textbook use for teaching

This textbook is intended to be used for senior undergraduate and/or junior graduate courses in IT/software project management.

This textbook is based on the "SE440 Software Project Management" course at JUST. The SE440 course has been in delivery as an online course since the second semester of 2020–2021.

To facilitate its online delivery, I have been using several tools, including the ZOOM Platform (e.g., *for delivering and recording all synchronous and asynchronous online lectures*), YouTube Platform (e.g., *for hosting all lectures' recordings*), JUST e-Learning Platform (e.g., *for hosting the course material*), JUST A+ Author Online Examination System (e.g., *for holding all midterm and final exams*), and a licensed video editing tool.

Supplemental materials

The website of this textbook can be reached via the author's website (https://www.just.edu.jo/~maradaideh). The indicated website will include the following:

- An executive summary of this textbook.
- Sample presentations (pdf format) that can be used by instructors to derive their own PowerPoint Presentations.
- A link to where this textbook can be ordered through the publisher.

Suggested course learning outcomes and their mapping to the IET competencies

	Course learning outcome	Weight	Mapped to IET competencies[1]
CLO1.	Understand and demonstrate the basic terms and concepts of software project management.	10%	C4
CLO2.	Understand and demonstrate the project management knowledge areas (e.g., *including their associated phases and processes along with their inputs, methods and tools, and outputs*) in accordance with the Project Management Body of Knowledge that was introduced years ago by the Project Management Institute (PMI).	30%	C6
CLO3.	Understand and demonstrate the software engineering knowledge areas including the software engineering management one (e.g., *including its associated processes*) in accordance with the Software Engineering Body of Knowledge (SWEBOK) that was introduced years ago by the IEEE Computer Society (IEEE-CS).	30%	C15 and C16
CLO4.	Understand and demonstrate the AGILE-SCRUM software development and software project management (e.g., *including their associated phases and processes along with their inputs, methods and tools, and outputs*) in accordance with the SCRUM Body of Knowledge (SBOK) that was introduced years ago by the SCRUMstudy.	30%	C15 and C16
	Total	**100%**	

1 Refer to Appendix A for the list of IET competencies.

Contents

List of figures

https://doi.org/10.1515/9783111206868-205

List of tables

https://doi.org/10.1515/9783111206868-206

Chapter 1
Introduction

1.1 Textbook background

This book is meant for use as a textbook for senior undergraduate and/or junior graduate IT/software project management courses. It was based on the "SE440 Software Project Management" course at Jordan University of Science and Technology (JUST), which has been already delivered by the author for 17 times.

The SE440 course has been in delivery as an online course since the second semester of 2020–2021. The author has been using several tools to facilitate the course's online delivery. These tools include the ZOOM platform for delivering and recording the synchronous and asynchronous online lectures, YouTube platform for hosting the lectures' recordings, JUST e-learning platform for hosting the course's material, JUST Author A+ Examining System for holding the midterm and final exams, and a licensed video editing tool for editing and enhancing the quality of the lectures' recordings before publishing them.

The most recent content (e.g., prior to the content of this textbook) of the SE440 course was composed of the following components:

- **Module I**: This module provides a generic PMI-flavored elaboration on project management in terms of its phases and processes. Also, it elaborates on the author's view of software project management in terms of its phases and processes. It elaborates as well on the International Standard for Systems and Software Life-Cycle Processes (ISO/IEC 12207).
- **Module II**: This module provides a generic elaboration on the software engineering knowledge areas, particularly the software engineering management knowledge area in accordance with the IEEE-CS Software Engineering Body of Knowledge (SWEBOK). These knowledge areas are the software requirements, software design, software construction, software testing, software maintenance, software configuration management, software engineering management, software models and methods, software quality, software engineering professional practice, software engineering economics, computing foundation, mathematical foundation, and engineering foundation.
- **Module III**: This module elaborates on the SCRUM software development and SCRUM project management approaches in terms of SCRUM's principles, aspects, phases, and processes.
- **Course project**: In addition to its status as a blended-online course, the SE440 course is also a project-based course. Students are expected to either carry on some research tasks that involve one or more of the 10 software project management knowledge areas, or derive detailed action plans for some case studies provided to them by the instructor.

https://doi.org/10.1515/9783111206868-001

1.2 Textbook organization

This textbook is organized in five chapters.

This chapter gives a short introduction.

Chapter 2 elaborates on the project management in terms of its history, definition, importance, standards, code of ethics, portfolios, programs, projects, operational environments, leadership, request for proposals, project manager essential skills, project management phases and processes, projects' contracts, project management approaches, projects' CRITICAL-SUCCESS-CRITERIA and factors, and a simplified model for project core action plans.

Chapter 3 provides an executive summary and elaboration on the project management 10 knowledge areas along with their associated processes. These knowledge areas are the project integration management, project scope management, project schedule management, project cost management, project quality management, project resource management, project communication management, project risk management, project procurement management, and project stakeholder management. This chapter also elaborates on the five phases of project management, which are the initiation phase, planning phase, execution phase, monitoring and control phase, and closure phase.

Chapter 4 provides an executive summary and elaboration on software engineering and its 15 knowledge areas in accordance with the IEEE-CS SWEBOK. These knowledge areas are the software requirements, software design, software construction, software testing, software maintenance, software configuration management, software engineering management, software engineering process, software engineering models and methods, software quality, software engineering professional practice, software engineering economics, computing foundation, mathematical foundation, and engineering foundation.

Chapter 5 provides an executive summary and elaboration on the:

- **Agile software project management**. This part elaborates on the various Agile methods such as the KANBAN, lean KANBAN, crystal methods, dynamic systems development methods, extreme programming method, feature-driven development method, test-driven development method, adaptive software development method, Agile unified process method, and domain-driven design method.
- **SCRUM project management method**. This part elaborates on the:
  - SCRUM principles of empirical process control, self-organization, collaboration, value-based prioritization, time-boxing, and iterative development;
  - SCRUM portfolios, programs, and projects hierarchy;
  - SCRUM core roles, which are the product owner, SCRUM master, and SCRUM team, and non-core roles such as the stakeholders;
  - SCRUM conflict management methods;

- SCRUM leadership styles;
- HR theories used in SCRUM projects;
- SCRUM quality;
- SCRUM change;
- SCRUM risks;
- The SCRUM initiate phase along with its its 6 processes;
- The SCRUM plan and estimate phase along with its 5 processes;
- The SCRUM implement phase along with its 3 processe;
- The SCRUM review and retrospect phase along with its 3 processes; and
- The SCRUM release phase along with its 2 processes.

Appendix A presents the IET competencies. Appendix B presents the banks of questions that can help the instructors prepare their quizzes and exams: Appendix B.1 presents chapter 2 bank of questions; Appendix B.2 presents chapter 3 bank of questions; Appendix B.3 presents chapter 4 bank of questions; and Appendix B.4 presents chapter 5 bank of questions.

Chapter 2
Project management background

This chapter presents and elaborates on the following topics, and some of which are based on the PMI's perspective of project management (PM) [1, 2]:

- The profession of PM in terms of its:
  - history, definition, importance, code of ethics, ANSI standard, and so forth;
  - portfolios, programs, and projects organizational hierarchy; and
  - operational environments and their impacting factors such as the enterprise environmental factors (EEFs), organizational process assets factors (OPAFs), organizational systems-related factors (OSFs), and organizational governance framework-related factors (OGFFs).
- The project managers, functional managers, operational managers, and leadership:
  - the roles of the project manager, functional manager, and operational manager;
  - the PMI project managers' Talent Triangle;
  - the project managers' circles of influence;
  - the project managers' qualities and skills;
  - the project managers' leadership classification;
  - the project managers' personalities and characters' aspects; and
  - the author's classification of the project managers' qualities and skills.
- The request for proposals (RFPs), project charters, contracts, and the common PM approaches.
- The author's view of the PM phases and their processes.
- The author's simplified change management process, simplified risk management process, and simplified model for deriving project core action plans.

The context of this chapter serves the course learning outcomes of CLO1 and CLO2 that were specified in the preface section at the beginning of this textbook.

2.1 Project management

2.1.1 Project management history

PM is not something new, but has been in use for tens of centuries in different ways of practice. If we look back in history, we can recall many legendary projects such as the Pyramids of Giza and those of Mexico, Great Wall of China, Olympic Games, Taj Mahal, and Panama Canal. Other historical projects include the marines, commercial jet airplanes, development of vaccinations, development of computer systems, human landing

https://doi.org/10.1515/9783111206868-002

on the Moon, development of the Global Positioning System, development and placement of the International Space Station in the Earth's orbit, and development of software systems. Each of these projects most likely was carried on differently in terms of their methods, practices, processes, planning methods, execution, monitoring and control, and evaluation. Thus, although PM has been in use since deep times in history, there has been no use of formal PM methods.

Prior to the 1950s, projects were managed directly by their engineers. They were managed using mostly Gantt charts and some other informal methods. During that time, two mathematical project-scheduling models were developed, which are the CRITICAL path method (CPM) that was developed as a joint venture between DuPont Corporation and Remington Rand Corporation; and the Program Evaluation and Review Technique (PERT) that was developed by the U.S. Navy Special Projects Office in collaboration with the Lockheed Corporation and Booz Allen Hamilton. Although PERT and CPM seem alike in terms of their approaches, they are different in terms of their assumptions. In other words, CPM is used for projects that assume predetermined times for carrying on the planned-for activities, while the PERT method is used for projects that assume uncertain or varied times for carrying on the activities.

During the 1950s, organizations started to use PM methods to manage complex projects, and then PM started to develop mainly from the civil and construction engineering projects:

- Henry Gantt, the father of planning and control methods, is known for his use of Gantt charts as a PM tool.
- Henri Fayol is known for creating the foundation of the Project Management Body of Knowledge.
- Gantt and Fayol were students of Frederick Winslow Taylor whose work lead to many of today's PM tools such as the work breakdown structure (WBS) and resource allocation.
- Project scheduling models and tools, project costing tools, cost management, and engineering economics tools started to evolve during the 1950s.
- In 1956, the Association for the Advancement of Cost Engineering (AACE International) was established by a Group of Project Management Practitioners. In 2006, the AACE International released their Total Cost Management Framework as the first integrated process for portfolio management, program management, and PM.
- After the 1950s, PM started to transform into a distinct profession.

2.1.2 Project management definition

PM is a profession that is based on having a team to initiate, plan, execute, monitor, and control, and close a set of tasks and activities (e.g., *a project*) in order to SUCCESSFULLY achieve a set of Specific, Measurable, Achievable, Realistic, and Time-bound

(S.M.A.R.T.) goals in accordance with predefined CRITERIA, within predefined schedules, and predetermined budgets.

While **software project management (SPM)** is a discipline of PM that is specific to software engineering management and software projects.

2.1.3 Project management importance

PM is the utilization of knowledge, methods, and tools in carrying on the tasks and activities of a project such that the project requirements are met and satisfied at the project end.

When applied effectively, PM leads to:
- meeting business objectives;
- satisfying shareholders' expectations;
- managing CHANGE-REQUESTS;
- managing and utilizing resources; and
- handling problems, issues, risks, and threats.

While poor PM leads to:
- disturbing the project schedule and missing its deadlines;
- disturbing the project expenditure and its allocated budget; and
- failing to meet the project predefined specifications.

2.1.4 Project management standard

The American National Standards Institute (ANSI) standard for PM was developed based on the concepts of consensus, openness, processes, and balance [1].

2.1.5 Project management code of ethics

The PM code of ethics involves the following four concepts in accordance with the PMI's view [2]:
- **Responsibility**. A project team must take full ownership and assume full responsibility for their decisions, actions, and consequences throughout the project life cycle, whether that project succeeds or fails at the end.
- **Respect**. A project team must show respect to themselves and to other involved individuals and parties. They also must take care of the resources entrusted to them such as people and money.

- **Fairness**. A project team must make decisions and act objectively without prejudgments.
- **Honesty**. A project team must act in a truthful manner in their communications and conduct.

2.2 Projects versus programs versus portfolio

2.2.1 What is a project?

A project can be referred to as a temporary mission that leads to a unique product, solution, and/or service with specific start and end dates, and within predefined schedule, budget, and specifications. The main challenge is to achieve the goals of that project given the constraints that are typically imposed on the project by the customer. Constraints include the scope, schedule times, budget and costs, quality, and so forth.

A project can also be referred to as a group of individuals assembled to carry on a set of tasks and activities related to predefined objectives and requirements that should be accomplished within a pre-estimated time, without exceeding a pre-estimated budget, and while its outcomes satisfy a predefined set of specifications.

There are many factors that can lead to initiating new projects such as:

- **Utilizing technological advancements**. Based on new technological advancements (e.g., *CPU speed and capacity, and memory/storage speed and capacity*), a firm may initiate a new project to create a new computer and/or software system using such advanced components to build that new system, improve/fix one of their existing computer and/or software systems, and/or implement a new or modify an existing business and/or technological strategy.
- **Remaining at the competitive edge**. If a competing product's price is lowered, then an affected firm may initiate a new project to implement a new or modify an existing business and/or technological strategy such that they lower the cost of their production in order to continue having their product at a competitive edge in the market.
- **Responding to changes in the market demand**. If a demand arises for a new product, then a concerned firm may initiate a new project to produce that new product.
- **Responding to customers' requests**. If a customer expresses their need for a new product, or for changing an existing one, then a concerned firm may initiate a new project in accordance with the specifications and requirements requested by that customer.

2.2.2 What is a program?

A program is a group of projects and probably subsidiary programs and activities that are managed together and belong to the same entity. An example would be the hospital program of projects at JUST. This can include all of the projects running for that hospital such as the hospital's building maintenance project, the hospital's information systems and infrastructure upgrade projects, and so forth.

2.2.3 What is a portfolio?

An organizational portfolio is a group of projects, programs, subsidiary portfolios, and other operational activities. It is typically managed by a Project Management Office (PMO) leveled at the top management of that organization. Program management and portfolio management are different from PM in terms of their objectives, focus, activities, life cycles, and benefits. However, portfolios, programs, projects, and operations share the same stakeholders, human resources, and physical resources.

2.2.4 Portfolios versus programs versus projects

Tables 2.1–2.7 provide comparative views of portfolios, programs, and projects from different aspects:
- Table 2.1 provides a PM organizational comparative view of portfolios, programs, and projects in terms of their definitions.
- Table 2.2 provides a PM organizational comparative view of portfolios, programs, and projects in terms of their scopes of work.
- Table 2.3 provides a PM organizational comparative view of portfolios, programs, and projects in terms of their planning methods.
- Table 2.4 provides a PM organizational comparative view of portfolios, programs, and projects in terms of their purposes.
- Table 2.5 provides a PM organizational comparative view of portfolios, programs, and projects in terms of their CHANGE-REQUESTS management.
- Table 2.6 provides a PM organizational comparative view of portfolios, programs, and projects in terms of their monitoring and control methods.
- Table 2.7 provides a PM organizational comparative view of portfolios, programs, and projects in terms of their SUCCESS CRITERIA.

Table 2.1: The definitions of projects, programs, and portfolios.

Projects	A project is a temporary mission that leads to a unique product, solution, and/or service with specific start date and end date.
Programs	A program is a group of projects and subsidiary programs and activities that are managed together and belong to the same entity.
Portfolios	A portfolio is a group of projects, programs, subsidiary portfolios, and operational activities. It is managed by a project management office (PMO) at the top management level of the concerned organization.

Table 2.2: The scopes of work of projects, programs, and portfolios.

Projects	A project scope of work composes the objectives and a high-level description of the requirements of that project along with a description of the tasks and activities that will be carried on throughout the life cycle of that project.
Programs	A program's scope of work composes the scopes of work of its integral projects, subsidiary programs, and operational activities.
Portfolios	A portfolio's scope of work composes the scopes of work of its integral programs, subsidiary portfolios, projects, and operational activities. It composes the overall scope of the concerned organization, which must be aligned with its strategic objectives.

Table 2.3: The planning methods of projects, programs, and portfolios.

Projects	A project planning function is used to transform the high-level information of the project objectives, scope of work, and requirements into a set of detailed action plans for managing, executing, monitoring, and controlling that project throughout its life cycle.
Programs	A program planning function is used to guide and monitor the progress of the plans of its integral projects, subsidiary programs, and operational activities; and to identify and watch their interdependencies.
Portfolios	A portfolio planning is used to guide and monitor the progress of the plans of its integral programs, subsidiary portfolios, and operational activities, and to identify and watch for their interdependencies.

Table 2.4: The purposes of portfolios, programs, and projects.

Projects	The purpose of managing a project is to ensure that its team members are executing the project plan carefully such that the project objectives and predefined requirements are accomplished as expected and as they were planned for.

Table 2.4 (continued)

Programs	The purpose of managing programs is to coordinate the activities of its integral projects, subsidiary programs, and operational activities.
Portfolios	The purpose of managing portfolios is to coordinate and supervise its integral programs, projects, and operational activities.

Table 2.5: The CHANGE-REQUESTS in projects, programs, and portfolios.

Projects	Managing CHANGE-REQUESTS is used to keep changes controlled by implementing the necessary change processes.
Programs	Program managers are expected to accept CHANGE-REQUESTS as necessary to facilitate the delivery of the components of their programs.
Portfolios	Portfolio managers are expected to monitor changes in the components of their portfolios.

Table 2.6: The monitoring and control methods in projects, programs, and portfolios.

Projects	Project managers are expected to conduct a series of quality assurance and quality control activities throughout the execution of their projects.
Programs	Program managers are expected to monitor the progress of the components of their programs. This is to ensure the satisfaction of the goals, schedules, budget, and benefits of their programs.
Portfolios	Portfolio managers are expected to monitor any strategic change across the components of their portfolios.

Table 2.7: The SUCCESS CRITERIA in projects, programs, and portfolios.

Projects	A project is considered SUCCESSFUL if and only it is completed: – **ON TIME.** Completed as per the predetermined schedule. – **WITHIN BUDGET.** Spendeture should not exceed the allocated budget. – **WHILE MEETING SPECS.** Product's quality is at the expected level. – **WHILE ENSURING THE CUSTOMER SATISFACTION.** Preserving the customer's satisfaction with the delivered product is at least as expected.
Programs	A program is considered SUCCESSFUL if and only if all of its integral components are SUCCESSFUL.
Portfolios	A portfolio is considered SUCCESSFUL if and only if all of its integral components are SUCCESSFUL.

2.3 Projects' operational environments

2.3.1 Projects' operational environmental influencing factors

In reference to Figure 2.1, projects are influenced by their surrounding environments. The main influencing environmental factors are EEFs, OPAFs, OSFs, and OGFFs.

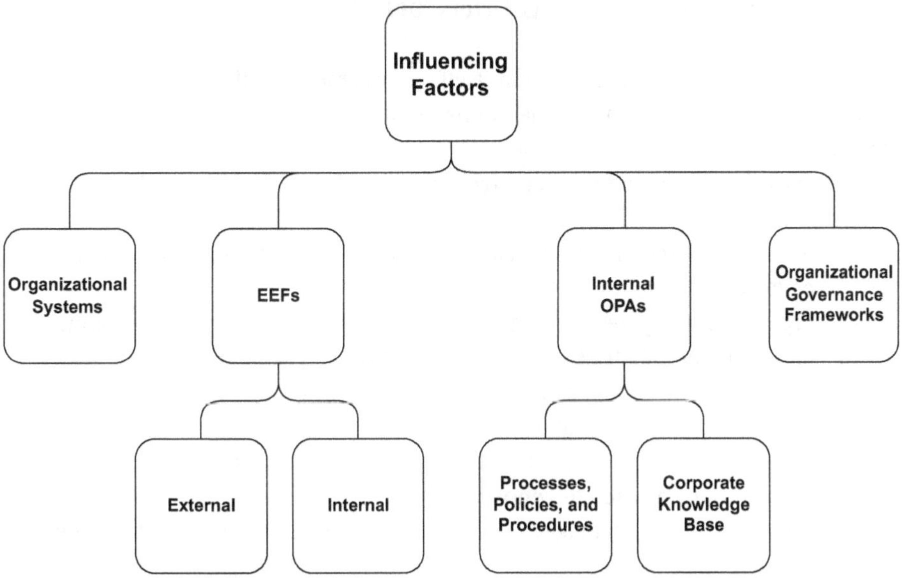

Figure 2.1: Projects environmental influencing factors.

2.3.2 Enterprise environmental factors (EEFs)

EEFs come from the environment around the project, mostly external to the organization, in which the project is being carried on. EEFs can be classified into internal and external factors as follows:

- **Internal to the organization EEFs**. These include the following:
  - **Organizational culture, structure, and governance**. This composes mainly the project vision, mission, values, culture, leadership style, and code of ethics
  - **Geographic distribution of facilities and resources**. This composes mainly the project locations, virtual teams, shared resources, and cloud computing
  - **Infrastructure**
  - **Resource capabilities and availabilities**

- **External to the organization EEFs.** These include the following:
  - Marketplace competition, recognition, and trademarks
  - Legal restrictions. This composes mainly the local regulations related to security, data, business conduct, and employment.
  - Governmental and industrial standards.

2.3.3 Organizational process assets factors (OPAFs)

OPAFs are internal to the organization itself. They come from the organization's portfolios, programs, projects, and/or operations. OPAFs can be classified into the following two sets:

- **Processes, policies, and procedures.** These include the organizational policies for managing:
  - human resources, health and safety, security and confidentiality, quality, and environment;
  - organizational procedures and methods for managing projects, estimation and estimation metrics, audits, and checklists;
  - organizational templates for PM plans, project documents, reports, contracts, and risk management; and
  - organizational procedures for conducting change control.
- **Organizational knowledge repositories.** These include the databases that contain the configuration management data, financial data, historical data, and issues management data.

2.3.4 Organizational systems-related factors (OSFs)

OSFs impact the people's ability to act within the organizational system. The OSFs include the people's power, influence, interests, competencies, and political capabilities.

2.3.5 Organizational governance frameworks-related factors (OGFFs)

OGFFs are used to manage the authority within the organization. They include the organizational rules, processes, policies, procedures, and systems. OGFFs influence projects in how to set and achieve the organization's objectives in alignment with their organizational objectives, monitor and assess the organization's potential risks, and optimize the organization's performance.

2.4 Project managers and leadership

2.4.1 The roles of project managers, functional managers, and operational managers

The role of a project manager is distinct from those of functional and/or operational managers:

- **The project manager role.** A project manager is assigned by his/her organization to lead a team that will be responsible for carrying on a project to achieve its objectives and requirements.
- **The functional manager role.** A functional manager manages a functional or a business unit.
- **The operational manager role.** An operational manager is responsible for ensuring that business operations go smooth, effective, and efficient.

2.4.2 The PMI's project managers' Talent Triangle

A project manager leads a team that will carry on a project and achieve its objectives and requirements.

A project manager is expected to balance the constraints on his/her project using the available resources, and to use his/her soft interpersonal skills to communicate with the stakeholders and/or balance their goals.

Figure 2.2: The PMI Talent Triangle.

In reference to Figure 2.2, the PMI Talent Triangle composes the following sets of skills that project managers should have:

- **Technical project management knowledge and skills.** This includes the knowledge and managerial and technical skills that are related to the specific business

domains of their projects, and to the project management, program management, and portfolio management.
- **Leadership knowledge and skills**. This includes the knowledge and skills needed to guide, motivate, and direct a team.
- **Strategic and business management knowledge and skills**. This includes the knowledge and expertise in the business industry to ensure the enhancement of the organization's performance.

2.4.3 A generic classification of the software project managers' essential skills

The following is a generic classification of the software project managers' essential skills:
- **Ability to organize and manage**. This implies that a software project manager must be able to:
  - handle the project complexities;
  - analyze the project artifacts;
  - set the project goals, scope, and implementation plan;
  - plan for the project budget;
  - staff for the project team;
  - control and solve problems; and
  - take effective actions.
- **Ability to cope and lead**. This implies that a software project manager must be able to:
  - handle CHANGE REQUESTS;
  - set directions;
  - provide vision;
  - delegate authority whenever there is a need for that;
  - be positive and energetic, flexible, creative, patient, and persistent throughout the project life cycle;
  - align people; and
  - take meaningful actions.
- **Ability to communicate and negotiate effectively**. This implies that a software project manager must be able to effectively:
  - communicate, negotiate, and make decisions; and
  - build teams and deal with people.
- **Hard and soft skills**. This implies that a software project manager must be able to:
  - understand his/her own organization as well as the customer's organization;
  - know enough about the targeted product and/or solution;
  - be well-experienced in the profession of PM and its relevant technological tools; and
  - be comfortable with handling CHANGE REQUESTS.

A project manager as a leader must be equipped with a set of qualities and skills such as:
- being a visionary, optimistic, positive, collaborative, respectful, courteous, friendly, kind, honest, trustworthy, loyal, ethical, lifelong learner, result/goal/action-oriented, culturally sensitive, courageous, problem-solver, service-oriented, and decisive;
- being able to develop personal and professional networks, manage relationships, and resolve conflicts by building trust and satisfying concerns; and
- being a CRITICAL thinker and using analytical methods to conclude decisions.

2.4.4 The author's classification of the software project managers' essential skills

The author classifies the essential skills for software project managers as follows:
- **PM skills**. This implies that a software project manager must be well-knowledgeable and experienced in the various PM principles, methods, processes, and tools.
- **Software project domain-specific skills**. This implies that a software project manager must be well-experienced in the domain of the project he/she will be managing.
- **Sociological skills**. This implies that a software project manager must be well-comfortable with the surrounding human sociological interactions, behaviors, and actions.
- **Psychological skills**. This implies that a software project manager must be well-comfortable with the surrounding human psychological views and actions.

2.4.5 Project managers' circles of influence

In reference to Figure 2.3, there are three layers through which a project manager can have influence on.

These layers are as follows:
- **Layer #1**. This layer includes the project manager's team members who are directly influenced by him/her. Also, it includes the peer managers and the resource managers such as the HR one.
- **Layer #2**. This layer includes the project sponsors, governing bodies, steering committee, and the PMO.
- **Layer #3**. This layer includes the project stakeholders, suppliers, customers, and users.

2.4.6 Project managers' leadership styles

Project managers lead their teams in different ways. Their leadership styles can change in accordance with the following factors:

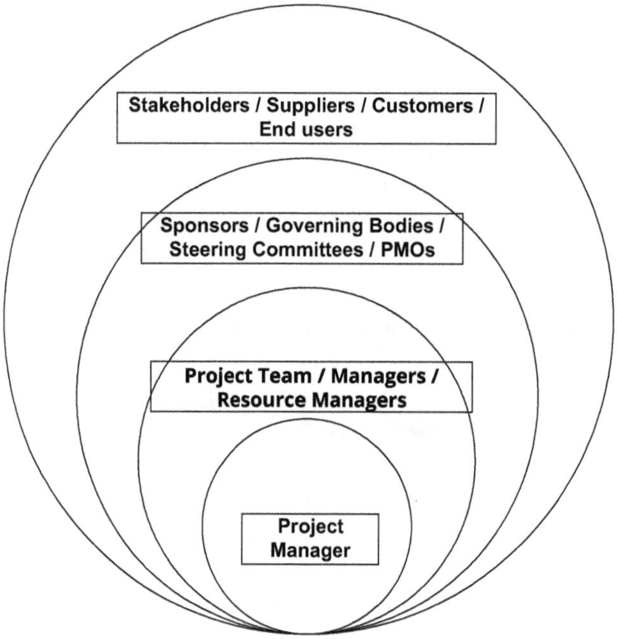

Figure 2.3: A sample project manager sphere of influence.

- **Leader characteristics**. This factor refers to the leader's attitude, mood, needs, values, and ethics.
- **Team members' characteristics**. This factor refers to the team members' attitudes, moods, needs, values, and ethics.
- **Organizational characteristics**. This factor refers to the organization's purpose, structure, and type of work performed.
- **Environmental characteristics**. This factor refers to the work environment's social situation, economic state, and political elements.

The following is a common classification of the project managers' leadership styles:
- **Laissez-faire leadership style**. With this leadership style, a project manager allows his/her team to make their own decisions and establish their own goals.
- **Transactional leadership style**. With this leadership style, a project manager focuses on the project goals, feedback, and accomplishments.
- **Servant-leader leadership style**. With this leadership style, a project manager commits to serve other people by focusing on their growth, learning, development, and autonomy.
- **Transformational leadership style**. With this leadership style, a project manager empowers his/her team through idealized behaviors, and inspirational motivation and encouragement.

- **Charismatic leadership style**. With this leadership style, a project manager inspires his/her teams by being high-energetic, self-confident, and enthusiastic.
- **Interactional leadership style**. With this style of leadership, the project manager should possess a combination of the transactional, transformational, and charismatic leadership styles.

The following is another generic classification of the project manager's leadership styles:
- **Autocratic leadership style**. This leadership style is no different from what is referred to as dictatorship. An autocratic project manager is needed to handle projects with customers who have autocratic work environments such as military and security services. An autocratic customer would not give the project manager enough room for discussion and negotiation. In other words, the customer simply issues instructions to the project manager who in turn obey and execute these instructions by simply acting autocratically with his/her team.
- **Bureaucratic leadership style**. This leadership style implies that a software project manager and his/her team are supposed to follow a set of predefined processes, procedures, guidelines, and/or regulations.
- **Charismatic leadership style**. This leadership style implies that a software project manager acts in accordance with the situation he/she is going through. He/she must spread hope and enthusiasm among his/her team, and encourage his/her team to move forward.
- **Democratic leadership style**. This leadership style implies that all of the software project team members are encouraged to participate in the decision-making process, and that their software project manager takes the final decision based on the outcomes of the team's discussions and suggestions.
- **Technocratic leadership style**. A technocratic leader is a bureaucratic leader, but he/she uses technological systems to quickly access the details of the needed bylaws and regulations. They also use the decision support systems.

2.4.7 Generic characteristics of the project managers' personalities

A project manager's personality refers to his/her way of thinking, feeling, and behaving. Examples of the characteristics of the project managers' personalities are as follows:
- **Authentic**. An authentic project manager accepts others for what and who they are.
- **Courteous**. A courteous project manager applies appropriate behaviors.
- **Creative**. A creative project manager thinks abstractly, sees things differently, and innovates.
- **Cultural**. A cultural project manager is sensitive to other cultures including values, norms, and beliefs.

- **Intellectual**. An intellectual project manager deals with human intelligence in multi-aptitudes.
- **Managerial**. A managerial project manager considers management practices.
- **Political**. A political project manager measures the political intelligence and make things happen.
- **Service-oriented**. A service-oriented project manager is willing to serve other people.
- **Social**. A social project manager understands and manages people.

2.5 Request for proposals (RFPs)

Before we step into the phases and their associated processes, it is important that we shed some light on RFPs and project charters.

An RFP is a document that is prepared by a customer for the purpose of specifying and declaring their interest in acquiring a product, a service, a solution, a system, and/or an asset. The RFPs issuing customer must make it clear to the service providers that their evaluation and selection procedures are well-structured.

2.5.1 A generic structure of an RFP document

The following are the main sections in an RFP document:

- The **project information section**. This section includes the following subsections:
  - The **issuing firm label subsection**. This includes the issuing firm's name, address, website, and phone.
  - The **issuing firm background subsection**. This includes the issuing firm's history, organizational structure, activities, and services.
  - The **project information subsection**. This includes the project title, issuance date, submission due date, questions time window, and business domain.
  - The **targeted categories of service providers' subsection**. This includes the list of categories and types of service providers that are targeted by RFPs.
  - The **intuitive budget and budget constraints subsection**. This includes the customer's suggestions with respect to the project budget, and the list of constraints that the customer is imposing on the project budget.
  - **The focal point of contact details subsection**. This includes the name, position, email, phone, and office address of the project focal point of contact.
- The **project description section**. This section presents an executive summary of the project context.
- The **project goals and objectives section**. This section presents the project goals and objectives. They must be well-articulated as S.M.A.R.T. ones (e.g., *Specific, Measurable, Achievable, Realistic, and Time-bounded*).

- The **project scope of work section**. This section includes the project formal statement of work that the to-be-selected service provider will be expected to honor and comply with.
- The **project requirements, terms and conditions, and constraints section**. This section includes:
  - The business and technical requirements that should be satisfied in the project outcomes (e.g., *product, solution, and/or service*);
  - The project terms and conditions, constraints, challenges, risks, threats, and so forth;
  - A preliminary architecture and design of the targeted outcome; and
  - The project targeted outcomes, deliverables, and milestones.
- The **project privacy, confidentiality, none disclosure, and intellectual property (IP) section**. This section presents:
  - The rules and guidelines that should be honored by the project parties with respect to the privacy, confidentiality, and non-disclosure of the project information throughout its life cycle; and
  - The ownership of any IPs that may be developed during the project.
- The **proposal submission guidelines Section**. This section presents a set of clear guidelines and requirements that relate to the proposal submission process.
- The **proposal evaluation CRITERIA section**. This section explains the CRITERIA used to evaluate and assess the to-be-received proposals.
- The **required service provider details section**. This section shows the service provider the details that must be provided in their submitted proposal. This includes at least the service provider's:
  - size, history, and domains of activities; and
  - profile of projects, SUCCESS stories and references, and focal point of contact (e.g., *Name, Email, Phone, and Office Address*).
- The **service providers qualifying CRITERIA section**. This section presents the CRITERIA that are used for qualifying the service providers.

2.5.2 Other forms of requests

The following are some other forms of requests other than the RFP document:
- **Request for information**. This request form is usually sent out by the customer to the pre-identified service providers to request information about their products and services. Such information benefits the RFP development process.
- **Request for quotation**. This request form is used to only ask for prices of certain products (e.g., *equipment, computers, laptops,* and *software solutions*).
- **Request for tender or invitation to tender**. This request form is similar to RFPs, but it is used by governmental organizations.

- **Request for qualifications or pre-qualification questionnaire**. This request form is used to invite potential service providers to complete a specific questionnaire for the purpose of becoming pre-qualified service providers in a specific domain of services.
- **Request for association, partnership, and/or alliance**. This request form is used to build consortiums to jointly carry on large-scale projects.

2.6 Project charters

A project charter is created by the service provider's team based on their initial review of the project RFPs. It provides an executive summary of the project opportunity and composes many sections such as:

- The **project vision section**. This section presents the project purpose.
- The **project objectives section**. This section presents the project objectives that should be S.M.A.R.T. ones (e.g., *Specific, Measurable, Achievable, Realistic*, and *Time-Bound*).
- The **project scope of work section**. This section presents high-level specifications of the project requirements, terms and conditions, and so forth.
- The **project deliverables and milestones section**. This section presents the project deliverables and milestones. Deliverables must be handed over to the customer during or at the end of the project. Milestones are check points on the project timeframe, at which the project manager should verify the accomplishment of the strategic goals associated with these milestones.
- The **project structure section**. This section presents the project structure in terms of its customers, stakeholders, shareholders, investors, roles, and responsibilities; and in terms of the dependencies and reporting hierarchy among its various roles.
- The **project proposed implementation plan section**. This section presents the preliminary WBS, schedule, resource allocation, risk management process, quality management process, and so forth.
- The **recommendations section**. This section presents the recommendations made by the team who conducted the initial evaluation of the project opportunity in accordance with the project RFPs.
- The **approval/disapproval section**. This section presents the decision made by the service provider to either proceed with the project opportunity or simply forget it.
- The **signatory section**. This section is where the customer and the service provider representatives inject their signatures.

2.7 The author's view of the software project management phases and their processes

This section presents the author's view of the SPM in terms of its phases and processes. This includes the initiation phase and its processes, planning phase and its processes, execution phase and its processes, monitoring and control phase and its processes, and closure phase and its processes [5–8].

2.7.1 The initiation phase and its processes

This group includes the set of processes and activities that are necessary to define and initiate new projects through determining their scopes of work.

Table 2.8 presents the activities that take place during the project initiation phase. While Table 2.9 presents the initiation phase processes.

Table 2.8: The author's view of the initiation phase activities.

Customer side (*project owner*)	The project owner would carry on the following activities: – Identifying and analyzing the business requirements. – Reviewing the organization's current operations. – Conducting a high-level benefits analysis. – Conducting a high-level budgeting. – Identifying the stakeholders and their expectations. – Developing a request for proposals (RFPs) that present a high level of the project requirements, terms and conditions, product's early design, and so forth. – Sharing the RFPs with the service providers.
Service providers' side	Each service provider would then carry on the following activities: – Developing a project charter that summarizes the project opportunity. This includes the project requirements, terms and conditions, cost, tasks, deliverables, milestones, and schedules. – Conducting a SWOT analysis to identify the strengths and weakness that would enable/disable the service provider to carry on the project, and to identify the opportunities and threats associated with the project.

Table 2.9: The author's view of the initiation phase processes.

1. **INProcess#1 Initialization** – An idea of a new system is initially suggested.
2. **INProcess#2 Endorsement**: If that idea is endorsed by the customer, a team of experts will then carefully study it.
3. **INProcess#3 RFP development**: The team of experts will carefully:
   - study that idea and its surroundings;
   - define the customer's objectives, scope of work, and requirements; and
   - put all of that into a request for proposal (RFP) document.

Table 2.9 (continued)

4.	**INProcess#4 RFP announcement**: The customer will then invite some of their prequalified service providers to submit proposals to carry on the intended project.
5.	**INProcess#5 RFPs acquirement**: Interested service providers will then obtain copies of RFPs.
6.	**INProcess#6 RFP review**: Each interested service provider will then review and evaluate RFPs.
7.	**INProcess#7 SWOT analysis**: Each interested service provider will then identify the strengths, weaknesses, opportunities, and threats of the project opportunity that is presented in RFPs.
8.	**INProcess#8 Project charter preparation**: Each interested service provider will then develop a project charter that elaborates on the project objectives, scope of work, requirements, work breakdown structure (WBS), deliverables and milestones, schedule, resources, and cost.
9.	**INProcess#9 Project charter presentation**: Each interested service provider will then present the project charter to their upper management to solicit their approval.
10.	**INProcess#10 YES/NO decision**: if the upper management endorses the project charter, then they need to sign that project charter for future reference.

2.7.2 The planning phase and its processes

This group includes the entire set of processes and activities that are necessary to define the scope of work for the new project and identify its requirements, and also to develop a detailed action plan for the project various phases of planning, execution, monitoring and control, quality assurance and control, closure, and so forth. The planning details are reached out by answering planning explicit as well as implicit questions. Table 2.10 presents the categories of the planning implicit questions along with some examples.

Table 2.11 presents the activities that take place during the planning phase, while Table 2.12 presents the planning phase processes.

Table 2.10: The author's view of the project planning implicit questions.

The WHAT questions	Examples of the WHAT implicit questions would be: – WHAT is the project scope of work? – WHAT is the project requirements? – WHAT are the tasks for each requirement? – WHAT are the deliverables and milestones?
The WHY questions	Examples of the WHY implicit questions would be: – WHY do we need this project? – WHY should we outsource the project or parts of it?
The WHEN questions	Examples of the WHEN implicit questions would be: – WHEN shall we start the project? – WHEN shall we close the project? – WHEN shall we start each of the project tasks and activities?

Table 2.10 (continued)

The WHO questions	Examples of the WHO implicit questions would be: – WHO should be the project manager? – WHO should be assigned to task x?
The HOW questions	Examples of the HOW implicit questions would be: – HOW should the software development be done? – HOW should the project be managed? – HOW should the testing cases be carried on?
The WHERE questions	Examples of the WHERE implicit questions would be: – WHERE should the software development environment be hosted? – WHERE should the software testing be hosted? – WHERE should the software operational be hosted? – WHERE should the project team members be seated?

Table 2.11: The author's view of the planning phase activities.

Customer side (e.g., project sponsor)	1. The customer reviews, and evaluates the proposals received from the service providers, and then creates a short list of the service providers of the best proposals (e.g., three to five *proposals*). 2. The customer then invites each of the shortlisted service providers to present and discuss their proposal. 3. Each service provider would then deliver a presentation on their proposal to the customer and negotiate it with them. 4. The proposal negotiation would involve: – the project scope of work and requirements, work breakdown structure (WBS), schedule, resources allocation; resources qualifications, and the proposed solution; – the proposed development and management methods; – the terms and conditions; and – the proposed price. 5. At the end of the service providers' presentations, the customer would select the best proposal and send a letter of awarding to their preferred service provider. 6. At the end of the planning phase, the customer and the selected service provider start the project contract negotiation that involves the same items as in the proposal negotiation. 7. A final version of the contract is then signed by the representatives of the customer and the service provider. 8. At this point, the planning phase is exited and the execution phase can then be started.

Table 2.12: The author's view of the planning phase processes.

1. **PLProcess#1 Requirements derivation**: Each interested service provider conducts a detailed study of the RFPs to extract the customer's requirements.
2. **PLProcess#2 Business and technical specifications derivation**: Each interested service provider then derives the business and technical specifications of the targeted software solution.
3. **PLProcess#3 High-level design derivation**: Each interested service provider then develops a design for the targeted software solution based on the technical specifications of PLProcess#2.
4. **PLProcess#4 Project core action plan derivation**: Each interested service provider then derives a project core action plan that incorporates the project work breakdown structure (WBS), schedule, task dependencies, resource allocation, and deliverables and milestones.
5. **PLProcess#5 Complete project action plan derivation**: Each interested service provider then adds to the project core action plan the needed processes such as the quality assurance and control process, risk management process, team building management process, change management process, and so forth.
6. **PLProcess#6 Proposal derivation**: Each interested service provider then prepares a proposal with technical and financial folds based on the complete project action plan of PLProcess#5.
7. **PLProcess#7 Proposal submission**: Each interested service provider then submits their proposal to the customer.
8. **PLProcess#8 Proposal assessment and shortlisting**: The customer then assigns a committee to review and assess all of the received proposals, then short-list the preferred ones.
9. **PLProcess#9 Proposal presentation and negotiation**: The customer then invites each of the shortlisted service providers to present, defend, and negotiate their proposal. At the end, the customer selects the service provider of the preferred proposal.
10. **PLProcess#10 Letter of awarding issuance**: The customer then issues a letter of awarding to the preferred service provider inviting them to start their contract negotiation.
11. **PLProcess#11 Contract negotiation**: The customer and the preferred service provider then negotiate the contract context including the scope of work, terms and conditions, price, and so forth, before a final version of the contract is concluded.
12. **PLProcess#12 Contract signatory**: The customer and the preferred service provider then sign the project contract.

2.7.3 The execution phase and its processes

This group includes the entire set of processes and activities that are necessary to execute the new project, complete its tasks and activities, produce its deliverables, and so forth.

Figure 2.4 presents the author's view of project staffing circles and options that he developed and used for many years, and proved to be SUCCESSFUL, particularly with small and mid-size projects, while Table 2.13 presents the details of this view.

Table 2.14 summarizes the author's view of the execution phase activities, while Table 2.15 presents the author's view of the execution phase processes.

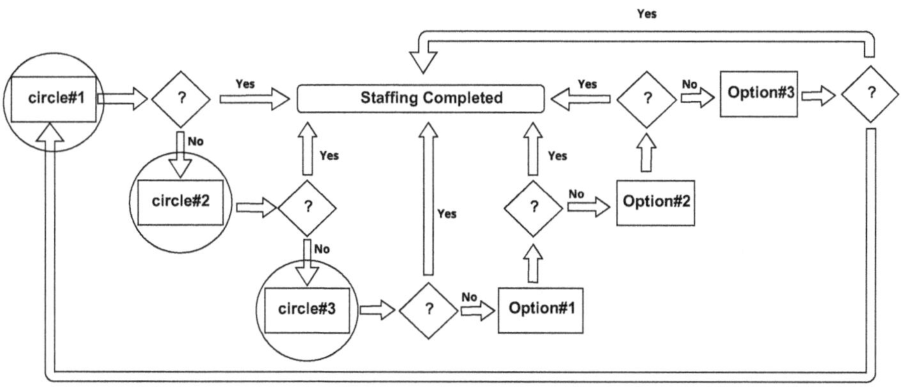

Figure 2.4: The author's view of project staffing circles and options.

Table 2.13: The author's view of project staffing circles and options.

(a) The author's view of staffing circles and options (Part#1).	
In-house recruits. This represents the first stage of the author's view of project staffing that composes three circles:	
Circle#1	A project manager attempts to hire staff from those who worked with them on previous projects.
Circle#2	If staffing remains incomplete, then the project manager seeks help from his/her colleague project managers to hire from those who worked with them on previous projects.
Circle#3	If staffing is still incomplete, then the project manager approaches the human resources department at his/her company to get access to the profiles of all available technical staff to fill for as many left roles as possible.
(b) The author's view of staffing circles and options (Part#2).	
External recruits. If there are still unfilled roles, then the project manager shall go through the following three options for external recruits in sequence:	
Option#1	The project manager advertises the unfilled roles searching for hirable independent external staff members. However, the human resources department should endorse hiring external staff on full-time bases. Otherwise, the project manager will have to consider Option#2.
Option#2	The project manager would offer contract-based employment to those who could not be hired on full-time bases in Option#1.
Option#3	If things do not work fine with Option#2, then the project manager approaches external employment agencies to get the needed staff.

Table 2.14: The author's view of the execution phase activities.

(a) The author's view of the execution phase activities (Part#1).

At the beginning of the execution phase, the project manager should check the following items to ensure that the contract is compliant with the RFPs:

- the project generic description;
- the project objectives and anticipated outcomes;
- the project terms and conditions, including the service-level agreement, if any;
- the team correspondence process for both of the horizontal and vertical communications;
- the team performance evaluation processes for both of the project manager and the functional manager;
- the contract management processes for both of the customer contract management, and the service provider contract management;
- the subcontractors' management process;
- the change management process as proposed by the service provider and accepted by the customer;
- the project accounting processes for both of the customer accounting, and the service provider accounting; and
- the contract governance and compliance process.

(b) The author's view of the execution phase activities (Part#2).

Then, the project manager must arrange for and take care of the following:

- holding a kick-off meeting to start the project;
- handling the logistical arrangements and the infrastructure setup;
- handling the team mobilization and resource allocations;
- managing and supervising the execution of the project action plan;
- compiling the progress and monitoring reports;
- compiling the accomplishments and milestone reports;
- compiling the issues and risk reports; and
- compiling the execution comprehensive report.

Table 2.15: The author's view of the execution phase processes.

1. EXProcess#1 Kick-off meeting.
2. EXProcess#2 Logistical arrangements.
3. EXProcess#3 Team build/mobilization/management.
4. EXProcess#4 Procurement and subcontracting.
5. EXProcess#5 Correspondence management.
6. EXProcess#6 Contract management (technical and financial).
7. EXProcess#7 Tasks execution.
8. EXProcess#8 Change management.
9. EXProcess#9 Quality assurance (e.g., *MCProcess#1 monitoring*).
10. EXProcess#10 Risk management.
11. EXProcesses#11-to-N All other Processes that may be needed for the Project.

2.7.4 The monitoring and control phase and its processes

This group includes the entire set of processes and activities that are necessary to track and regulate the project progress and performance throughout its life cycle. The project performance is observed and measured repeatedly during the project execution, and accordingly, corrective actions are taken to resolve any issue.

Table 2.16 presents the author's view of the monitoring and control phase activities, whereas Table 2.17 presents the author's view of the monitoring and control phase processes.

Table 2.16: The author's view of the monitoring and control phase activities.

During the monitoring and control phase, many monitoring activities are carried on as follows:

- Assessing and measuring the ongoing activities throughout the project life cycle
- Monitoring the various factors including the project cost, effort, and scope against its plan and performance baseline
- Identifying the corrective actions needed to resolve the issues observed during the measurement and monitoring activities
- Establishing a communication channel to share and discuss issues throughout the project life cycle
- Establishing a key performance indicator (KPI) system for the project
- Conducting functional testing to verify the correctness and preciseness of the various functions, and nonfunctional testing to test the performance, availability, reliability, scalability, maintainability, and testability of the targeted solution.

Table 2.17: The author's view of the monitoring and control phase processes.

1. MCProcess#1 Assessing the ongoing project activities – monitoring and quality assurance (e.g., *EXProcess#9 Quality Assurance*)
2. MCProcess#2 Testing the product operation factors including the correctness, reliability, efficiency, integrity, and usability
3. MCProcess#3 Testing the product revision factors including the maintainability, testability, and flexibility
4. MCProcess#4 Testing the product transition factors including the portability, reusability, and interoperability
5. MCProcess#5 Monitoring the project variables including the cost, effort, scope, schedule, and resources against the project targeted performance baseline
6. MCProcess#6 Identifying necessary corrective actions to resolve and fix any issue that arose during the testing
7. MCProcess#7 Managing changes

2.7.5 The closure phase and its processes

This group includes the entire set of processes and activities that are necessary to complete and close on the project and its contract. The project closure phase is used to obtain from the customer their formal acceptance of the project deliverables, and then ending the project. The project documents must be archived, including the project management plan, implementation plan, testing plan, risk management plan, quality assurance and control plan, requirements documentation, analysis reports, testing reports, design documentation, and so forth.

Table 2.18 presents the closure phase activities and Table 2.19 presents the closure phase processes.

Table 2.18: The author's view of the closure phase Activities.

The project closure phase involves the following activities:
- A post-implementation review. Reviewing things that succeeded and analyzing things that went the other way to come up with lessons learned
- Closure of the project contacts. Complete all contracts related to the project
- The project closure. Complete all activities across all groups of processes throughout the project life cycle.

Table 2.19: The author's view of the closure phase processes.

1. CLProcess#1 Checklist preparation.
2. CLProcess#2 Deliverables handover.
3. CLProcess#3 Customer's users training.
4. CLProcess#4 Warranty management.
5. CLProcess#5 Financial settlement.
6. CLProcess#6 Project signoff.

2.8 Contracts

A contract is a legally binding agreement between two or more parties to fulfill the terms and conditions outlined in that agreement.

Contract management is used to manage contracts made with customers, vendors, partners, and/or employees.

Contract management includes the negotiation of the contracts' terms and conditions, while ensuring the compliance of these contracts with their terms and conditions. It includes also documenting and agreeing on any changes or amendments that may arise during the contracts' execution.

2.8.1 Types of contracts

Contract types are as follows:

A) **Sales contracts.** A sales contract composes several sections such as:
   - the details of the involved parties;
   - the list of products, solutions, and/or services that the representative will be authorized to sell out;
   - the selling prices;
   - the territory throughout which the representative can operate;
   - the promotional discounts; and
   - the commission percentage that the representative will receive when he/she closes a deal.

   A sales contract can be with a sales representative or a sales agency as follows:
   - **Sales representative contracts.** By signing this type of contracts, a sales representative agrees to promote and sell the products to the customers, and then gets paid a base salary plus sales commission.
   - **Agency (agent) contract agreements.** By signing this type of contracts, an agency agrees to promote and sell a predefined list of products on behalf of the seller to customers in a predetermined region and for predefined prices, and then gets paid a percentage of these sales.

B) **Purchasing contracts.** A purchasing contract is a contract between a customer (e.g., *a buyer*) and a service provider (e.g., *a seller*) who is promising to sell products, solutions, and/or services in accordance with a set of terms and conditions. The customer is required to acknowledge the reception of these products, solutions, and/or services and pay for them.

C) **Performance-based contracts.** A performance-based contract is judged by the performance of its products, solutions, and/or services. In other words:
   - If the performance is in full compliance with the customer's predefined performance expectations, then the service provider will get paid the same amount of money as agreed upon in the project contract.
   - Else-if the performance is behind the customer's expectations, then either the outcomes are rejected by the customer and in that case no money will be paid to the service provider, or the service provider will get paid an amount of money that is less than the amount agreed upon in the project contract. The deducted amount is this case would be governed by the terms and conditions of the project contract.
   - Else-if the performance degree exceeds the customer's expectations, then the service provider would be paid some bonus money on the top of the agreed-upon amount as in the project contract. The bonus amount in this case would be governed by the terms and conditions of the project contract.

D) **Partnership agreements**. A partnership agreement is a contract that establishes the terms for a partnership between at least two parties.

E) **Subcontractor contracts**. A subcontractor is an individual or a business that is hired by a prime contractor to carry on part of the work that is defined in their contract with their customer. The incentive to hire a subcontractor is either to reduce the project cost, improve its quality, or mitigate its risks. A subcontractor can be one of the following categories:
   - **Domestic and interdisciplinary subcontractors**. A domestic subcontractor signs a subcontract with the prime contractor to supply material or to execute part of the prime contract. Subcontractors of this category are typically not specialized, but can work across many areas.
   - **Nominated subcontractors**. A nominated subcontractor is directly contracted. Subcontractors of this category are nominated to join the project team because of their specialization related to the part of the project that they are needed to carry on.
   - **Named subcontractors**. A named subcontractor signs a contract with the prime contractor to supply materials or to execute part of the work of the prime contract. Subcontractors of this category are listed on the prime contractor's lists of prequalified subcontractors.

2.8.2 Contract management phases

The three phases of contract management are as follows:

- **The pre-contract phase**. During this contract management phase, the project contract is planned for, articulated, and then endorsed and signed by its parties.

  In terms of PM, this contract management phase is lined up with the PM phases of initiation and planning.
- **The contract execution phase**. During this contract management phase, the project execution, monitoring and control, and closure phases' activities are carried on.

  In terms of PM, this contract management phase is lined up with the PM phases of execution, monitoring and control, and closure.
- **The contract post-execution phase**. During this contract management phase, the contract is checked to ensure its compliance with its governing standards, and to ensure having adequate support to the delivered products solutions and/or services.

  In terms of PM, this contract management phase is lined up with the PM phases of monitoring and control, and closure. Also, it is lined up with the additional PM phase of support (e.g., *either as an additional phase to the five phases of PM, or as a separate support contract*).

2.9 The author's simplified change management process

The change management process is a process for dealing with CHANGE-REQUESTS received throughout the project life cycle.

The types of CHANGE-REQUESTS are as follows:

- **CHANGE-REQUESTS received from the customer.**[2] This refers to the CHANGE-REQUESTS received verbally or in writing from the customer throughout the project life cycle.
  - If a change request is received verbally from the customer, then a verbal or written response should be sent to the customer regardless whether the service provider accepted it or turned it down. If accepted, then a verbal agreement or a contract amendment must take place to legitimize the endorsement of that change request.
  - Else if that change request is received in writing from the customer, then a written response should be sent to the customer regardless whether the service provider accepted it or turned it down. If accepted, then a written amendment to the contract must take place to legitimize the endorsement of that change request.
- **CHANGE-REQUESTS received from the project team.** This refers to the CHANGE-REQUESTS received from one or more of the project team members. Such CHANGE-REQUESTS can slightly impact the project schedule and/or specifications. They are mostly handled internally without involving the customer.
- **CHANGE-REQUESTS received from the project accountant.** This refers to the CHANGE-REQUESTS received from the project accountant and/or financial advisor. Such CHANGE-REQUESTS can slightly impact the project budget distribution across various components of the project. They are mostly handled internally without involving the customer.

2.10 The author's simplified risk management process

The author's simplified risk management process goes through the following steps:

- **Step#1: Risk identification.** This step is to identify all potential risks and threats and their anticipated impacts along with the probabilities of their potential occurrences.

2 **Note**: It is important to note that a contract amendment may alter one or more articles in the original contract. It may alter the scope of work, requirements, work breakdown structure (WBS), schedule, deliverables, milestones, and/or any of the processes associated with the project detailed action plan.

- **Step#2: Risk classification**. This step is to classify the identified potential risks into three categories, which are the risks with low-impact category, risks with moderate-impact category, and risks with high-impact category.
- **Step#3: Responses to risks**. In reference to Table 2.20, risks can be dealt with as follows:
  - In the case of low-impact risks, all what is needed is to proactively identify these risks. However, if it happens that a low-impact risk occurs during the project's life cycle, then the project manager should reactively devise a plan and immediately implement it to counterpart that risk.
  - In the case of moderate-impact risks, all what is needed is to proactively identify and plan for these risks. However, if it happens that a moderate-impact risk occurs during the project life cycle, then the project manager should reactively implement the pre-prepared plan to counterpart that risk.
  - In the case of high-impact risks, all what is needed is to proactively identify, plan for encountering these risks, then immediately implement these plans.

Table 2.20: The author's simplified risk management process.

Category	Identify	Plan	Implement
Low-impact risks	Proactively	Reactively	Reactively
Moderate-impact risks	Proactively	Proactively	Reactively
High-impact risks	Proactively	Proactively	Proactively

2.11 Project management approaches

As of the 1990s, there has been many PM approaches. In reference to Figure 2.5, PM approaches include the phase-based PM (PBPM), lean-based PM (LBPM), iterative-and-incremental-based PM (IIBPM), critical-chain-based PM (CCBPM), product-and-production-based PM (PPBPM), process-based PM (PRBPM), and benefits-and-earned-value-based PM (BEVBPM).

Regardless of the PM approach that is used for a given project, the project manager must take into consideration the objectives, scope, timeline, cost, and the roles and responsibilities of team and stakeholders of that project. Around the 2000s, the Agile PM methods started to evolve. Agile PM approaches, particularly the SCRUM one, are addressed in detail in Chapter 5.

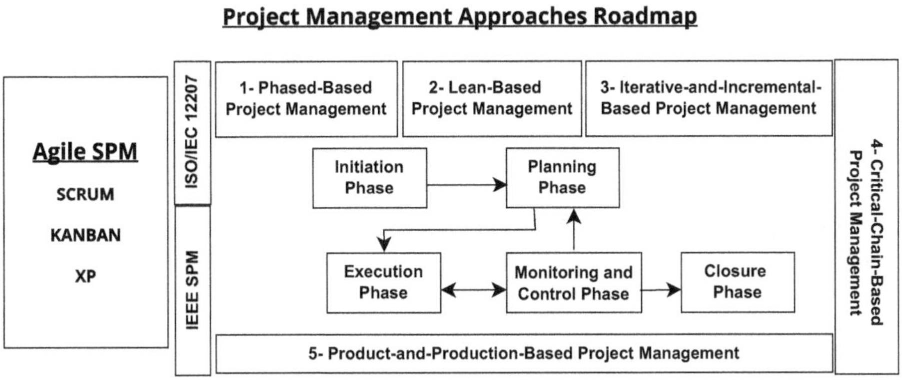

Figure 2.5: Project management approaches' overview.

2.11.1 Phase-based project management (PBPM)

The PBPM approach carries on the project predefined work through the following five phases:

- **Initiation phase**. This stage is meant to:
  - initiate the project idea; and
  - intuitively define the project objectives, scope of work, requirements, milestones, and deliverables.
- **Planning phase**. This stage is meant to:
  - extract the project objectives, scope, requirements, deliverables, and milestones from RFPs;
  - estimate the project tasks, activities, WBS, and schedule;
  - allocate resources; and
  - devise a preliminary design for the targeted products, solutions, and/or services.
- **Execution phase**. This stage is meant to:
  - carry on the tasks and activities according to WBS; and
  - build and construct the targeted products, solutions, and/or services.
- **Monitoring and control phase**. This stage is meant to carry on a predefined set of quality assurance and quality control tasks and activities.
- **Closure phase**. This stage is meant to carry on a predefined set of tasks and activities needed to complete and close the project.

2.11.2 Lean-based project management (LBPM)

The LBPM approach varies from the PBPM approach. Although it goes through the same five phases of PBPM approach, it is meant to produce deliverables in less times, and with less waste by RE-CYCLING components from previous products, solutions, and/or services.

2.11.3 Iterative-and-incremental-based project management (IIBPM)

The PBPM approach is not suitable for large-size projects with fast-changing requirements that are not clearly predefined. It is not suitable as well for projects that have much risks, uncertainties, and/or dependencies. Several IIBPM methods have evolved to handle such complexities. It is important to notice that the IIBPM methods are considered part of the Agile methods.

2.11.4 Critical-chain-based project management (CCBPM)

The CCBPM approach is designed to deal with the uncertainties in managing projects such as having limited availability of the needed human resources and/or physical resources. Constrained tasks and activities are given higher priorities during the project planning, resource allocation, and execution.

2.11.5 Product-and-production-based project management (PPBPM)

The PPBPM approach is a structured project management approach that identifies the project's targeted products, and then produces the targeted outcome products, solutions, and/or services using the materialized and informational inputs at a predefined production volume for each of these outcomes.

2.12 Projects' SUCCESS

In reference to the following sections, we can think of many ways to conclude whether a project is SUCCESSFUL or not.

2.12.1 The four pillars for project management

Table 2.21 presents and describes the four pillars used to declare the project as SUCCESSFUL or not. A project is considered SUCCESSFUL if and only if these four pillars are met and satisfied by the project outcomes.

Table 2.21: The four pillars for projects' SUCCESS.

ON-TIME	This pillar refers to completing the project on-time in accordance with its predefined schedule.
WINTHIN-BUDGET	This pillar refers to completing the project without exceeding the pre-allocated budget. In other words, its overall spendeture should not exceed that budget.
MEET-SPECS	This pillar refers to completing the project with its outcome product's specifications meeting the project's predefined specifications. In other words, the specifications of the project outcomes should be compliant with its governing baseline specifications.
SATISFY-CUSTOMER	This pillar refers to having the customer to accept the project deliverables. In other words, the deliverables should pass the customer testing and ACCEPTANCE-CRITERIA.

2.12.2 The four Ps for project management

The project manager should have four Ps as described in Table 2.22 in order to ensure the SUCCESS of his/her project. With these four Ps, a project manager may proceed with confidence toward a great story of SUCCESS.

Table 2.22: The four Ps for project's SUCCESS.

PLAN	This **P** refers to the project core action plan that should be developed first. A core action plan composes an estimated work breakdown structure (WBS), an estimated schedule with careful considerations of the task's interdependencies, a carefully estimated allocation of resources, and a set of carefully defined deliverables and milestones.
PROCESSES	This **P** refers to the set of project management processes needed besides the project core action plan to create the project detailed action plan. The set of all necessary processes must be developed, and then added to the project core action plan to form the project detailed action plan.
PEOPLE	This **P** refers to allocating the right staff members as part of the project team. The project manager should choose the right staff members to carry on the right sets of tasks and activities. In other words, the right people should be assigned the right roles and responsibilities.
POWER	This **P** refers to the necessary level of power and authority that must be given to the project manager and the project team in order to enable them to carry on their duties with less issues.

2.12.3 Projects' SUCCESS factors

The main project's essential SUCCESS factors are:

- **Definitions**. This refers to the necessity of having the project goals, scope, and requirements well-defined before the project starts.
- **Upper management support**. For a project to succeed, it is essential to have a high level of support from the upper management at both sides of the customer and the service provider. Also, it is essential to have support from the project stakeholders and shareholders.
- **User involvement**. It is essential to involve the end users from the customer side as early as possible in the project life cycle.
- **Project manager competency**. The project manager must be competent in terms of:
  - **Being a professional project manager**. He/she must be very-well experienced in PM.
  - **Having adequate knowledge of the project subject domain**. For example, if the project involves telecommunications, then the project manager should have adequate knowledge and experience in the domain of telecommunications.
  - **Being able to handle sociological and psychological issues**. The project manager must be able to handle any sociological and/or psychological issues that may arise among the project team or while dealing with the customer's representative throughout the project life cycle.
  - **Having technological skills**. The project manager must be able to use and utilize the necessary technological tools to plan for and track the project execution.
  - **Being able to manage**. The project manager must be able to handle any complexity in the project, devise a reasonable budget for the project, build and manage the project team, control and solve problems that may arise during the project life cycle, and take effective actions and decisions.
  - **Being able to lead**. The project manager must be able to handle and promote changes; give directions and provide vision; align the team members in a way that would prevent any inter-team conflicts; motivate and inspire the team members; take meaningful actions; delegate authority whenever that is necessary; remain positive, energetic, flexible, creative, persistent, and patient throughout the project life cycle; and conduct careful planning and analysis for the project.
- **WBS, schedule, and resource allocations**. As part of the planning phase, the project manager must be able to transform the requirements into lists of tasks and activities such that each list belongs to one requirement.
- **Deliverables and milestones**. As part of the planning phase, the project manager should find out the required deliverables that should be handed over to the customer throughout or at the end of the project. Also, he/she should devise a set of

milestones to monitor and watch the achievements of the project strategic objectives throughout the project life cycle.
- **Needed processes**. As part of the planning for the project detailed action plan, the project manager should identify the list of processes that must be used throughout the project life cycle. Together with the project core action plan (e.g., *WBS, schedule, inter-task dependencies, resource allocations, deliverables,* and *milestones*), these processes compose the project detailed action plan.
- **Project team quality**. As part of the resource allocation activities, the project manager should select the right resources to be assigned to work on his/her project.
- **Resource availability**. If resources with the needed competencies are not available in the local market, then the project manager will face hard times while staffing for his/her project.
- **TEST-CASES quality and comprehensiveness**. In collaboration with the project team and the project quality assurance officer, the project manager should devise a set of TEST-CASES that is comprehensive and leads to high coverage of the solution's functions (e.g., *black-box TEST-CASES as well as white-box TEST-CASES for both functional and nonfunctional TEST-CASES*).
- **Users' training quality and comprehensiveness**. During the closure phase, the project manager should facilitate an adequate set of training sessions for the end users at the customer's side.
- **Estimation accuracy**. The project manager should do the best he/she can to secure correctness and preciseness when he/she estimates for tasks, schedule, and resource allocations.

2.12.4 Best practices, de factos, and software standards

The journey from practices to standards goes through the following steps:
- **Step #1**: A large set of generic practices is identified.
- **Step #2**: The large set is then narrowed down to a subset of practices that can be referred to as BEST-PRACTICES. A practice can be identified as a best practice based on the reputation of the project manager who used that practice to carry on his/her projects with positive impacts on his/her customers.
- **Step #3**: The BEST-PRACTICES are then tried for few years with many real projects, and based on the feedback from their project managers, these BEST-PRACTICES would diverge into one or more enhanced practice(s). Each enhanced practice can then be referred to as a de facto.
- **Step #4**: A de facto is then tried for many years with many real projects, and based on the feedback from their project managers, that de facto can be announced as a ready-for-standardization de facto.

- **Step #5**: The concerned organization can then initiate the standardization process with the concerned standardization authority (e.g., ISO and IEEE) to standardize their ready-for-standardization de facto.
- **Step #6**: The concerned standardization authority then reviews the submitted de facto, adjusts it if necessary in coordination with the owning organization, arranges to use it with additional real projects if needed, discusses it carefully and thoroughly until a solid conclusion is reached before that de facto is declared officially as a new standard.

2.13 The author's simplified model for the project core action plan

2.13.1 The simplified model

Table 2.23 presents the author's simplified model that can be used to derive a WBS for a project along with its schedule, resource allocation, and cost.

Table 2.23: The author's simplified model for project work breakdown structure (WBS), schedule, resource allocations, and cost.

TID	Categories/tasks		Start date	End date
C_0	**Project management** (e.g., *a project manager has been assigned to the project on full-time basis*)		P_{Start}	P_{End}
C_1	**Category #1** (e.g., *Requirement #1 has been transformed into (n) tasks*)		C_1.Start	C_1.End
	$C_1.T_1$	Task #1 of Category #1	C_1T_1.Start	C_1T_1.End
	$C_1.T_2$	Task #2 of Category #1	C_1T_2.Start	C_1T_2.End
		....		
	$C_1.T_n$	Task #n of Category #1	C_1T_n.Start	C_1T_n.End
C_2	**Category #2** (e.g., *Requirement #2 has been transformed into (m) tasks*)		C_2.Start	C_2.End
	$C_2.T_1$	Task #1 of Category #2	C_2T_1.Start	C_2T_1.End
	$C_2.T_2$	Task #2 of Category #2	C_2T_2.Start	C_2T_2.End
		....		
	$C_2.T_m$	Task #m of Category #2	C_2T_m.Start	C_2T_m.End
C_3	**Category #3** (e.g., *Requirement #3 has been transformed into (x) tasks*)		C_3.Start	C_3.End
	$C_3.T_1$	Task #1 of Category #3	C_3T_1.Start	C_3T_1.End

Table 2.23 (continued)

TID	Categories/tasks		Start date	End date
	$C_3.T_2$	Task #2 of Category #3	C_3T_2.Start	C_3T_2.End
		. . . .		
	$C_3.T_x$	Task #x of Category #3	C_3T_x.Start	C_3T_x.End
		. . . .		
C_z	**Category #z** (e.g., *Requirement #z has been transformed into (y) tasks*)		C_z.Start	C_z.End
	$C_z.T_1$	Task #1 of Category #z	C_zT_1.Start	C_zT_1.End
	$C_z.T_2$	Task #2 of Category #z	C_zT_2.Start	C_zT_2.End
		. . . .		
	$C_z.T_y$	Task #y of Category #z	C_zT_y.Start	C_zT_y.End

Legend:

P_{Start}	Project start date. This is the same as the start date of the project manager. Also, it is the same as the start date of the first task across the categories to be started.
P_{End}	Project end date. This is the same as the end date of the project manager. Also, it is the same as the end date of the last task across the categories to be completed/ended.
C_i	The ID of the ith Category.
C_i.Start	The start date of the ith category. This is the same as the start date of the first task from the ith category to be started.
C_i.End	The end date of the ith category. This is the same as the end date of the last task from the ith category to be completed/ended.
$C_i.T_j$	The ID of the jth task in the ith category list of tasks.
C_i.Start	The start date of the ith category.
C_i.End	The end date of the ith category.
C_iT_j.Start	The start date of the jth task in the ith category list of tasks.
C_iT_j.End	The end date of the jth task in the ith category list of tasks.

2.13.2 A simplified formula for the project duration

Project elapsed days = $P_{e\text{-}days} = P_{end} - P_{start}$

Project working days ($P_{w\text{-}days}$) is calculated using a programming script as follows:

```
// edays: elapsed days --- wdays: working days
wdays = 0;
while (edays ≥ 7)
{
    wdays = wdays + 5;
    edays = edays – 7;
}
wdays = wdays + edays;
```

2.13.3 A simplified formula for resource allocations and costing

Assume the following:

- The project manager has a pool of resources to use to assign resources on various tasks in the project plan.
- The $\mathbf{C_i.T_j}$ (e.g., *the jth task in the ith category*) is scheduled to be completed within $\mathbf{WD_{ij}}$ working days.
- The $\mathbf{C_i.T_j}$ (e.g., *the jth task in the ith category*) is assigned by the project manager to a resource that can be referred to as the $\mathbf{R_{ij}}$ Resource.
- The daily rate of that resource is $\mathbf{R_{ij}.DRT}$.

Then

- The cost of task $\mathbf{C_i.T_j}$ (e.g., is referred to in this model as $\mathbf{C_iT_j.Cost}$) can be calculated using this formula: $\mathbf{C_iT_j.Cost = WD_{ij} * R_{ij}.DRT}$.

Assume the following:

- The cost of task $\mathbf{C_i.T_j}$ (e.g., *the jth task in the ith category*) is $\mathbf{C_iT_j.Cost}$.
- The number of tasks in the **ith** category is $\mathbf{N_i}$.
- The number of categories is **z**.

Then

- The overall cost of the **ith** category is $\mathbf{C_i.Cost}$, and the project's overall cost ($\mathbf{P_{Cost}}$) will be the sum of the costs of the individual categories:

$$\mathbf{Ci.\,Cost} = \sum_{j=1}^{Ni} \mathbf{CiTj.\,Cost}$$

$$\mathbf{P_{Cost}} = \sum_{i=1}^{z} \mathbf{Ci.\,Cost}$$

Chapter 3
Project management knowledge areas and their processes: the PMI perspective

This chapter presents and elaborates on the inputs, methods, and outputs of the following project management knowledge area processes [1]:
- **Project integration management processes**. These processes are the Develop Project Charter, Develop Project Management Plan, Direct and Manage Project Work, Manage Project Knowledge, Monitor and Control Project Work, Perform Integrated Change Control, and Close Project or Phase.
- **Project scope management processes**. These processes are the Plan Scope Management, Collect Requirements, Define Scope, Create WBS (work breakdown structure), Validate Scope, and Control Scope.
- **Project schedule management processes**. These processes are the Plan Schedule Management, Define Activities, Sequence Activities, Estimate Activity Durations, Develop Schedule, and Control Schedule.
- **Project cost management processes**. These processes are the Plan Cost Management, Estimate Costs, Determine Budget, and Control Costs.
- **Project quality management processes**. These processes are the Plan Quality Management, Manage Quality, and Control Quality.
- **Project resource management processes**. These processes are the Plan Resource Management, Estimate Activity Resources, Acquire Resources, Develop Team, Manage Team, and Control Resources.
- **Project communications management processes**. These processes are the Plan Communications Management, Manage Communications, and Monitor Communications.
- **Project risk management processes**. These processes are the Plan Risk Management, Identify Risks, Perform Qualitative Risk Analysis, Perform Quantitative Risk Analysis, Plan Risk Responses, Implement Risk Responses, and Monitor Risks.
- **Project procurement management processes**. These processes are the Plan Procurement Management, Conduct Procurements, and Control Procurements.
- **Project stakeholder management processes**. These processes are Identify Stakeholders, Plan Stakeholder Engagement, Manage Stakeholder Engagement, and Monitor Stakeholder Engagement.

Similar to the previous chapter, the context of this chapter serves the course learning outcomes of CLO1 and CLO2 as specified in the Preface at the beginning of this textbook.

https://doi.org/10.1515/9783111206868-003

3.1 Project integration management processes

Project integration management includes the processes that are used to identify, define, combine, unify, and coordinate the various project management activities.

In reference to Figure 3.1, the project integration management processes are the Develop Project Charter, Develop Project Management Plan, Direct and Manage Project Work, Manage Project Knowledge, Monitor and Control Project Work, Perform Integrated Change Control, and Close Project or Phase.

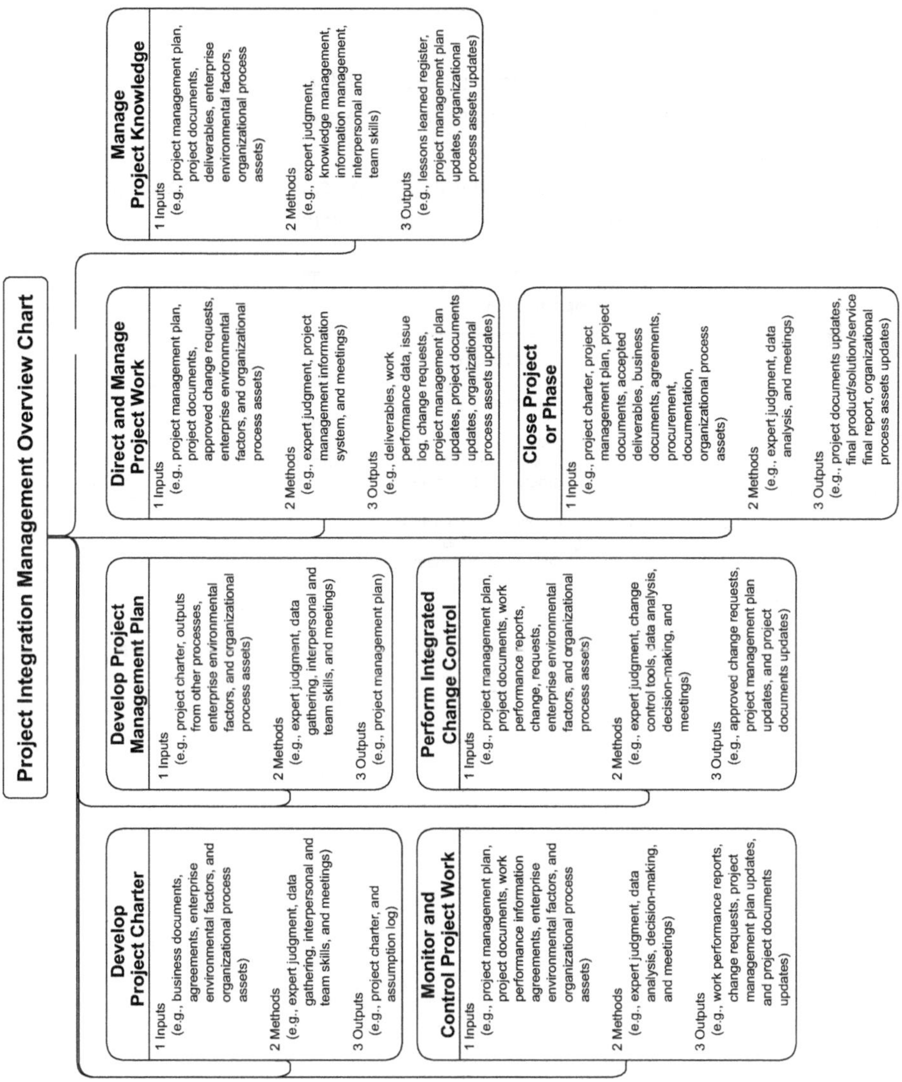

Figure 3.1: Project integration management overview chart.

3.1.1 The "Develop Project Charter" process

The "Develop Project Charter" process is used to develop a project charter that facilitates the existence of a project and enables the project manager to utilize the organizational resources.

In reference to Figures 3.2 and 3.3, the main inputs, methods, and outputs of this process are detailed in Tables 3.1, 3.2, and 3.3, respectively.

The "Develop Project Charter" Process		
Inputs	**Methods**	**Outputs**
1 Business documents (e.g., business case, and benefits management plan) 2 Agreements 3 Enterprise environmental factors 4 Organizational process assets	1 Expert judgment 2 Data gathering (e.g., brainstorming, focus groups, and interviews) 3 Interpersonal and team skills (e.g., conflict management, facilitation, and meeting management) 4 Meetings	1 Project charter 2 Assumption log

Figure 3.2: The "Develop Project Charter" process.

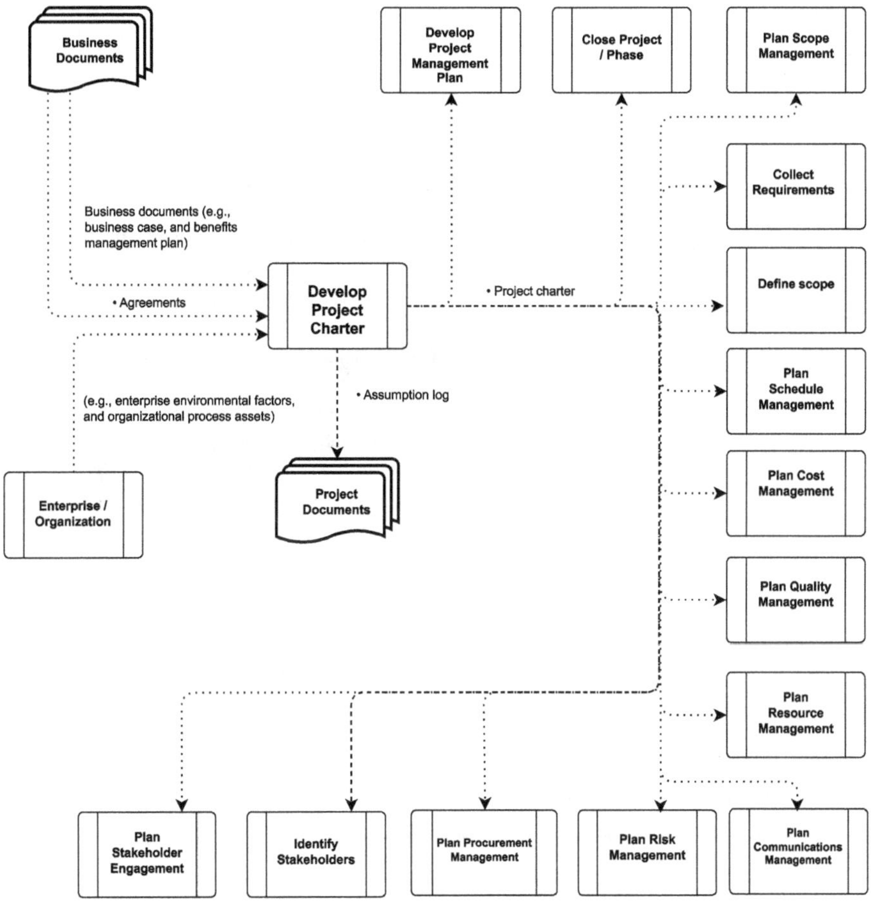

Figure 3.3: The "Develop Project Charter" process data flow diagram.

Table 3.1: Inputs to the "Develop Project Charter" process.

Input	Details
Business documents	The main business documents used as inputs to this process are the business case, market demand report, customer requests, organizational needs, technological advances, legal requirements, ecological impacts, and social needs.

Table 3.1 (continued)

Input	Details
Agreements and contracts	The main agreements used as inputs to this process are the project contract including all of their amendments, memorandum of understanding, alliance agreements, association agreements, and subcontracting agreements. A contract is an agreement that: – obligates the service provider to provide the customer with the agreed-upon products, solutions, and/or services; – obligates the customer to compensate the service provider for their provision; and – legalizes the contractual relationship between the service provider and the customer. The main components of an agreement are the procurement statement of work, deliverables, schedule, milestones, performance reporting, pricing and payment terms, inspection, quality, ACCEPTANCE-CRITERIA, warranty and support, incentives and penalties, insurance and performance bonds, subcontractors' approvals, terms and conditions, CHANGE-REQUESTS handling, termination clause, and dispute resolution methods.
Enterprise environmental factors (EEFs)	The main EEFs used as inputs to this process are the governmental standards, industrial standards, legal and regulatory requirements, marketplace conditions, organizational culture, political climate, organizational governance framework, and stakeholder requirements.

Table 3.2: Methods used during the "Develop Project Charter" process.

Method	Details
Expert judgment	This refers to all judgments and decisions made while developing the project charter. Such judgments require the project manager and his/her team to have adequate level of the required relevant skills and experience.
Data gathering approaches	The main data gathering methods used during this process are the brainstorming, checklists, focus groups, interviews, and questionnaires.
Interpersonal and team skills	This refers to the skills utilized in the project charter development process.
Meetings	This refers to the meetings with stakeholders as part of this process to identify the project objectives, SUCCESS-CRITERIA, deliverables, requirements, milestones, and so forth.

Table 3.3: Outputs of the "Develop Project Charter" process.

Output	Details
Project charter	The project charter presents high-level information on the project and on its targeted products, solutions, and/or services. The main items that compose the project charter are the project description, purpose, objectives, SUCCESS-CRITERIA, requirements description, deliverables, terms and conditions, risks, milestones, schedule, stakeholders, financial requirements, APPROVAL-CRITERIA, CLOSURE-CRITERIA, project manager, and project sponsor.
Assumption log	The assumption log presents the high-level strategic and operational assumptions and constraints throughout the project life cycle.

3.1.2 The "Develop Project Management Plan" process

The "Develop Project Management Plan" process is used to define, prepare, coordinate, and consolidate the needed components into a project management plan. This process produces a comprehensive project management plan that defines the basis of the project work activities and how they will be performed. In other words, the project management plan defines how the project will be executed, monitored, controlled, and closed.

In reference to Figures 3.4 and 3.5, the main inputs, methods, and outputs of this process are detailed in Tables 3.4, 3.5, and 3.6, respectively.

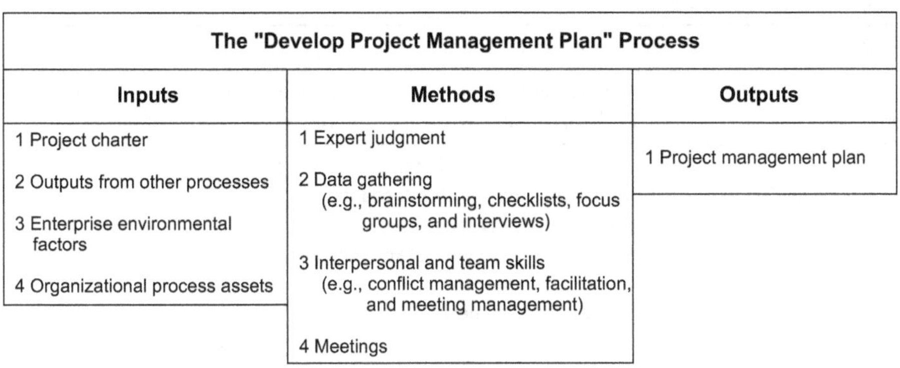

Figure 3.4: The "Develop Project Management Plan" process.

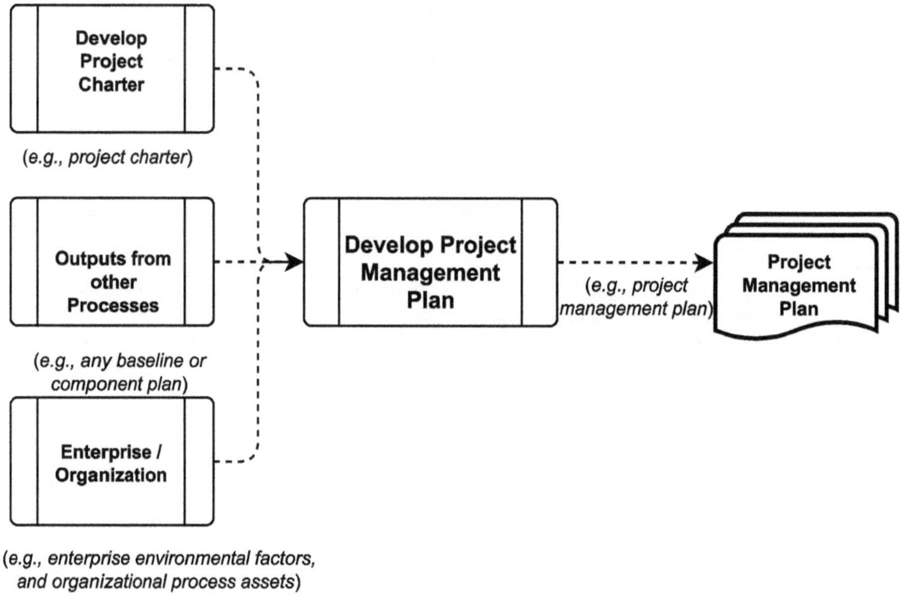

Figure 3.5: The "Develop Project Management Plan" process data flow diagram.

Table 3.4: Inputs to the "Develop Project Management Plan" process.

Input	Details
Project charter	As described earlier in Table 3.3.
Outputs from other processes	The project manager and his/her team may use from the outputs of other processes. Such outputs come into this process in the form of subsidiary plans.
Enterprise environmental factors (EEFs)	The main EEFs used as inputs to this process are the: – Governmental and industry standards – Legal and regulatory requirements – Project management body of knowledge – Focus areas and focus groups – Operational environment – Safety – Risks – Software development methods – Organizational structure, culture, and management practices – Organizational governance framework that controls, directs, and coordinates people, policies, and processes – Organizational infrastructure

Table 3.4 (continued)

Input	Details
Organizational process asset factors (OPAFs)	The main OPAFs used as inputs to this process are: – Organizational standard policies, processes, and Procedures – Project management plan templates – Change control procedures – Monitoring and reporting methods – Risks control procedures – Communication requirements – Information from similar projects

Table 3.5: Methods used during the "Develop Project Management Plan" process.

Method	Details
Expert judgment	This refers to all judgments and decisions made while developing the project management plan. Such judgments require the project manager and his/her team to have adequate level of the required relevant skills and experience.
Data gathering approaches	The main data gathering methods used during this process are the brainstorming, checklists, focus groups, interviews, and questionnaires.
Interpersonal and Team skills	This refers to the skills that are utilized in the Develop Project Management Plan process.
Meetings	This refers to the meetings that: – discuss the project management approach; – determine how to execute the project work to accomplish the project objectives; and – how to monitor and control the project. The project kick-off meeting is associated with the end of the project planning phase and the start of the execution phase.

Table 3.6: Outputs of the "Develop Project Management Plan" process.

Output	Details
Project management plan	The project management plan describes how the project will be executed, monitored, controlled, and closed. The main components of project management plans are a set of subsidiary plans and a set of baseline documents. The subsidiary project plans are the scope management plan, requirements management plan, schedule management plan, cost management plan, quality management plan, resource management plan, communications management plan, risk management plan, procurement management plan, stakeholders' engagement plan, change management plan, and configuration management plan. The subsidiary project baseline documents are the scope baseline, schedule baseline, cost baseline, and performance measurement baseline.
Project life cycle description	This document presents a description of the project's entire life cycle.
Development approach	This document illustrates the approach used to produce the project outcomes: product, solution, or service.
Project documents	This includes the project documents of the activity attributes, activity list, assumption log, basis of estimates, change log, cost estimates, cost forecasts, duration estimates, issue log, lessons learned register, milestone list, physical resource assignments, project calendars, project communications, project schedule, project schedule network diagram, project scope statement, project team assignments, quality control measurements, quality metrics, quality report, requirements documentation, requirements traceability matrix, resource breakdown structure, resource calendars, resource requirements, risk register, risk report, schedule forecasts, stakeholder register, team charter, and test and evaluation documents.

3.1.3 The "Direct and Manage Project Work" process

The "Direct and Manage Project Work" process is used to lead the work defined in the project management plan, and then implement the approved CHANGE-REQUESTS to achieve the project objectives.

In this process, the project activities are carried on throughout the project life cycle, and the project deliverables are developed and shipped to the customer.

In reference to Figures 3.6 and 3.7, the main inputs, methods, and outputs of this process are detailed in Tables 3.7, 3.8, and 3.9, respectively.

The "Direct and Manage Project Work" Process		
Inputs	**Methods**	**Outputs**
1 Project management plan (e.g., all components)	1 Expert judgment	1 Deliverables
2 Project documents (e.g., change log, lessons learned register, milestone list, project communications, project schedule, requirements traceability matrix, risk register, and risk report)	2 Project management information system	2 Work performance data
3 Approved change requests	3 Meetings	3 Issue log
4 Enterprise environmental factors		4 Change requests
5 Organizational process assets		5 Project management plan updates (e.g., any components)
		6 Project documents updates (e.g., activity list, assumption log, lessons learned register, requirements documentation, risk register, and stakeholder register)
		7 Organizational process assets updates

Figure 3.6: The "Direct and Manage Project Work" process.

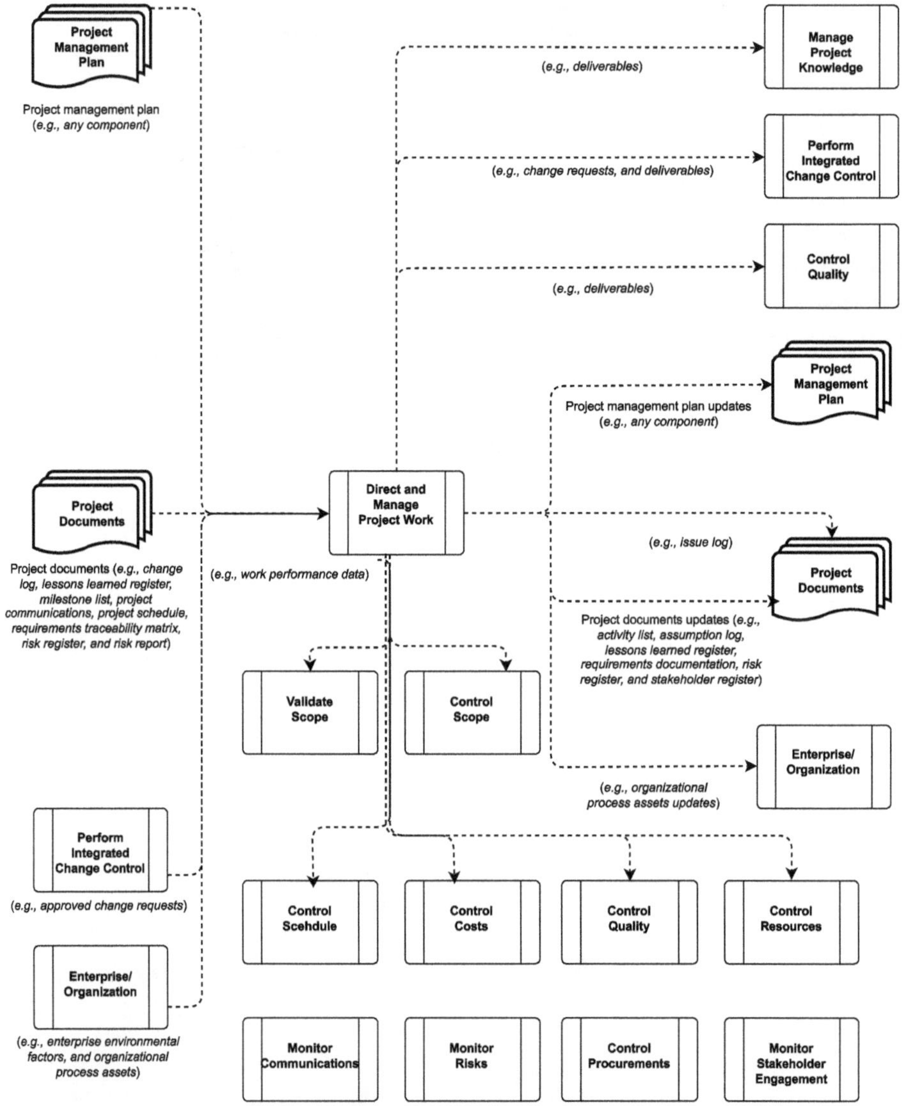

Figure 3.7: The “Direct and Manage Project Work” process data flow diagram.

Table 3.7: Inputs to the "Direct and Manage Project Work" process.

Input	Details
Project management plan	As described earlier in Table 3.6.
Project documents	The main documents used as inputs to this process are the change log, lessons learned register, milestone list, project communications, project schedule, requirements traceability matrix, risk register, and risk report.
Approved CHANGE-REQUESTS	This includes all CHANGE-REQUESTS approved during the perform integrated change control process.
Enterprise environmental factors (EEFs)	The main EEFs used as inputs to this process are the organizational structure, culture, management practices, and infrastructure; and the stakeholders risk thresholds.

Table 3.8: Methods used during the "Direct and Manage Project Work" process.

Method	Details
Expert judgment	This refers to all judgments and decisions made while directing and managing the project work. Such judgments require the project manager and his/her team to have adequate level of the required relevant skills and experience.
Project management information system (PMIS)	The PMIS provides access to information technology tools such as scheduling software tools, work authorization systems, configuration management systems, information collection and distribution systems, electronic project management software tools, and meeting and virtual office support software.
Interpersonal and team skills	This refers to the skills that are utilized during the project management plan development.
Meetings	This refers to the meetings throughout the execution of the project management plan and its subsidiary plans.

Table 3.9: Outputs of the "Direct and Manage Project Work" process.

Output	Details
Deliverables	Deliverables are typically the outcomes of the project as defined in the project management plan.
Project work performance data	Work performance data is the observations and measurements identified while performing the project activities. This includes the project status details such as the authorized/incurred/invoiced/paid costs; project scope, schedule, budget, and quality performance metrics; project status data such as the implemented risk responses, occurred risks, active and closed risks; seller's data such as technical performance; started/in-progress/completed activities; and incurred and committed costs.
Issue log	The issue log is the project document, in which all issues are recorded and tracked.
CHANGE-REQUESTS	This includes all CHANGE-REQUESTS received from the project sponsor and stakeholders.
Project management plan updates	This includes all updates made in the project management plan and/or its subsidiary plans and documents.
Project documents' updates	The main documents that may be updated during this process are the activity list, assumption log, lessons learned register, requirements documentation, risk register, and stakeholder register.

3.1.4 The "Manage Project Knowledge" process

The "Manage Project Knowledge" process is used to utilize the existing knowledge and extend that to achieve the project objectives and to contribute to the organizational learning.

In reference to Figures 3.8 and 3.9, the main inputs, methods, and outputs of this process are detailed in Tables 3.10, 3.11, and 3.12, respectively.

The "Manage Project Knowledge" Process		
Inputs	**Methods**	**Outputs**
1 Project management plan (e.g., all components). 2 Project documents (e.g., lessons learned register, project team assignments, resource breakdown structure, source selection criteria, and stakeholder register). 3 Deliverables. 4 Enterprise environmental factors. 5 Organizational process assets.	1 Expert judgment. 2 Knowledge management. 3 Information management. 4 Interpersonal and team skills (e.g., active listening, facilitation, leadership, networking, and political awareness).	1 Lessons learned register . 2 Project management plan updates (e.g., any component). 3 Organizational process assets updates.

Figure 3.8: The "Manage Project Knowledge" process.

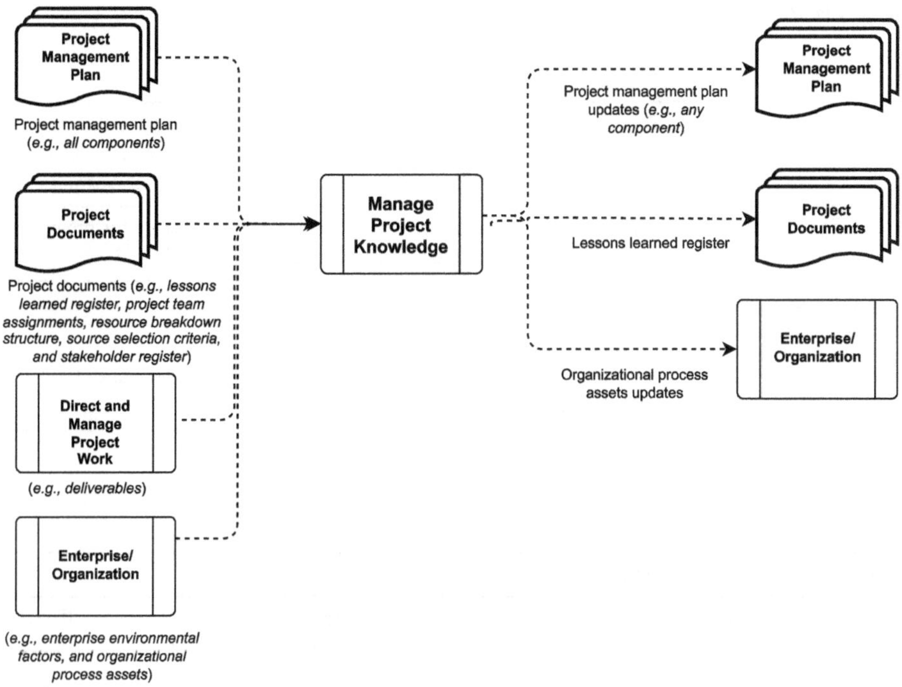

Figure 3.9: The "Manage Project Knowledge" process data flow diagram.

Table 3.10: Inputs to the "Manage Project Knowledge" process.

Input	Details
Project management plan	As described earlier in Table 3.6.
Project documents	The main documents used as inputs to this process are the lessons learned register, project team assignments, resource breakdown structure, and stakeholder register.
Deliverables	As described earlier in Table 3.9.
Enterprise environmental factors (EEFs)	The main EEFs used as inputs to this process are the Organizational Factors, Stakeholders, customer culture, facilities and resources geographic distribution, knowledge experts, and legal and regularity requirements.
Organizational process asset factors (OPAFs)	The main OPAFs used as inputs to this process are the organizational standard policies, processes, procedures, human resources management, and communication requirements.

Table 3.11: Methods used during the "Manage Project Knowledge" process.

Method	Details
Expert judgment	This refers to all judgments and decisions made while managing the project knowledge. Such judgments require that the project manager and his/her team to have adequate level of the required relevant skills and experience.
Knowledge management tools	This includes all necessary technological tools used to manage the project knowledge throughout the project life cycle.
Project management information system (PMIS)	As described earlier in Table 3.8.
Interpersonal and team skills	This includes active listening, facilitation, leadership, networking, and political awareness.

Table 3.12: Outputs of the "Manage Project Knowledge" process.

Output	Details
Lessons learned register	This includes the necessary recommendations and proposed actions. It also includes all challenges, problems, risks, and opportunities.
Project management plan updates	This includes all updates made in the project management plan and/or its subsidiary plans and documents.
Project document updates	This includes all updates made in various project documents.

3.1.5 The "Monitor and Control Project Work" process

The "Monitor and Control Project Work" process is used to track, review, and report the project progress. It allows stakeholders to understand the project status, to recognize any corrective actions taken to resolve any performance issues, and to realize the future project status and its cost and schedule forecasts.

In reference to Figures 3.10 and 3.11, the main inputs, methods, and outputs of this process are detailed in Tables 3.13, 3.14, and 3.15, respectively.

The "Monitor and Control Project Work" Process		
Inputs	**Methods**	**Outputs**
1 Project management plan (e.g., any component)	1 Expert judgment	1 Work performance reports
2 Project documents (e.g., assumption log, basis of estimates, cost forecasts, issue log, lessons learned register, milestone list, quality reports, risk register, risk report, schedule forecasts)	2 Data analysis (e.g., alternatives analysis, cost-benefit analysis, earned value analysis, root cause analysis, trend analysis, and variance analysis)	2 Change requests
3 Work performance information	3 Decision-making	3 Project management plan updates (e.g., any component)
4 Agreements	4 Meetings	4 Project documents updates (e.g., cost forecasts, issue log, lessons learned register, risk register, and schedule forecasts)
5 Enterprise environmental factors		
6 Organizational process assets		

Figure 3.10: The "Monitor and Control Project Work" process.

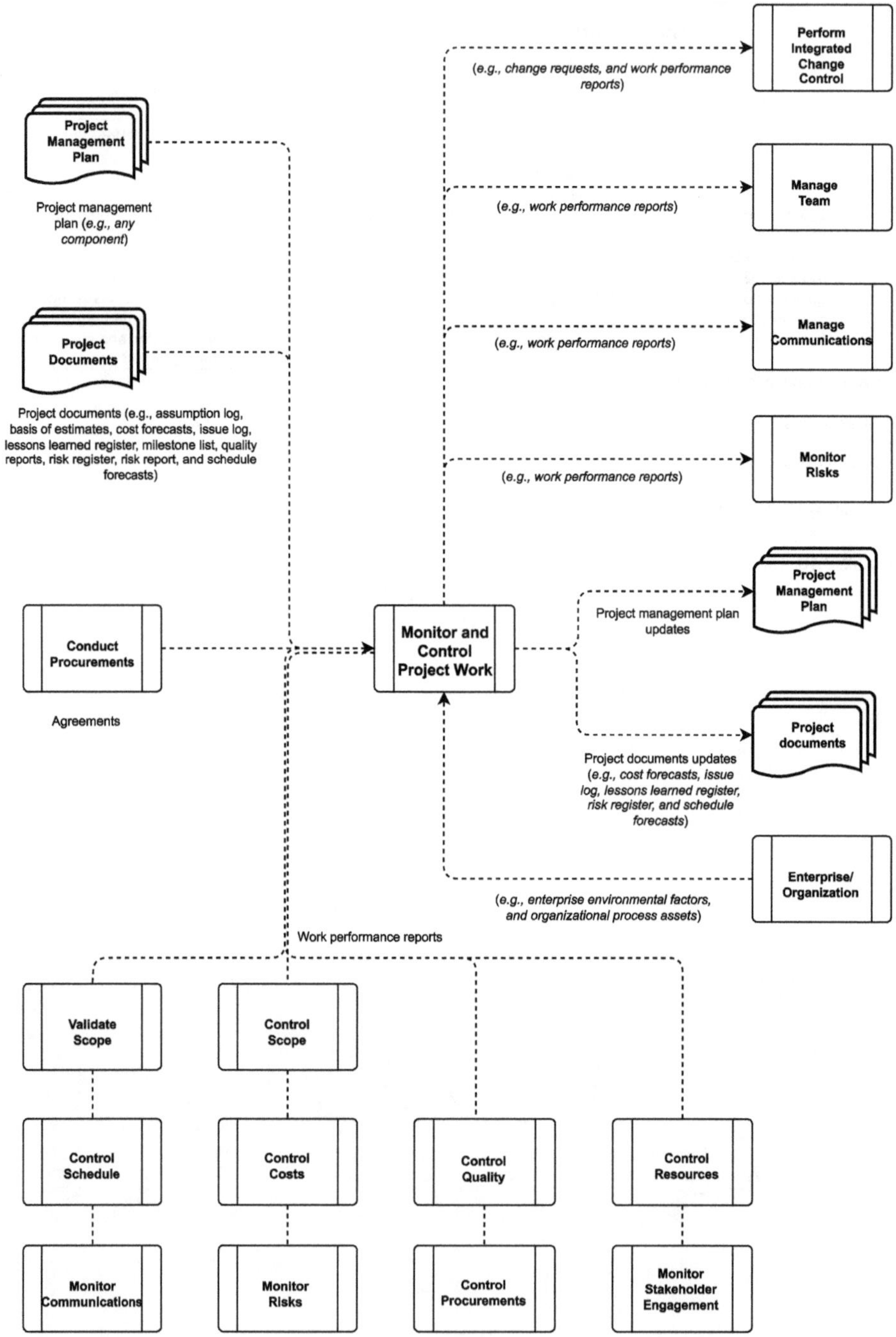

Figure 3.11: The "Monitor and Control Project Work" process data flow diagram.

Table 3.13: Inputs to the "Monitor and Control Project Work" process.

Input	Details
Project management plan	As described earlier in Table 3.6.
Project documents	The main documents used as inputs to this process are the assumption log, basis of estimates, cost forecasts, issue log, lessons learned register, milestone list, quality reports, risk register, risk report, and schedule forecasts.
Project work performance data	As described earlier in Table 3.9.
Agreements	As described earlier in Table 3.1.
Enterprise environmental factors (EEFs)	The main EEFs used as inputs to this process are the scheduling, cost, performance management information systems, infrastructure, stakeholders' expectations, stakeholders' risk thresholds, and governmental and industry standards.
Organizational process asset factors (OPAFs)	The main OPAFs used as inputs to this process are the organizational standard policies, processes, and procedures; financial control procedure; and monitoring and reporting methods.

Table 3.14: Methods used during the "Monitor and Control Project Work" process.

Method	Details
Expert judgment	This refers to all judgments and decisions made while monitoring and controlling the project work. Such judgments require the project manager and his/her team to have adequate level of the required relevant skills and experience.
Data analysis	The main data analysis methods that are used during this process are the alternatives' analysis, cost-benefit analysis, earned-value analysis, root-cause analysis, trend analysis, and variance analysis.
Decision-making	Voting is the main method used by this process in support to decision-making.
Meetings	As described earlier in Table 3.2.

Table 3.15: Outputs of the "Monitor and Control Project Work" process.

Output	Details
Project work performance reports	The project work performance reports include the project status and progress reports. They are circulated to the stakeholders as defined in the communications management plan. They contain earned-value graphs, forecasts, defect histograms, contract performance information, risks, and so forth.
CHANGE-REQUESTS	As described earlier in Table 3.9.

Table 3.15 (continued)

Output	Details
Project management plan updates	This includes all updates made in various components of the project management plan.
Project documents' updates	The main of the documents that may be updated during this process are the cost forecasts, issue log, lessons learned register, risk register, and schedule forecasts.

3.1.6 The "Perform Integrated Change Control" process

The "Perform Integrated Change Control" process is used to manage and review all CHANGE-REQUESTS, and approve and apply some of them to the deliverables, project documents, and project management plan.

In reference to Figures 3.12 and 3.13, the main inputs, methods, and outputs of this process are detailed in Tables 3.16, 3.17, and 3.18, respectively.

The "Perform Integrated Change Control" Process		
Inputs	**Methods**	**Outputs**
1 Project management plan (e.g., change management plan, and configuration management plan, scope baseline, schedule baseline, and cost baseline)	1 Expert judgment	1 Approved change requests
2 Project documents (e.g., basis of estimates, requirements traceability matrix, and risk report)	2 Change control tools	2 Project management plan updates
3 Work performance reports	3 Data analysis (e.g., alternatives analysis, and cost-benefit analysis)	3 Project documents updates (e.g., change log)
4 Change requests	4 Decision-making (e.g., voting, autocratic decision-making, and multicriteria decision analysis)	
5 Enterprise environmental factors	5 Meetings	
6 Organizational process assets		

Figure 3.12: The "Perform Integrated Change Control" process.

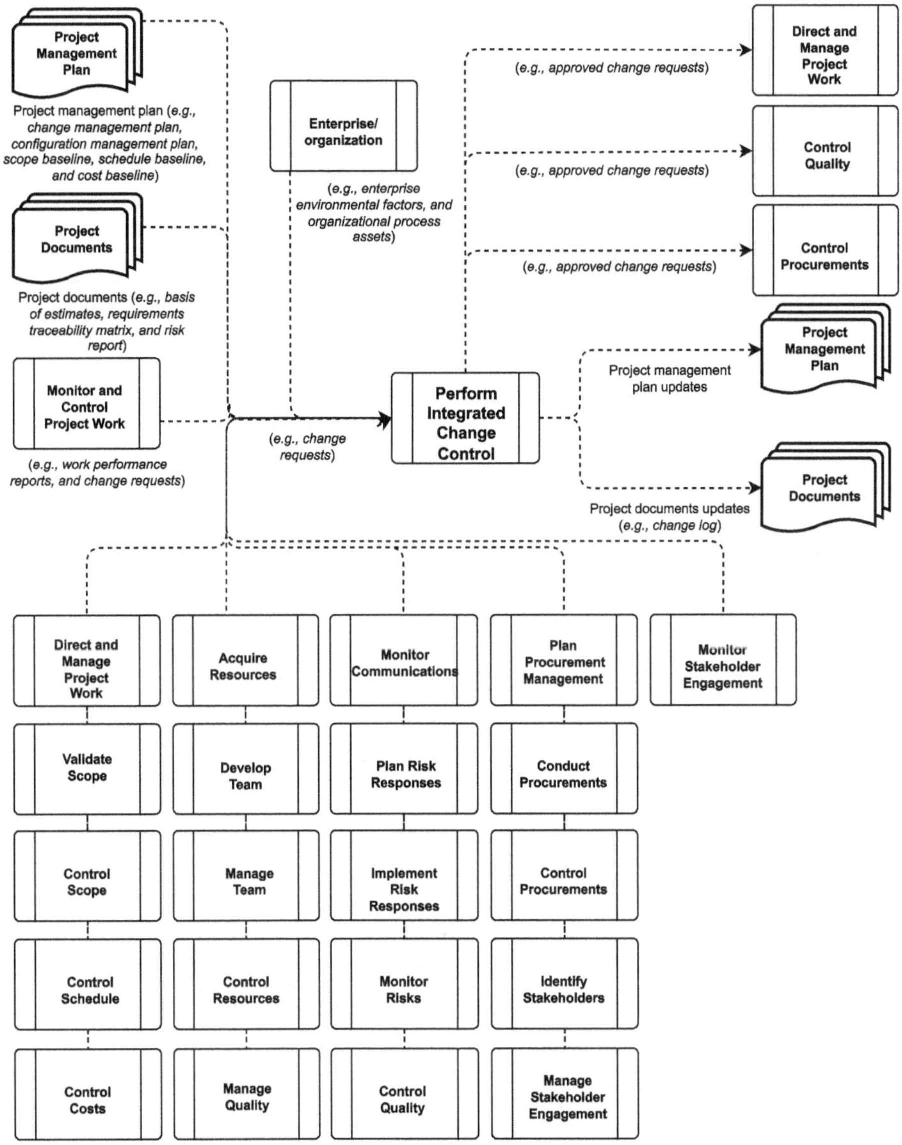

Figure 3.13: The “Perform Integrated Change Control” process data flow diagram.

Table 3.16: Inputs to the "Perform Integrated Change Control" process.

Input	Details
Project management plan	The components of the project management plan used as inputs to this process are the change management plan, configuration management plan, scope baseline, schedule baseline, and cost baseline.
Project documents	The main documents used as inputs to this process are the basis of estimates, requirements traceability matrix, and risk report.
Project work performance data	As described earlier in Table 3.9.
CHANGE-REQUESTS	As described earlier in Table 3.9.
Enterprise environmental factors (EEFs)	The main EEFs used as inputs to this process are the legal restrictions, governmental and industry standards, legal and regularity requirements, and contracting and purchasing constraints.
Organizational process asset factors (OPAFs)	The main OPAFs used as inputs to this process are the change control procedures, approval procedures, and configuration management knowledge base.

Table 3.17: Methods used during the "Perform Integrated Change Control" process.

Method	Details
Expert judgment	This refers to all Judgments made while performing the perform integrated change control process. Such judgments require the project manager and his/her team to have adequate level of the relevant skills and experience.
Data analysis	The main data analysis methods during this process are the alternatives' analysis, and cost-benefit analysis.
Decision-making	The main decision-making methods used during this process are the voting, automatic decision-making, and multi-CRITERIA decision analysis.
Meetings	As described earlier in Table 3.2.
Change control tools	This includes identifying the configuration team; recording and reporting the items' statuses; performing the configuration items, verifications, and audits; and identifying, documenting, deciding on, and tracking changes.

Table 3.18: Outputs of the "Perform Integrated Change Control" process.

Output	Details
Project work performance reports	As described earlier in Table 3.15.
Approved CHANGE-REQUESTS	As described earlier in Table 3.7.
Project management plan updates	This includes all updates made in various components of the project management plan.
Project documents' updates	This includes all updates made in various project documents.

3.1.7 The "Close Project or Phase" process

The "Close Project or Phase" process is used to:
- finalize all of the project activities such as ensuring that all documents and deliverables are up-to-date and that all issues are resolved;
- ensure that all costs are charged to the project, confirming the delivery and the formal acceptance of deliverables by the customer;
- measure the stakeholders' satisfaction;
- reassign human resources, dealing with excess project materials, and reallocating the project facilities and equipment;
- prepare the project reports in accordance with the organizational policies;
- manage knowledge sharing and transfer;
- identify lessons learned;
- transfer the project products, solutions, and/or services to the next phase or to production/operations;
- collect suggestions for improving the organizational policies and procedures;
- archive the project information for future use by the organization; and
- close the project accounts.

In reference to Figures 3.14 and 3.15, the inputs, methods, and outputs of this process are detailed in Tables 3.19, 3.20, and 3.21, respectively.

The "Close Project or Phase" Process		
Inputs	**Methods**	**Outputs**
1 Project charter 2 Project management plan (e.g., All components) 3 Project documents (e.g., Assumption log, basis of estimates, change log, issue log, lessons learned register, milestone list, project communications, quality control measurements, quality reports, requirements documentation, risk register, and risk report) 4 Accepted deliverables 5 Business documents (e.g., Business case, and benefits management plan) 6 Agreements 7 Procurement documentation 8 Organizational process assets	1 Expert judgment 2 Data analysis (e.g., Document analysis, regression analysis, trend analysis, and variance analysis) 3 Meetings	1 Project documents updates (e.g., Lessons learned register) 2 Final product, service, or result transition 3 Final report 4 Organizational process assets updates

Figure 3.14: The "Close Project or Phase" process.

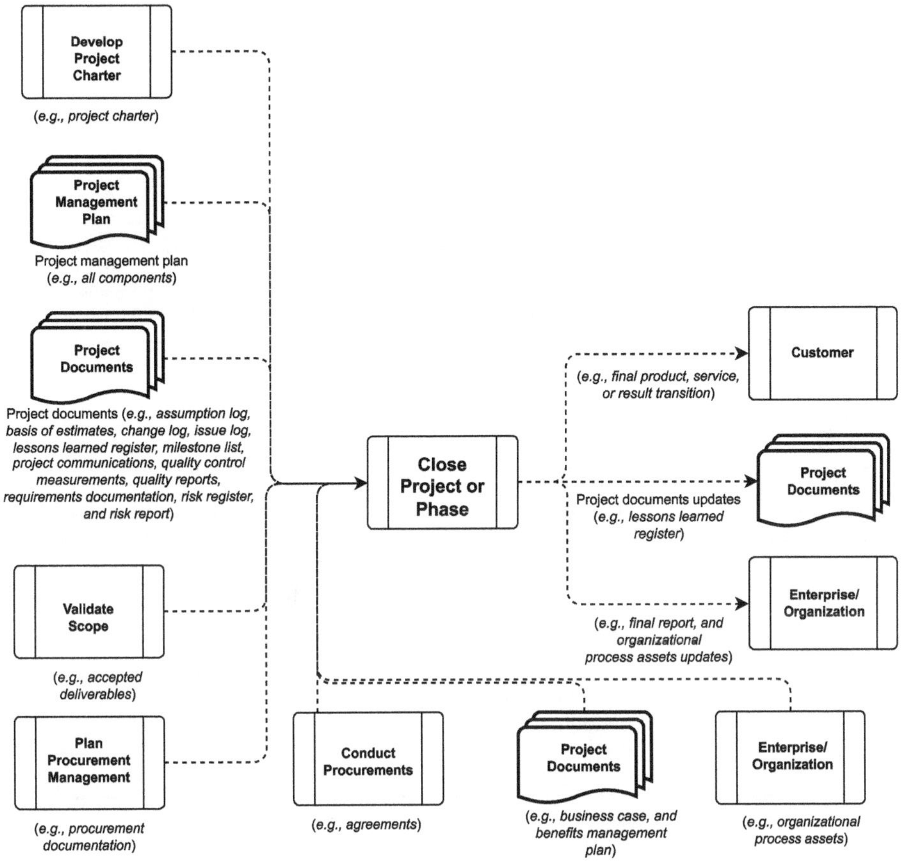

Figure 3.15: The "Close Project or Phase" process data flow diagram.

Table 3.19: Inputs to the "Close Project or Phase" process.

Input	Details
Project charter	As described earlier in Table 3.3.
Project management plan	As described earlier in Table 3.6.
Project documents	The main documents that are used as inputs to this process are the assumption log, basis of estimates, change log, issue log, lessons learned register, milestone list, project communications, quality control measurements, quality report, requirements documentation, risk register, and risk report.
Accepted deliverables	As described earlier in Table 3.9.

Table 3.19 (continued)

Input	Details
Business documents	This includes the business case and the benefits management plan documents.
Agreements	As described earlier in Table 3.1.
Procurement documentation	This includes all documentation of the procurement processes.
Organizational process asset factors (OPAFs)	This includes the project guidelines and/or requirements, and the configuration management knowledge base.

Table 3.20: Methods used during the "Close Project or Phase" process.

Method	Details
Expert judgment	This refers to all judgments made while closing on the project or one of its phases. Such judgments require the project manager and his/her team to have adequate level of the required relevant skills and experience.
Data analysis	The main data analysis methods used during this process are the documents analysis, regression analysis, trend analysis, and variance analysis.
Meetings	As described earlier in Table 3.2.

Table 3.21: Outputs of the "Close Project or Phase" process.

Output	Details
Final product/solution/service transition	At the end of this process, the final product, solution, or service will be handed over to the customer.
Project documents' updates	Most of the project documents may be updated and derived into final versions during this process, particularly the lessons learned register.
Final report	The project final report presents an executive summary of the project performance, including the project description, scope, objectives, scope evaluation CRITERIA, quality objectives, product quality evaluation CRITERIA, cost objectives, validation information, schedule objective, and so forth.
Organizational process asset factor (OPAF) updates	This includes the project documents, operational and support documents, project or phase closure documents, and lessons learned repository.

3.2 Project scope management

Project scope management includes the processes that are necessary to ensure that the project includes only the predefined work that is required to complete the project SUCCESSFULLY.

The term "**scope**" refers to:

- The **project scope** that includes the work that is carried-on throughout the project life cycle to deliver a product, service, and/or solution while at the same time satisfying a predefined set of specifications in terms of its functionalities.
- The **product scope** that includes the functionalities that should be satisfied by the delivered product, service, and/or solution.

In reference to Figure 3.16, the project scope management processes are the Plan Scope Management, Collect Requirements, Define Scope, Create WBS, Validate Scope, and Control Scope Processes.

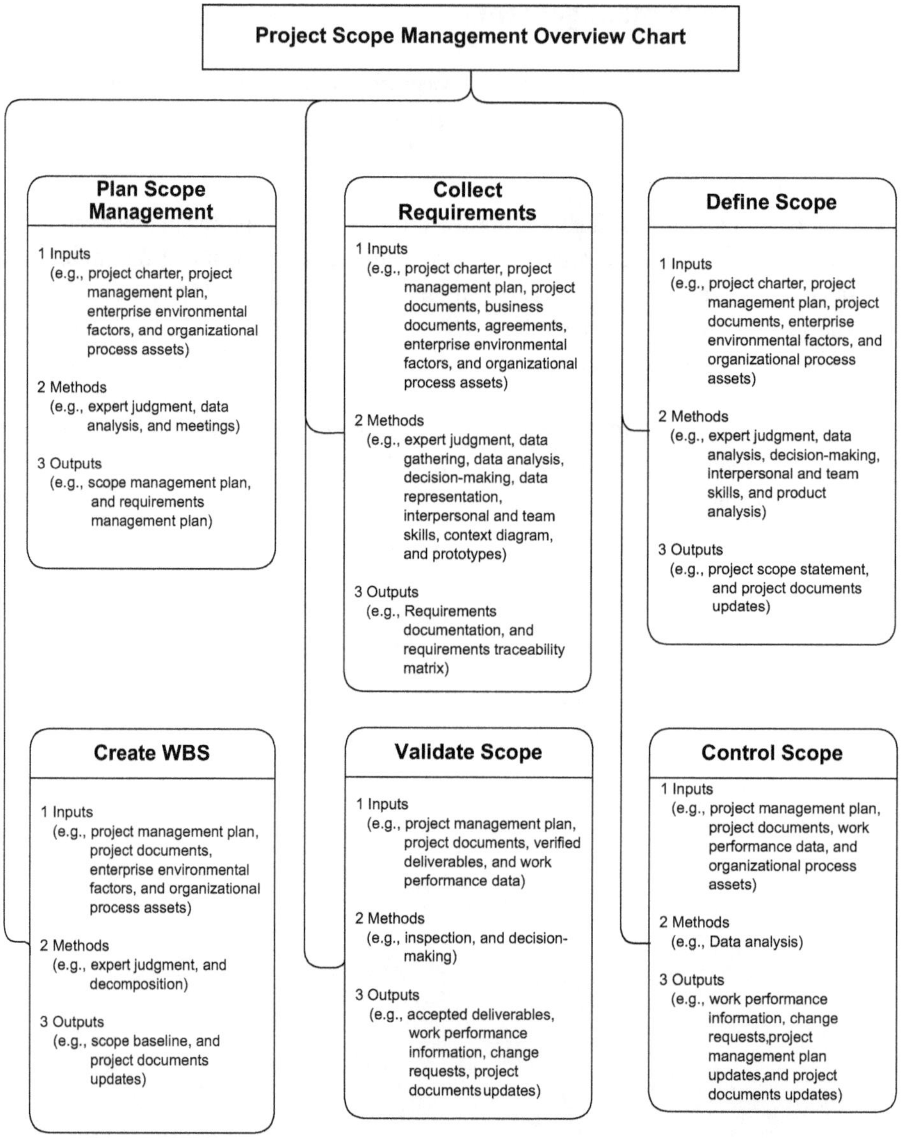

Figure 3.16: Project scope management overview chart.

3.2.1 The "Plan Scope Management" process

The "Plan Scope Management" process is used to create a scope management plan that shows how to define, validate, and control the project and the product scopes. It shows how to manage the project and the product scopes throughout the project life cycle.

In reference to Figures 3.17 and 3.18, the main inputs, methods, and outputs of this process are detailed in Tables 3.22, 3.23, and 3.24, respectively.

The "Plan Scope Management" Process		
Inputs	**Methods**	**Outputs**
1 Project charter	1 Expert judgment	1 Scope management plan
2 Project management plan (e.g., quality management plan, project life cycle description, and development approach)	2 Data analysis (e.g., Alternatives analysis)	2 Requirements management plan
	3 Meetings	
3 Enterprise environmental factors		
4 Organizational process assets		

Figure 3.17: The "Plan Scope Management" process.

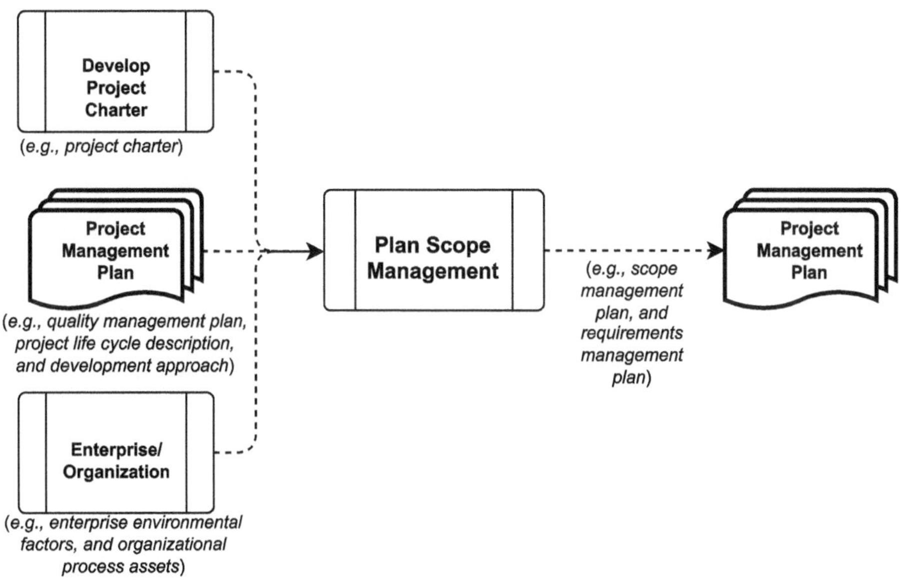

Figure 3.18: The "Plan Scope Management" process data flow diagram.

Table 3.22: Inputs to the "Plan Scope Management" process.

Input	Details
Project charter	As described earlier in Table 3.3.
Project management plan	The main components of the project management plan used as inputs to this process are the quality management plan, project life-cycle description, and development approach.
Organizational process asset factors (OPAFs)	The main OPAFs used as inputs to this process are the organizational policies and procedures, and lessons learned repositories.
Enterprise environmental factors (EEFs)	The main EEFs used as inputs to this process are the organizational culture, infrastructure, human resources management, and marketplace conditions.

Table 3.23: Methods used during the "Plan Scope Management" process.

Method	Details
Expert judgment	This refers to all judgments and decisions made while planning for the scope management. Such judgments require the project manager and his/her team to have adequate level of the required relevant skills and experience.
Data analysis	The main data analysis method used during this process is the alternatives' analysis.
Meetings	As described earlier in Table 3.2.

Table 3.24: Outputs of the "Plan Scope Management" process.

Output	Details
Scope management plan	The scope management plan is an integral component of the project management plan that describes how to define, monitor, control, and validate the project and product scopes. It includes processes for preparing the project scope statement, creating the work breakdown structure, approving and maintaining the scope baseline, and obtaining/accepting the deliverables.
Requirements management plan	The requirements management plan is an integral component of the project management plan that describes how to analyze, document, and manage the project and product requirements. It includes processes for planning, tracking, and reporting activities; analyzing their impacts; tracing, tracking, and reporting them; and for prioritizing the requirements.

3.2.2 The "Collect Requirements" process

The "Collect Requirements" process is used to extract and organize the requirements from the formal documents such as request for proposals.

In reference to Figures 3.19 and 3.20, the main inputs, methods, and outputs of this process are detailed in Tables 3.25, 3.26, and 3.27, respectively.

The "Collect Requirements" Process		
Inputs	**Methods**	**Outputs**
1 Project charter	1 Expert judgment	1 Requirements documentation
2 Project management plan (e.g., scope management plan, requirements management plan, and stakeholder engagement plan)	2 Data gathering (e.g., brainstorming, interviews, focus groups, questionnaires and surveys, and benchmarking)	2 Requirements traceability matrix
3 Project documents (e.g., assumption lo, lessons learned register, and stakeholder register)	3 Data analysis (e.g., document analysis)	
4 Business documents (e.g., business case)	4 Decision-making (e.g., voting, and multicriteria decision analysis)	
5 Agreements	5 Data representation (e.g., affinity diagrams, mind mapping)	
6 Enterprise environmental factors	6 Interpersonal and team skills (e.g., nominal group technique, observation/conversation, and facilitation)	
7 Organizational process assets	7 Context diagram	

Figure 3.19: The "Collect Requirements" process.

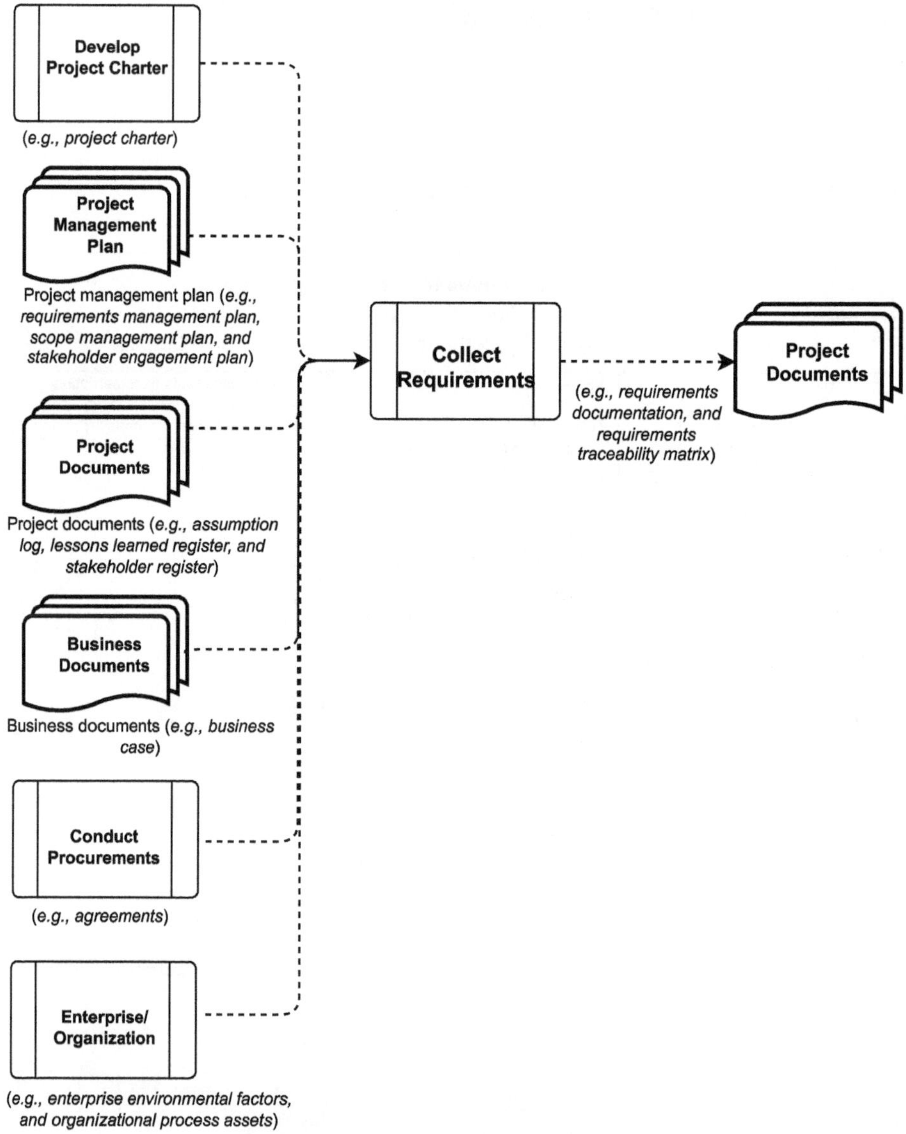

Figure 3.20: The “Collect Requirements” process data flow diagram.

Table 3.25: Inputs to the "Collect Requirements" process.

Input	Details
Project charter	As described earlier in Table 3.3.
Project management plan	The main components of project management plan used as inputs to this process are the scope management plan, requirements management plan, and stakeholders' engagement plan.
Project documents	The main documents used as inputs to this process are the assumption log, lessons learned register, and stakeholder register.
Business documents	As described earlier in Table 3.19.
Agreements	As described earlier in Table 3.1.
Organizational process asset factors (OPAFs)	The main OPAFs used as inputs to this process are the organizational policies and procedures, and lessons learned repositories.
Enterprise environmental factors (EEFs)	The main EEFs used as inputs to this process are the organizational culture, infrastructure, human resource management, and marketplace conditions.

Table 3.26: Methods used during the "Collect Requirements" process.

Method	Details
Expert judgment	This refers to all judgments and decisions made while collecting the project requirements. Such judgments require the project manager and his/her team to have adequate level of the required relevant skills and experience.
Data gathering	The main data gathering methods used during this process are brainstorming, interviews, focus groups, questionnaires and surveys, and benchmarking.
Data analysis	The main documents that are analyzed during this process are the agreements, business plans, business rules repositories, interface document, issue log, policies, procedures, and use cases.
Decision-making	The main decision-making methods used during this process are voting, automatic decision-making, and multi-CRITERIA decision analysis.
Data representation	The main data representation methods used during this process are the affinity diagrams (e.g., *classifying large number of ideas into reviewable and analyzable groups*), and mind mapping (e.g., *consolidating ideas from brainstorming sessions into a single map to determine their similarities and differences and generate new ideas accordingly*).

Table 3.26 (continued)

Method	Details
Interpersonal and team skills	The main interpersonal and team skills' techniques used during this process are the nominal group techniques (e.g., giving useful ideas higher priorities and pushing them for further brainstorming), observation and conversation, and facilitation.
Context diagrams	Context diagrams depict the product scope in terms of its business systems (e.g., *process* and *computer system*).
Prototypes	Prototypes are used to demonstrate parts of the targeted system at early stages of the project.

Table 3.27: Outputs of the "Collect Requirements" process.

Output	Details
Requirements documentation	The requirements documentation includes the following: – Business requirements. They describe the business issues or opportunities, and why the project has been carried on. – Stakeholder requirements. They describe the stakeholders' needs. – Solution requirements. They describe the targeted solution's functional and nonfunctional requirements. – Transition and readiness requirements. They describe the data conversion and training requirements. – Project requirements. They describe the actions and processes that should be satisfied by the project. – Quality requirements. They describe the CRITERIA for validating the SUCCESSFUL completion of a project.
Requirements traceability matrix	The requirements traceability matrix includes the business needs, opportunities, project objectives, scope, deliverables, solution design, development method, testing strategy, and testing scenario.

3.2.3 The "Define Scope" process

The "Define Scope" process is used to develop a detailed statement describing the project and the product scopes. A project scope of work is intuitively developed using the project deliverables and constraints that are defined during the project initiation phase, then it is detailed during the project planning phase.

In reference to Figures 3.21 and 3.22, the main inputs, methods, and outputs of this process are detailed in Tables 3.28, 3.29, and 3.30, respectively.

The "Define Scope" process		
Inputs	**Methods**	**Outputs**
1 Project charter	1 Expert judgment	1 Project scope statement
2 Project management plan (e.g., Scope management plan)	2 Data analysis (e.g., Alternatives analysis)	2 Project documents updates (e.g., Assumption log, requirements documentation, requirements traceability matrix, stakeholder register)
3 Project documents (e.g., Assumption log, requirements documentation, and risk register)	3 Decision-making (e.g., Multicriteria decision analysis)	
4 Enterprise environmental factors	4 Interpersonal and team skills (e.g., Facilitation)	
5 Organizational process assets	5 Product analysis	

Figure 3.21: The "Define Scope" process.

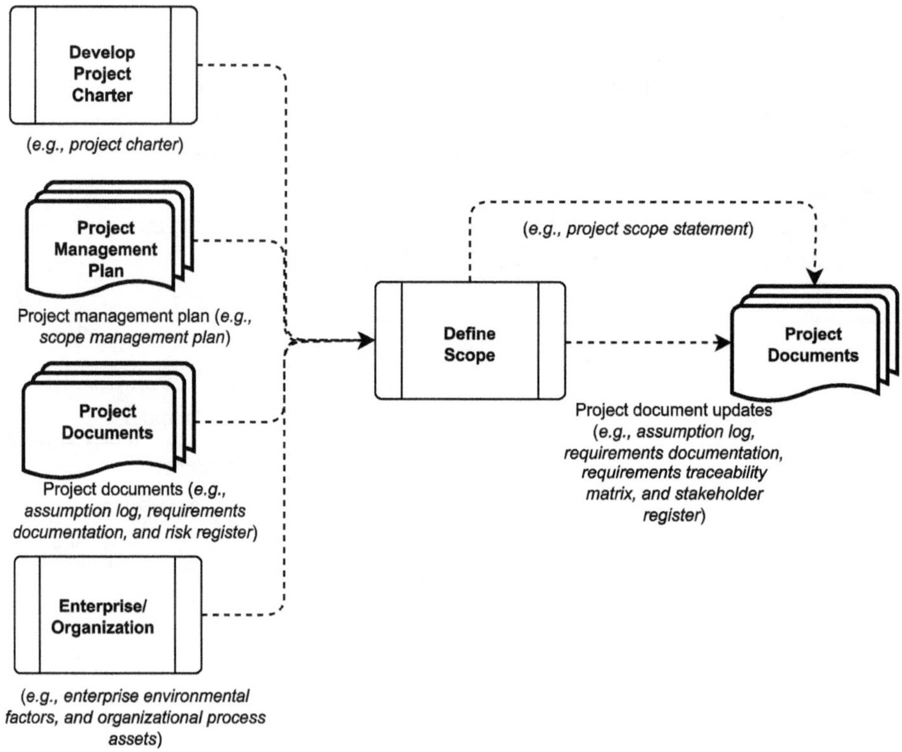

Figure 3.22: The "Define Scope" process data flow diagram.

Table 3.28: Inputs to the "Define Scope" process.

Input	Details
Project charter	As described earlier in Table 3.3.
Project management plan	As described earlier in Table 3.6.
Project documents	The main documents that are used as inputs to this process are the assumption log, requirements documentation, and risk register.
Business documents	As described earlier in Table 3.19.
Agreements	As described earlier in Table 3.1.
Organizational process asset factors (OPAFs)	The main OPAFs used as inputs to this process are the relevant organizational policies, procedures, and templates; project files; and lessons learned from previous projects.
Enterprise environmental factors (EEFs)	The main EEFs used as inputs to this process are the organizational culture, infrastructure, human resources management, and marketplace conditions.

Table 3.29: Methods used during the "Define Scope" process.

Method	Details
Expert judgment	This refers to all judgments and decisions made while defining the project and product scopes. Such judgments require the project manager and his/her team to have adequate level of the required relevant experience.
Data analysis	The main data analysis method used during this process is the alternatives' analysis.
Decision-making	The main decision-making method used during this process is the multi-CRITERIA decision analysis.
Interpersonal and team skills	The main interpersonal and team skill used during this process is the facilitation.
Product analysis	The main product analysis methods used as inputs to this process are the product breakdown, requirements analysis, systems analysis and engineering, and value analysis and engineering.

Table 3.30: Outputs of the "Define Scope" process.

<table>
<tr><th>Output</th><th>Details</th></tr>
<tr><td>Project scope statement</td><td>This statement presents a detailed description of the project and the product scopes, along with the project deliverables, constraints, and ACCEPTANCE-CRITERIA.
The following is a high-level comparison between the components of the project charter and the components of the project scope statement:
<table>
<tr><th>Project charter elements</th><th>Project scope statement elements</th></tr>
<tr><td>Project purpose</td><td rowspan="2">Project scope description</td></tr>
<tr><td>SMART project objectives and their related SUCCESS-CRITERIA</td></tr>
<tr><td>High-level requirements</td><td rowspan="2">Project deliverables</td></tr>
<tr><td>High-level project description, boundaries, and key deliverables</td></tr>
<tr><td>Overall project risk</td><td rowspan="2">ACCEPTANCE-CRITERIA</td></tr>
<tr><td>Summary of the milestone schedule</td></tr>
<tr><td>Preapproved financial resources</td><td rowspan="2">Project exclusions</td></tr>
<tr><td>Key stakeholder list</td></tr>
<tr><td>Project approval requirements</td><td></td></tr>
<tr><td>Project exit CRITERIA</td><td></td></tr>
<tr><td>Assigned project manager, responsibility, and authority level</td><td></td></tr>
<tr><td>Name and authority of the project sponsor</td><td></td></tr>
</table></td></tr>
<tr><td>Project documents updates</td><td>The main documents that may be updated during this process are the assumption log, requirements documentation, requirements traceability matrix, and stakeholder register.</td></tr>
</table>

3.2.4 The "Create WBS" process

The "Create WBS" process subdivides the project work into small components that are easy to manage. The WBS is a hierarchical breakdown and decomposition of the work to be carried on by the project team to accomplish the project objectives and to create its deliverables.

In reference to Figures 3.23 and 3.24, the main inputs, methods, and outputs of this process are detailed in Tables 3.31, 3.32, and 3.33, respectively.

The "Create WBS" Process		
Inputs	**Methods**	**Outputs**
1 Project management plan (e.g., Scope management plan)	1 Expert judgment	1 Scope baseline
2 Project documents (e.g., Project scope statement, and requirements documentation)	2 Decomposition	2 Project documents updates (e.g., Assumption log, requirements documentation)
3 Enterprise environmental factors		
4 Organizational process assets		

Figure 3.23: The "Create WBS" process.

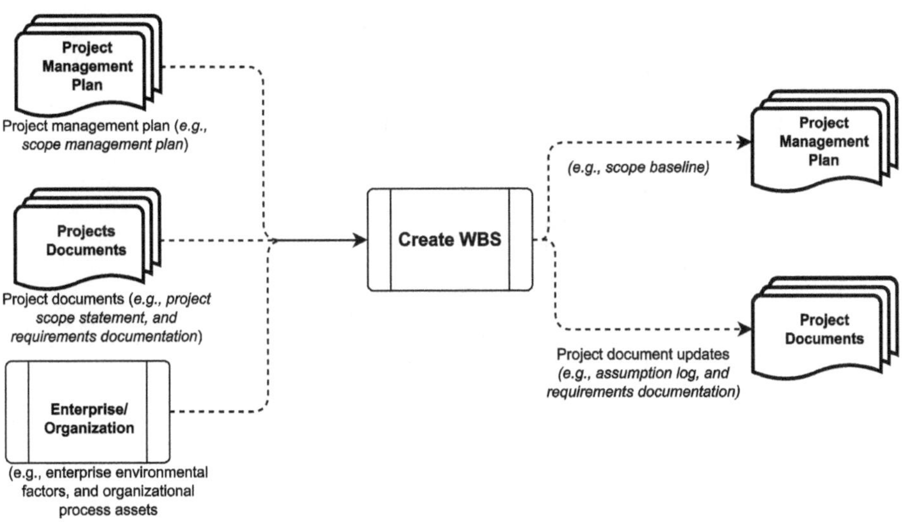

Figure 3.24: The "Create WBS" process data flow diagram.

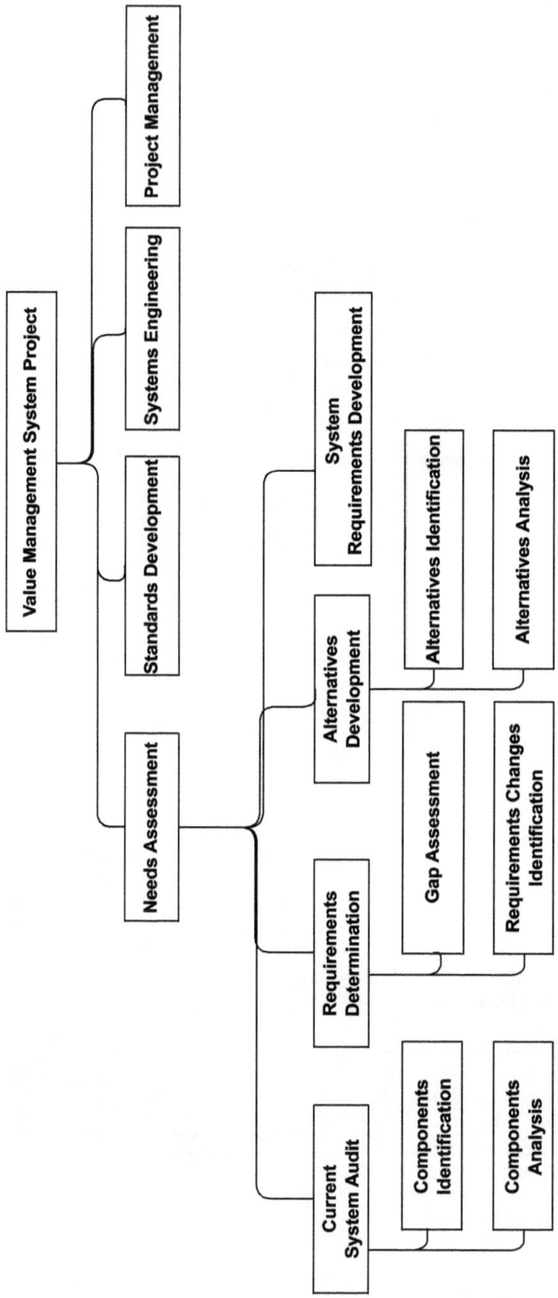

Figure 3.25: A sample WBS decomposition.

Table 3.31: Inputs to the "Create WBS" process.

Input	Details
Project management plan	As described earlier in Table 3.6.
Project documents	The main documents used as inputs to this process are the project scope statement, and requirements documentation.
Organizational process asset factors (OPAFs)	The main OPAFs used as inputs to this process are the organizational policies, procedures, and templates; project files; and lessons learned from previous projects.
Enterprise environmental factors (EEFs)	The main EEFs used as inputs to this process are the industrial work breakdown structure standards.

Table 3.32: Methods used during the "Create WBS" process.

Method	Details
Expert judgment	This refers to all judgments and decisions made while creating the project work breakdown structure. Such judgments require the project manager and his/her team to have adequate level of the required relevant skills and experience.
Decomposition	This technique is used to divide the project work into small and manageable components. Figure 3.25 shows a sample decomposition.

Table 3.33: Outputs of the "Create WBS" process.

Output	Details
Scope baseline	This includes the approved scope statement; WBS; and WBS dictionary including an identifier, work description, constraints, schedule milestones, cost estimates, and so forth.
Project documents' updates	The main documents that may be updated during this process are the assumption log and requirements documentation.

3.2.5 The "Validate Scope" process

The "Validate Scope" process formalizes the project deliverables ACCEPTANCE-CRITERIA.

In reference to Figures 3.26 and 3.27, the main inputs, methods, and outputs of this process are detailed in Tables 3.34, 3.35, and 3.36, respectively.

The "Validate Scope" Process		
Inputs	**Methods**	**Outputs**
1 Project management plan (e.g., scope management plan, requirements management plan, and scope baseline) 2 Project documents (e.g., lessons learned register, quality reports, requirements documentation, and requirements traceability matrix) 3 Verified deliverables 4 Work performance data	1 Inspection 2 Decision-making (e.g., Voting)	1 Accepted deliverables 2 Work performance information 3 Change requests 4 Project document updates (e.g., lessons learned register, requirements documentation, and requirements traceability matrix)

Figure 3.26: The "Validate Scope" process.

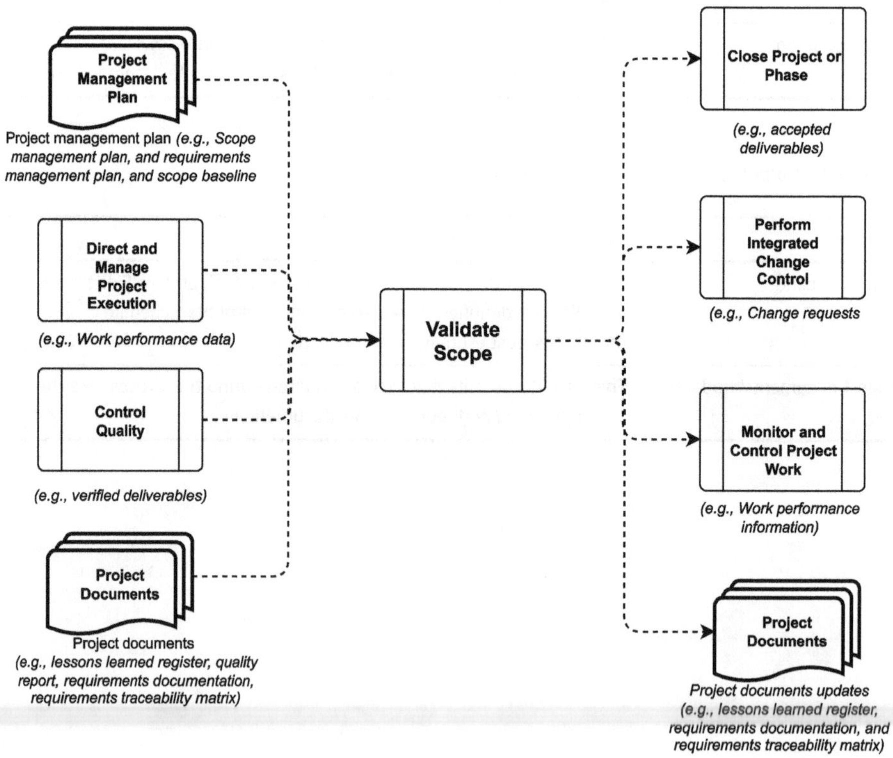

Figure 3.27: The "Validate Scope" process data flow diagram.

Table 3.34: Inputs to the "Validate Scope" process.

Input	Details
Project management plan	The main components of the project management plan used as inputs to this process are the scope management plan, requirements management plan, and scope baseline.
Project documents	The main documents used as inputs to this process are the lessons learned register, quality reports, requirements documentation, and requirements traceability matrix.
Verified deliverables	This includes all deliverables that will be handed over to the customer, but after they are verified by the project team and approved by the product owner.
Work performance data	As described earlier in Table 3.9.

Table 3.35: Methods used during the "Validate Scope" process.

Method	Details
Inspection	This includes measuring, examining, and validating the project work and deliverables in order to satisfy the requirements and pass the product ACCEPTANCE-CRITERIA.
Decision-making	The main decision-making method used during this process is voting.

Table 3.36: Outputs of the "Validate Scope" process.

Output	Details
Accepted deliverables	This includes all deliverables that are accepted by the project sponsor.
Project documents' updates	The main documents that may be updated during this process are the lessons learned register, requirements documentation, and requirements traceability matrix.
Work performance data	As described earlier in Table 3.9.
CHANGE-REQUESTS	As described earlier in Table 3.9.

3.2.6 The "Control Scope" process

The "Control Scope" process is used to monitor the project status and the project and product scopes, and manage changes to the scope baseline. The "Control Scope" process ensures that all CHANGE-REQUESTS and their corrective actions are processed through the "Perform Integrated Change Control" process.

In reference to Figures 3.28 and 3.29, the main inputs, methods, and outputs of this process are detailed in Tables 3.37, 3.38, and 3.39, respectively.

The "Control Scope" Process		
Inputs	**Methods**	**Outputs**
1 Project management plan (e.g., scope management plan, requirements management plan, change management plan, configuration management plan, scope baseline, and performance measurement baseline)	1 data analysis (e.g., variance analysis, and trend analysis)	1 Work performance information
2 Project documents (e.g. Lessons learned register, requirements documentation, and requirements traceability matrix)		2 Change requests
3 Work performance data		3 Project management plan updates (e.g., scope management plan, scope baseline, schedule basel, cost baseline, and performance measurement baseline)
4 Organizational process assets		4 Project documents updates (e.g., lessons learned register, requirements documentation, and requirements traceability matrix)

Figure 3.28: The "Control Scope" process.

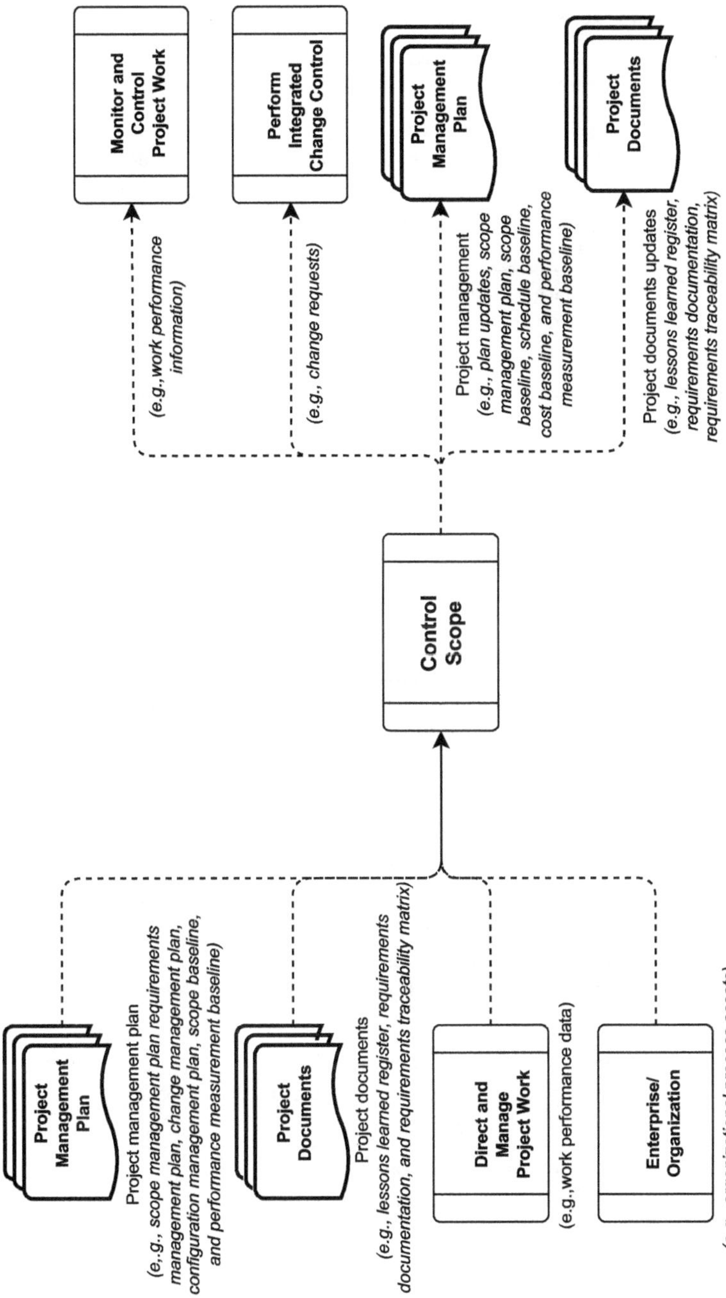

Figure 3.29: The "Control Scope" process data flow diagram.

Table 3.37: Inputs to the "Control Scope" process.

Input	Details
Project management plan	The main components of the project management plan used as inputs to this process are the scope management plan, requirements management plan, change management plan, configuration management plan, scope baseline, and performance measurement baseline.
Project Documents	The main documents used as inputs to this process are the lessons learned register, requirements documentation, and requirements traceability matrix.
Organizational process asset factors (OPAFs)	The main OPAFs used as inputs to this process are the existing formal and informal scope; relevant policies, procedures, and guidelines; and monitoring and reporting templates and methods.
Work performance data	As described earlier in Table 3.9.

Table 3.38: Methods used during the "Control Scope" process.

Method	Details
Data analysis	The main data analysis methods used during this process are the variance analysis, and trend analysis.

Table 3.39: Outputs of the "Control Scope" process.

Output	Details
Project management plan updates	The main components of the project management plan that may be updated during this process are the scope management plan, scope baseline, schedule baseline, cost baseline, and performance measurement baseline.
Project documents' updates	The main documents that may be updated during this process are the lessons learned register, requirements documentation, and requirements traceability matrix.
Work performance data	As described earlier in Table 3.9.
CHANGE-REQUESTS	As described earlier in Table 3.9.

3.3 Project schedule management

Project scope management includes the processes necessary to ensure that the project is completed on time per its predefined schedule.

In reference to Figure 3.30, the project schedule management has six processes: Plan Schedule Management, Define Activities, Sequence Activities, Estimate Activity Durations, Develop Schedule, and Control Schedule.

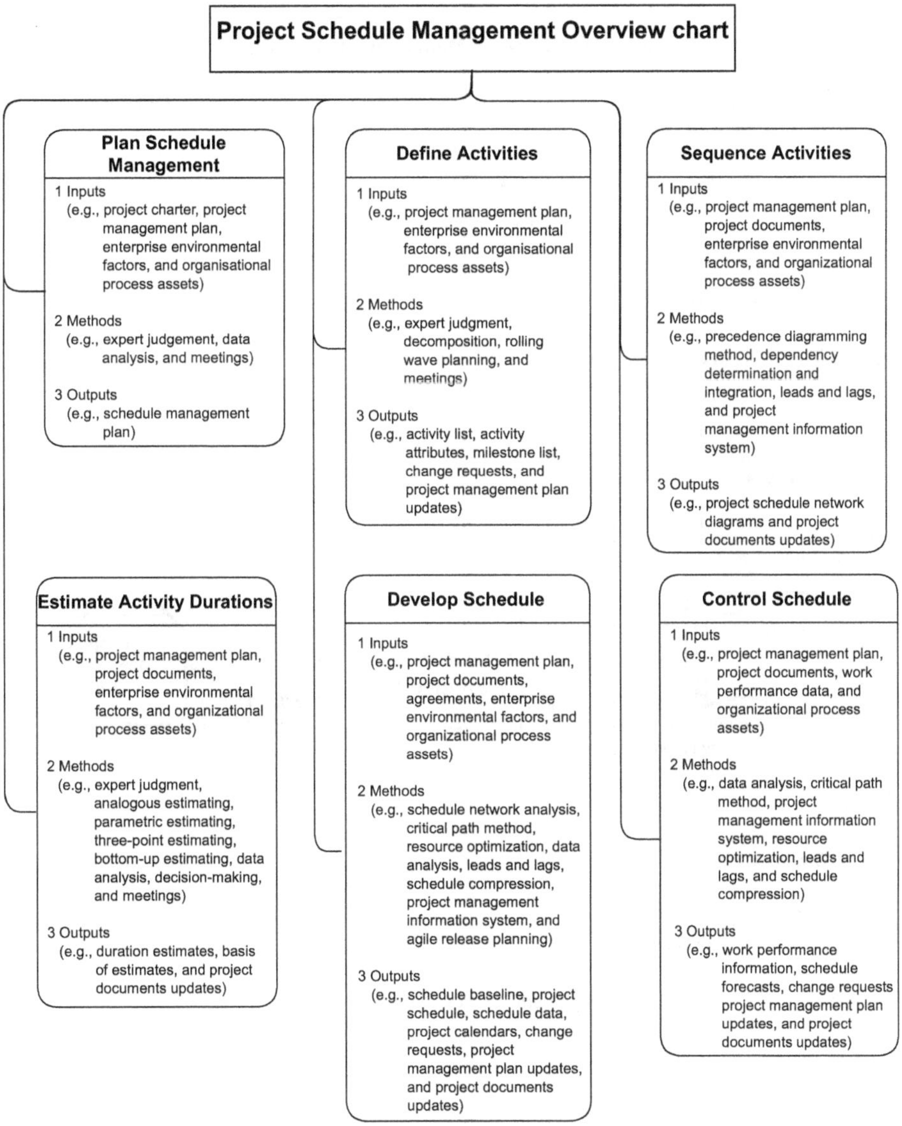

Figure 3.30: Project schedule management overview chart.

3.3.1 The "Plan Schedule Management" process

The "Plan Schedule Management" process establishes the policies, procedures, and documents that are required to plan, develop, manage, execute, and control the project schedule.

In reference to Figures 3.31 and 3.32, the main inputs, methods, and outputs of this process are detailed in Tables 3.40, 3.41, and 3.42, respectively.

The "Plan Schedule Management" Process		
Inputs	**Methods**	**Outputs**
1 Project charter	1 Expert judgment	1 Schedule management plan
2 Project management plan (e.g., scope management plan, and development approach)	2 Data analysis	
	3 Meetings	
3 Enterprise environmental factors		
4 Organizational process assets		

Figure 3.31: The "Plan Schedule Management" process.

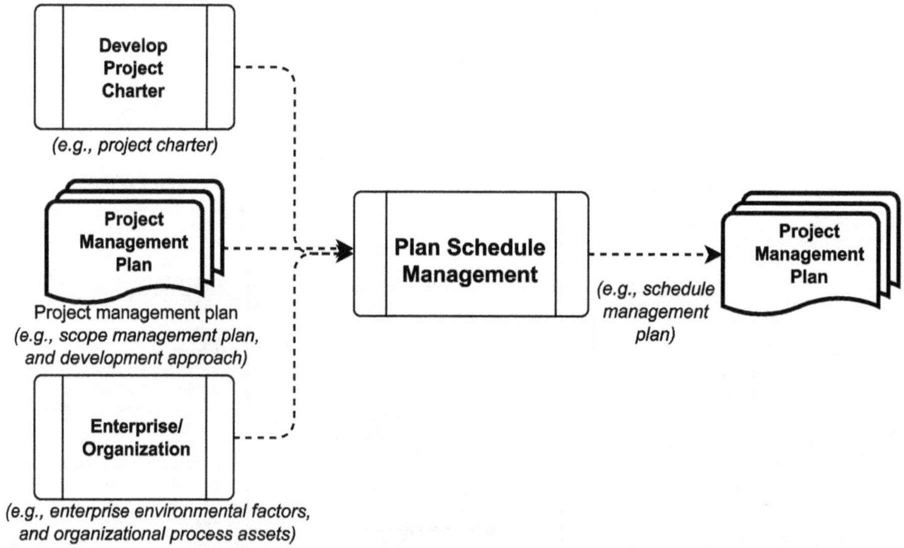

Figure 3.32: The "Plan Schedule Management" process data flow diagram.

Table 3.40: Inputs to the "Plan Schedule Management" process.

Input	Details
Project charter	As described earlier in Table 3.3.
Project management plan	The main components of the project management plan used as inputs to this process are the scope management plan and development approach.
Enterprise environmental factors (EEFs)	The main EEFs used as inputs to this process are the organizational culture and structure, availability of scheduling tools, availability of commercial databases, and availability of skilled human resources and physical resources.
Organizational process asset factors (OPAFs)	The main OPAFs used as inputs to this process are the historical information and lessons learned repositories; availability of scheduling/management/control policies, procedures and guidelines; availability of relevant templates and forms; and availability of monitoring/control/reporting tools.

Table 3.41: Methods used during the "Plan Schedule Management" process.

Method	Details
Expert judgment	This refers to all judgments and decisions made while planning for the project schedule management. Such judgments require the project manager and his/her team to have adequate level of the required relevant skills and experience.
Data analysis	The main data analysis method used during this process is the alternatives' analysis.
Meetings	This refers to all meetings that are concerned with the planning for the schedule management.

Table 3.42: Outputs of the "Plan Schedule Management" process.

Output	Details
Schedule management plan	The schedule management plan is an integral component of the project management plan. It establishes the CRITERIA and the activities that are required to develop, monitor, and control the project schedule. It also establishes the project schedule model development, release and iteration length, accuracy level, measurement units, organizational procedure links, schedule model maintenance, control thresholds, performance measurement rules, and reporting formats.

3.3.2 The "Define Activities" process

The "Define Activities" process is used to identify and document the required to produce the project deliverables.

In reference to Figures 3.33 and 3.34, the main inputs, methods, and outputs of this process are detailed in Tables 3.43, 3.44, and 3.45, respectively.

The "Define Activities" Process		
Inputs	**Methods**	**Outputs**
1 Project management plan (e.g., schedule management plan, and scope baseline)	1 Expert judgment	1 Activity list
	2 Decomposition	2 Activity attributes
2 Enterprise environmental factors	3 Rolling wave planning	3 Milestone list
3 Organizational process assets	4 Meetings	4 Change requests
		5 Project management plan updates (e.g., schedule management plan, and scope baseline)

Figure 3.33: The "Define Activities" process.

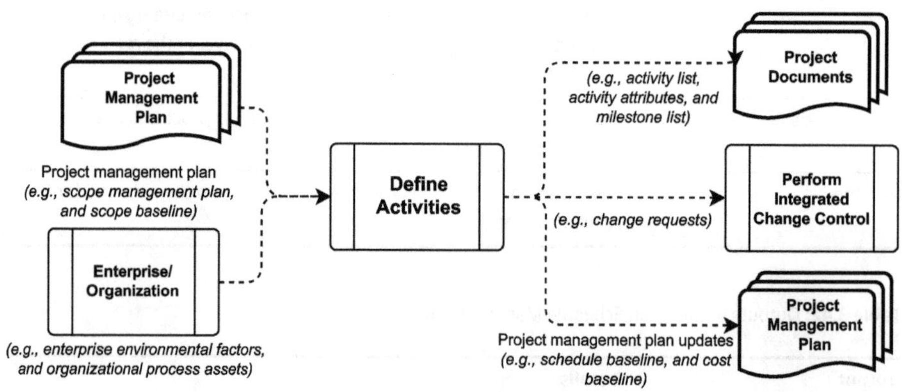

Figure 3.34: The "Define Activities" process data flow diagram.

Table 3.43: Inputs to the "Define Activities" process.

Input	Details
Project management plan	The main components of the project management plan used as inputs to this process are the schedule management plan, and the scope baseline.
Enterprise environmental factors (EEFs)	The main EEFs used as inputs to this process are the organizational culture and structure, availability of commercial databases, and availability of a project management information system (PMIS).
Organizational process asset factors (OPAFs)	The main OPAFs used as inputs to this process are the lessons learned repositories, standard processes, and so forth.

Table 3.44: Methods used during the "Define Activities" process.

Method	Details
Expert judgment	This refers to all judgments and decisions made while defining the project work activities. Such judgments require the project manager and his/her team to have adequate level of the required relevant skills and experience.
WBS decomposition	WBS decomposition is a method used to divide and subdivide the project scope and deliverables into smaller, but more manageable parts.
Meetings	This refers to all meetings that are concerned with defining the project work activities.

Table 3.45: Outputs of the "Define Activities" process.

Output	Details
Activity list	This includes all activities that are scheduled for the project.
Activity attributes	This includes the attributes of each activity such as activity ID and duration.
Milestone list	This document presents the list of the milestones defined during the project planning phase.
CHANGE-REQUESTS	As described earlier in Table 3.9.
Project management plan updates	The main components of the project management plan that may be updated during this process are the schedule baseline, and cost baseline.

3.3.3 The "Sequence Activities" process

The "Sequence Activities" process identifies and documents the project's inter-activity relationships.

In reference to Figures 3.35 and 3.36, the main inputs, methods, and outputs of this process are detailed in Tables 3.46, 3.47, and 3.48, respectively.

The "Sequence Activities" Process		
Inputs	**Methods**	**Outputs**
1 Project management plan (e.g., schedule management plan, and scope baseline) 2 Project documents (e.g., activity attributes, activity list, assumption log, and milestone list 3 Enterprise environmental factors 4 Organisational process assets	1 Precedence diagramming method 2 Dependency determination and integration 3 Leads and lags 4 Project management information system	1 Project schedule network diagrams 2 Project documents updates (e.g., activity attributes, activity list, assumption log, and milestone list)

Figure 3.35: The "Sequence Activities" process.

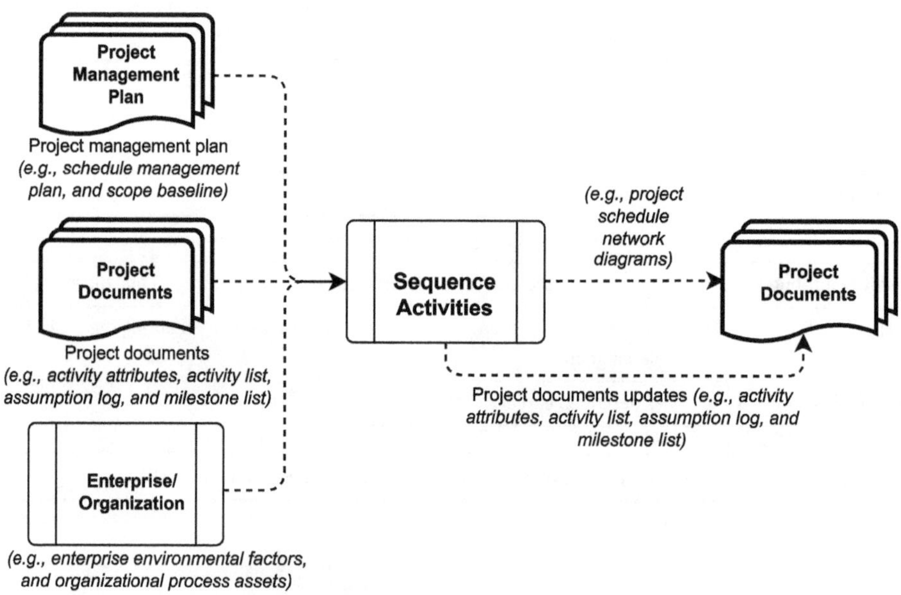

Figure 3.36: The "Sequence Activities" process data flow diagram.

Table 3.46: Inputs to the "Sequence Activities" process.

Input	Details
Project management plan	The main components of the project management plan used as inputs to this process are the schedule management plan and scope baseline.
Project documents	The main documents used as inputs to this process are the activity attributes, activity list, assumption log, and milestone list.
Enterprise environmental factors (EEFs)	The main EEFs used as inputs to this process are the governmental and industry standards, the project management information system (PMIS), and availability of scheduling tools.
Organizational process asset factors (OPAFs)	The main OPAFs used as inputs to this process are the portfolio and program plans, project dependencies and relationships; existing planning-related policies, procedures, and guidelines; and lessons learned repository.

Table 3.47: Methods used during the "Sequence Activities" process.

Method	Details
Precedence diagramming method (PDM)	The PDM is used to construct a schedule model. Activities are represented by nodes and graphically interlinked by logical relationships to show the sequence in which the activities are to be performed. These relationships are of four types: – **Finish-to-start (FS).** With this relation, a successor activity cannot start before its predecessor activity is completed. – **Finish-to-finish (FF).** With this relation, both of a successor activity and its predecessor activity must be completed at the same time. – **Start-to-start (SS).** With this relation, a successor activity and its predecessor activity must be started at the same time. – **Start-to-finish (SF).** With this relation, a successor activity cannot be completed before its predecessor activity is started.
Dependency determination and integration	This includes the dependencies among project activities that impose some restrictions on the activities sequencing in the schedule.
Leads and lags	A lead is the time period that a successor activity can proceed in respect of its predecessor activity. In other words, when a successor activity does not need an input from its predecessor when it starts, but rather it will need that input after some amount of time (e.g., *lead time*), then the successor activity can proceed in parallel with its predecessor for that lead time.
Project management information system (PMIS)	As described earlier in Table 3.8.

Table 3.48: Outputs of the "Sequence Activities" process.

Output	Details
Project documents' updates	The main documents that may be updated during this process are the activity attributes including all of the project's scheduled activities; activity attributes including the attributes of each activity such as the activity Id and duration; and milestone list as defined during the project planning phase.
Project schedule network diagram	This is a graphical representation of the project inter-activity relations.
Project documents' updates	The project documents that may be updated during this process are the activity attributes, activity list, assumption log, and milestone list.

3.3.4 The "Estimate Activity Durations" process

The "Estimate Activity Durations" process estimates the work periods needed to complete the project activities.

In reference to Figures 3.37 and 3.38, the main inputs, methods, and outputs of this process are detailed in Tables 3.49, 3.50, and 3.51, respectively.

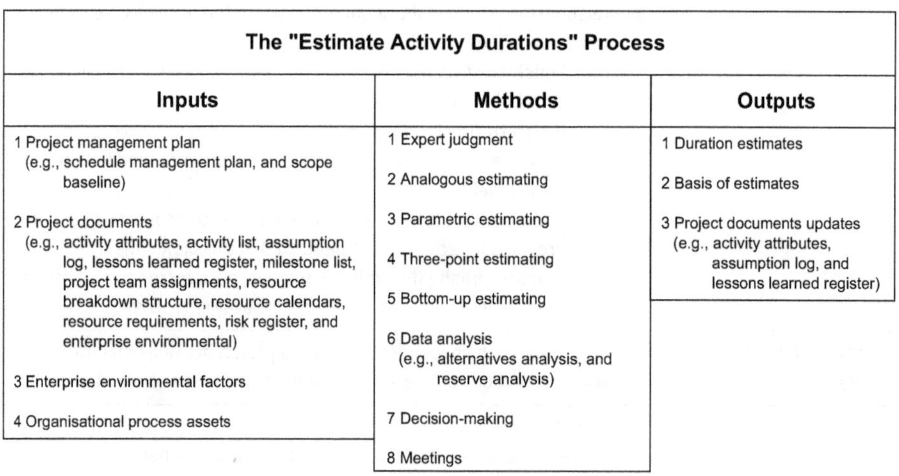

Figure 3.37: The "Estimate Activity Durations" process.

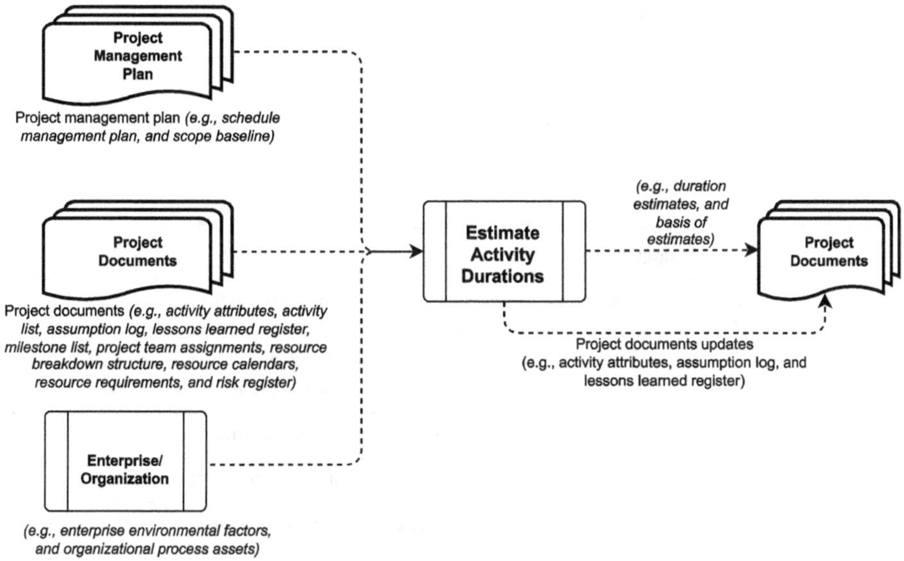

Figure 3.38: The "Estimate Activity Durations" process data flow diagram.

Table 3.49: Inputs to the "Estimate Activity Durations" process.

Input	Details
Project management plan	The main components of the project management plan used as inputs to this process are the schedule management plan and scope baseline.
Project documents	The main documents used as inputs to this process are the activity attributes, activity list, assumption log, lessons learned register, milestone list, project team assignments, resource requirements, risk register, resource calendars, and resource breakdown structure.
Enterprise environmental factors (EEFs)	The main EEFs used as inputs to this process are the duration estimating databases, productivity metrics, and team members' locations.
Organizational process asset factors (OPAFs)	The main OPAFs used as inputs to this process are the historical duration information, scheduling method, estimating policies, project calendars, and lessons learned repository.

Table 3.50: Methods used during the "Estimate Activity Durations" process.

<table>
<tr><th>Method</th><th colspan="2">Details</th></tr>
<tr><td>Expert judgment</td><td colspan="2">This refers to all judgments and decisions made while estimating the project activities' durations. Such judgments require the project manager and his/her team to have adequate level of the required relevant skills and experience.</td></tr>
<tr><td>Analogous estimating</td><td colspan="2">This uses from the attributes of similar projects. This includes the scope, cost, budget, and duration.</td></tr>
<tr><td>Parametric estimating</td><td colspan="2">This method is algorithm-based, which estimates the activities' durations and cost based on some of the project parameters and on historical data from similar projects. It uses a statistical relation between historical data and other variables to calculate a duration and cost estimates for the project work activities.</td></tr>
<tr><td rowspan="4">Three-point estimating</td><td colspan="2">This method helps define approximated activities' durations using three points:</td></tr>
<tr><td>Most likely</td><td>Highly accurate duration estimates. Best-case estimate</td></tr>
<tr><td>Optimistic</td><td>Moderately accurate duration estimate.</td></tr>
<tr><td>Pessimistic</td><td>Poorly accurate duration estimates. Worst-case estimate</td></tr>
<tr><td>Bottom-up estimating</td><td colspan="2">This method is used to estimate project duration and/or cost by aggregating the estimates of the lower level of the WSB, and take it then upward.</td></tr>
<tr><td>Data analysis</td><td colspan="2">The data analysis methods used during this process are the alternatives' analysis and reserve analysis.</td></tr>
<tr><td>Decision-making</td><td colspan="2">The main decision-making method used during this process is the voting.</td></tr>
<tr><td>Meetings</td><td colspan="2">This refers to all meetings that are concerned with estimating durations for the project work activities.</td></tr>
</table>

Table 3.51: Outputs of the "Estimate Activity Durations" process.

Output	Details
Project documents' updates	The main documents that may be updated during this process are the activities' attributes, assumption log, and lessons learned register.
Duration estimates	This includes all duration estimates of the project activities.
Basis of estimates	This includes the assumptions used to estimate the project activities' durations and/or cost.

3.3.5 The "Develop Schedule" process

The "Develop Schedule" process is used to analyze the activities' sequences, durations, resource requirements, and schedule constraints to create a project schedule model.

In reference to Figures 3.39 and 3.40, the main inputs, methods, and outputs of this process are detailed in Tables 3.52, 3.53, and 3.54, respectively.

The "Develop Schedule" Process		
Inputs	**Methods**	**Outputs**
1 Project management plan (e.g., schedule management plan, scope baseline) 2 Project documents (e.g., activity attributes, activity list, assumption log, basis of estimates, duration estimates, lessons learned register, milestone list, project schedule network diagrams, project team assignments, resource calendars, resource requirements, risk register) 3 Agreements 4 Enterprise environmental factors 5 Organizational process assets	1 Schedule network analysis 2 Critical path method 3 Resource optimization 4 Data analysis (e.g., what-if scenario analysis, simulation) 5 Leads and lags 6 Schedule compression 7 Project management information system 8 Agile release planning	1 Schedule baseline 2 Project schedule 3 Schedule data 4 Project calendars 5 Change requests 6 Project management plan updates (e.g., schedule management plan, cost baseline) 7 Project documents updates (e.g., activity attributes, assumption log, duration estimates, lessons learned register, resource requirements, risk register)

Figure 3.39: The "Develop Schedule" process.

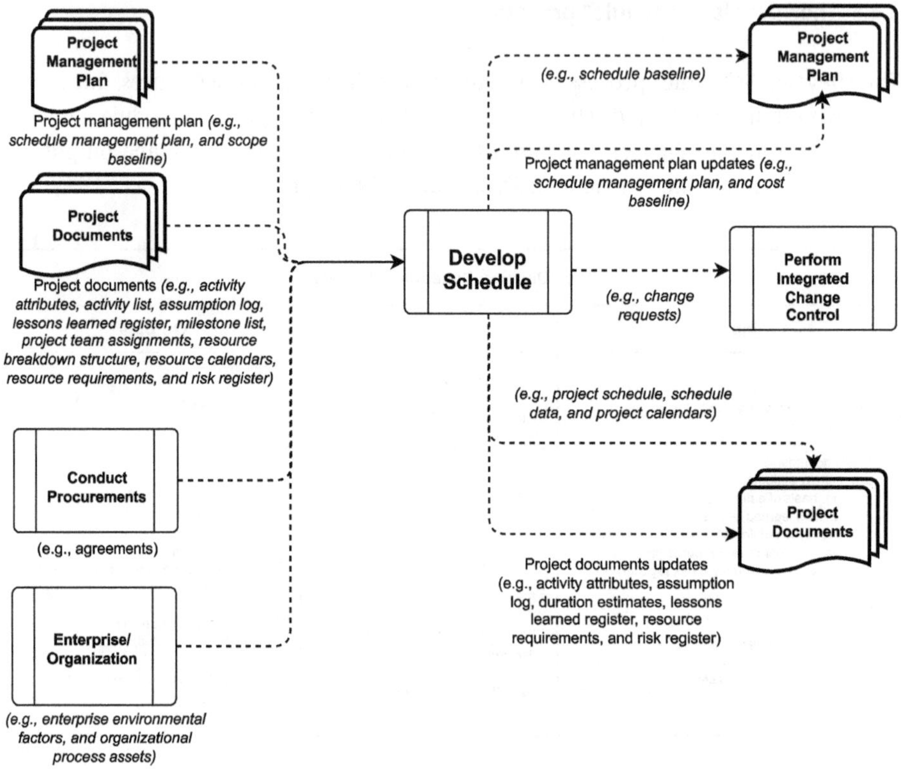

Figure 3.40: The "Develop Schedule" process data flow diagram.

Table 3.52: Inputs to the "Develop Schedule" process.

Input	Details
Project management plan	The main components of the project management plan used as inputs to this process are the schedule management plan and scope baseline.
Project documents	The main documents used as inputs to this process are the activities' attributes, activity list, assumption log, basis of estimates, duration estimates, lessons learned register, milestone list, project schedule network diagrams, project team assignments, resource calendars, resource requirements, and risk register.
Agreements	As described earlier in Table 3.1.
Enterprise environmental factors (EEFs)	The main EEFs used as inputs to this process are the governmental and industry standards, and communication channels.
Organizational process asset factors (OPAFs)	The main OPAFs used as inputs to this process are the scheduling method including the policies that govern the schedule and its maintenance, and the project calendars.

Table 3.53: Methods used during the "Develop Schedule" process.

Method	Details
Scheduling network analysis	This method is used to generate the project schedule by several methods such as the CRITICAL path method, resource optimization techniques, and modeling techniques.
CRITICAL path method	This method is used to estimate the minimum project duration and determine the schedule flexibility on the different network paths in the schedule model.
Resource optimization	This method is used to adjust the start and finish dates of the project activities such that the planned resource use is adjusted to become equal to or less than resource availability.
Data analysis	For data analysis, this process uses the what-if scenario analysis and/or simulation.
Leads and lags	As described earlier in Table 3.47.
Schedule compression	This method is used to shorten the schedule duration without reducing the project scope.

Table 3.54: Outputs of the "Develop Schedule" process.

Output	Details
Schedule baseline	This is the approved schedule model that can be changed only through formal change control procedures. During monitoring and controlling, the approved baseline dates are compared to the actual start and finish dates to determine if variances have occurred.
Project schedule	A project schedule model can be presented in tabular and graphical format. Graphical formats include bar charts, milestone charts, and project schedule network diagram.
Schedule data	This is the collection of information for describing and controlling the schedule. The schedule data includes the schedule milestones, schedule activities, activity attributes, and assumptions and constraints documentation.
Project calendars	A project calendar identifies the working days for scheduled activities.
CHANGE-REQUESTS	As described earlier in Table 3.9.
Project management plan updates	The main components of the project management plan that may be updated during this process are the schedule management plan and cost baseline.
Project documents' updates	The main documents that may be updated during this process are the activity attributes, assumption log, duration estimates, lessons learned register, resource requirements, and risk register.

3.3.6 The "Control Schedule" process

The "Control Schedule" process is used to monitor the project status and update its schedule and manage changes to the schedule baseline.

In reference to Figures 3.41 and 3.42, the main inputs, methods, and outputs of this process are detailed in Tables 3.55, 3.56, and 3.57, respectively.

The "Control Schedule" Process		
Inputs	**Methods**	**Outputs**
1 Project management plan (e.g., schedule management plan, schedule baseline, scope baseline, and performance measurement baseline)	1 Data analysis (e.g., earned value analysis, iteration burndown chart, performance reviews, trend analysis, variance analysis, and what-if scenario analysis)	1 Work performance information
2 Project documents (e.g., lessons learned register, project calendars, project schedule, resource calendars, and schedule data)	2 Critical path method	2 Schedule forecasts
3 Work performance data	3 Project management information system	3 Change requests
4 Organisational process assets	4 Resource optimization	4 Project management plan updates (e.g., schedule management plan, schedule baseline, cost baseline, and performance measurement baseline)
	6 Leads and lags	5 Project documents updates (e.g., assumption log, basis of estimates, lessons learned register, project schedule,
	7 Schedule compression	

Figure 3.41: The "Control Schedule" process.

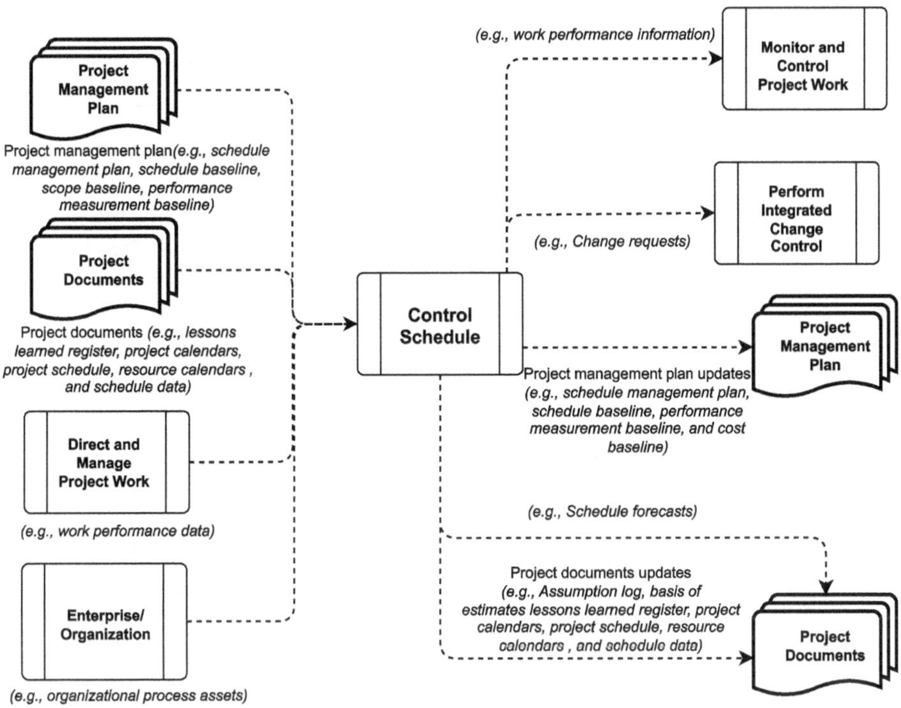

Figure 3.42: The “Control Schedule” process data flow diagram.

Table 3.55: Inputs to the “Control Schedule” process.

Input	Details
Project management plan	The components of the project management plan used as inputs to this process are the schedule management plan, schedule baseline, the scope baseline, and the performance measurement baseline.
Project documents	The project documents used as inputs to this process are the lessons learned register, project calendars, project schedule, resource calendars, and schedule data.
Project work performance data	As described earlier in Table 3.9.
Organizational process asset factors (OPAFs)	The main OPAFs used as inputs to this process are the existing schedule control-related policies, procedures, and guidelines; and available schedule controlling, monitoring, and reporting tools.

Table 3.56: Methods used during the "Control Schedule" process.

Method	Details
CRITICAL path method	As described earlier in Table 3.53.
Resource optimization	As described earlier in Table 3.53.
Data analysis	The data analysis techniques used during this process are the earned-value analysis, iteration BURNDOWN-CHART, performance reviews, trend analysis, variance analysis, and what-if scenario analysis.
Leads and lags	As described earlier in Table 3.47.
Schedule compression	As described earlier in Table 3.53.

Table 3.57: Outputs of the "Control Schedule" process.

Output	Details
Project work performance data	As described earlier in Table 3.9.
Schedule forecasts	This includes the predictions of conditions and events in the project life cycle that may affect its schedule.
CHANGE-REQUESTS	As described earlier in Table 3.9.
Project management plan updates	The main components of the project management plan that may be updated during this process are the schedule management plan, schedule baseline, cost baseline, and performance measurement baseline.
Project documents' updates	The project documents that may be updated during this process are the assumption log, basis of estimates, lessons learned register, project schedule, resource calendars, risk register, and schedule data.

3.4 Project cost management

Project cost management includes the processes that are necessary to plan, estimate, manage, monitor, and control the project cost and budget.

In reference to Figure 3.43, the project cost management processes are the Plan Cost Management, Estimate Costs, Determine Budget, and Control Costs.

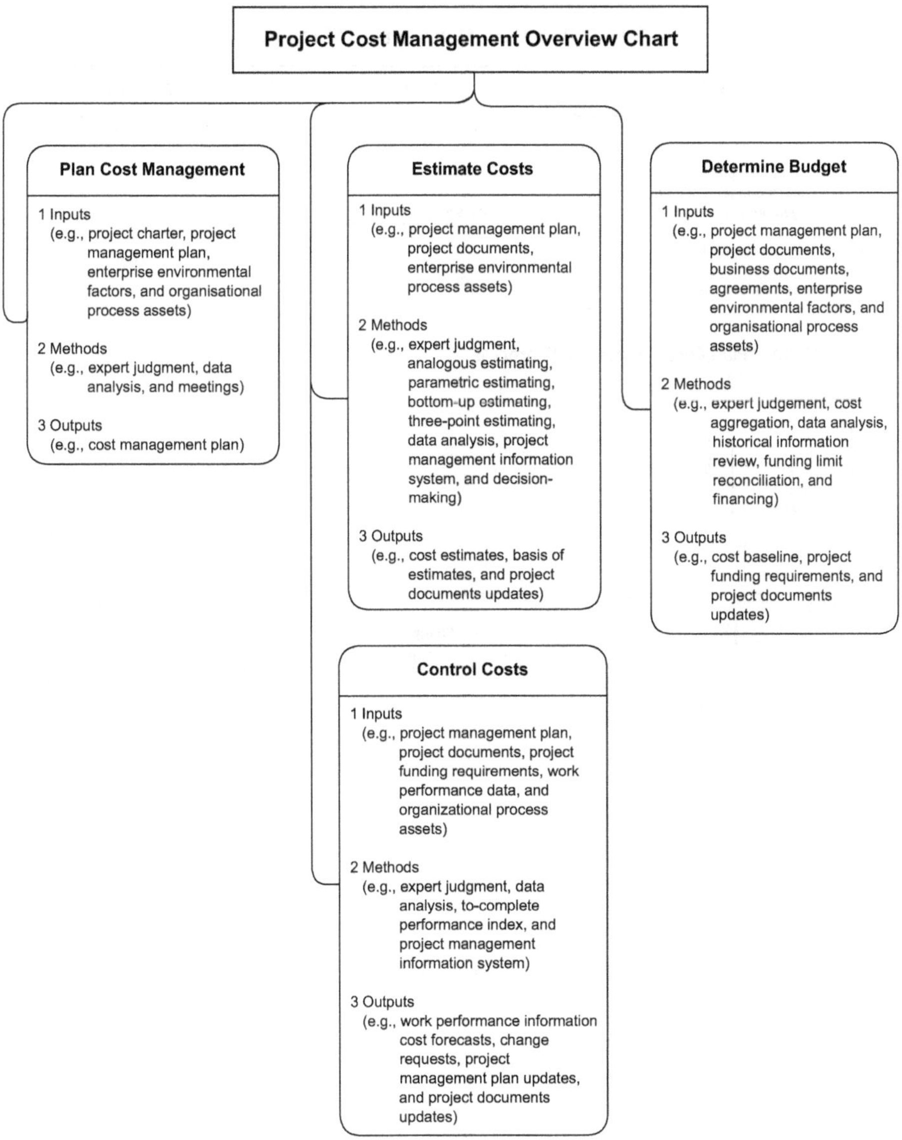

Figure 3.43: Project cost management overview chart.

3.4.1 The “Plan Cost Management” process

The “Plan Cost Management” process defines how the project costs should be estimated, budgeted, managed, monitored, and controlled.

In reference to Figures 3.44 and 3.45, the main inputs, methods, and outputs of this process are detailed in Tables 3.58, 3.59, and 3.60, respectively.

The "Plan Cost Management" Process		
Inputs	**Methods**	**Outputs**
1 Project charter	1 Expert judgment	1 Cost management plan
2 Project management plan (e.g., schedule management plan, and risk management plan)	2 Data analysis	
3 Enterprise environmental factors	3 Meetings	
4 Organizational process assets		

Figure 3.44: The “Plan Cost Management” process.

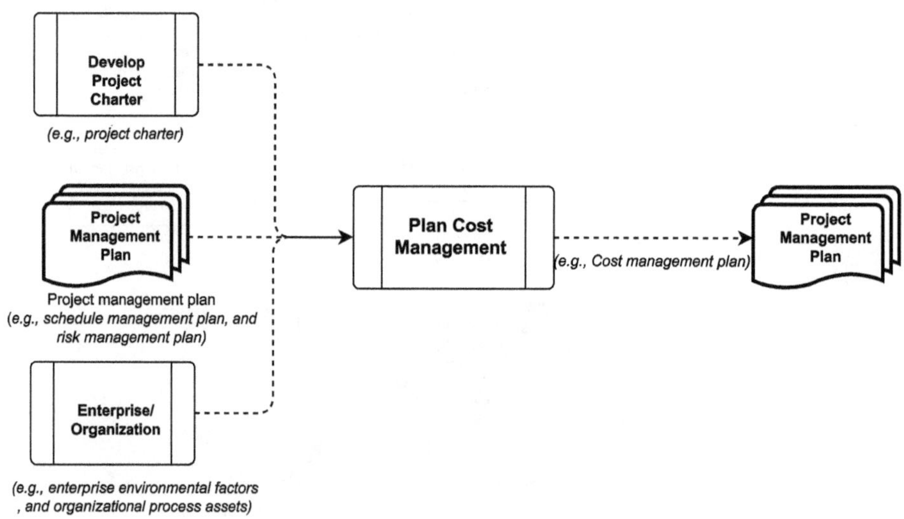

Figure 3.45: The “Plan Cost Management” process data flow diagram.

Table 3.58: Inputs to the "Plan Cost Management" process.

Input	Details
Project charter	As described earlier in Table 3.3.
Project management plan	The main components of the project management plan used as inputs to this process are the schedule management plan and risk management plan.
Enterprise environmental factors (EEFs)	The main EEFs used as inputs to this process are the organizational culture and structure, marketplace conditions, currency exchange rates, and project management information system (PMIS).
Organizational process asset factors (OPAFs)	The main OPAFs used as inputs to this process are the financial control procedures, financial databases, and lessons learned repository.

Table 3.59: Methods used during the "Plan Cost Management" process.

Method	Details
Expert judgment	This refers to all judgments and decisions made while planning for the project cost management. Such judgments require the project manager and his/her team to have adequate level of the required relevant skills and experience.
Data analysis	The alternatives' analysis is the main data analysis method used during this process.
Meetings	As described earlier in Table 3.5.

Table 3.60: Outputs of the "Plan Cost Management" process.

Output	Details
Cost management plan	The cost management plan establishes the cost units of measure, cost levels of precision and accuracy, organizational procedure links, cost control thresholds, rules of performance measurement, and the reporting forms.

3.4.2 The "Estimate Costs" process

The "Estimate Costs" process estimates the cost of resources required to carry on and complete the project work.

In reference to Figures 3.46 and 3.47, the main inputs, methods, and outputs of this process are detailed in Tables 3.61, 3.62, and 3.63, respectively.

The "Estimate Costs" process		
Inputs	**Methods**	**Outputs**
1 Project management plan (e.g., cost management plan, quality management plan, and scope baseline)	1 Expert judgment	1 Cost estimates
2 Project documents (e.g., lessons learned register, project schedule, resources requirements, and risk register)	2 Analogous estimating	2 Basis of estimates
3 Enterprise environmental factors	3 Parametric estimating	3 Project documents updates (e.g., assumption log, lessons learned register, and risk register)
4 Organizational process assets	4 Bottom-up estimating	
	5 Three-point estimating	
	6 Data analysis (e.g., alternatives analysis, reserve analysis, and cost of quality)	
	7 Project management information system	
	8 Decision-making (e.g., voting)	

Figure 3.46: The "Estimate Costs" process.

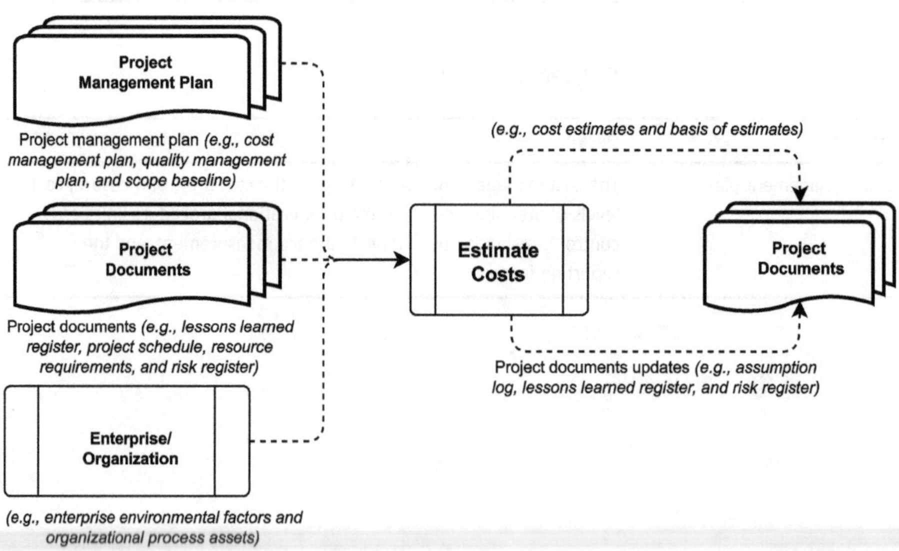

Figure 3.47: The "Estimate Costs" process data flow diagram.

Table 3.61: Inputs to the "Estimate Costs" process.

Input	Details
Project charter	As described earlier in Table 3.3.
Project management plan	The main components of the project management plan used as inputs to this process are the cost management plan; quality management plan; and scope baseline including the project scope statement, work breakdown structure (WBS), and WBS directory.
Project documents	The main documents used as inputs to this process are the lessons learned register, project schedule, resource requirements, and risk register.
Enterprise environmental factors (EEFs)	The main EEFs used as inputs to this process are the marketplace conditions, published commercial information, and exchange rates and inflation.
Organizational process asset factors (OPAFs)	The main OPAFs used as inputs to this process are the cost estimation policies and templates, and lessons learned register.

Table 3.62: Methods used during the "Estimate Costs" process.

Method	Details
Expert judgment	This refers to all judgments and decisions made while estimating the project ingredients' costs. Such judgments require the project manager and his/her team to have adequate level of the required relevant skills and experience.
Data analysis	The main methods used for data analysis are the alternatives' analysis, reserve analysis, and the cost of quality analysis.
Analogous estimating	As described earlier in Table 3.50.
Parametric estimating	As described earlier in Table 3.50.
Bottom-up estimating	As described earlier in Table 3.50.
Three-point estimating	The accuracy of single-point cost estimates uncertainty and risk and using three estimates to define an approximate range for an activity's cost: – **Most likely**. The cost of the activity based on a realistic effort assessment for the required work. – **Optimistic**. The cost based on analysis of the best-case scenario for the activity. – **Pessimistic**. The cost based on analysis of the worst-case scenario for the activity.
Project management information system (PMIS)	As described earlier in Table 3.8.
Decision-making	The main decision-making methods used in this process are the alternatives' analysis, reserve analysis, and cost of quality.

Table 3.63: Outputs of the "Estimate Costs" process.

Output	Details
Cost estimates	This includes quantitative assessments of the costs required to complete the project work, risk contingency fund, and management reserve to cover unplanned work.
Basis of estimates	The main documents that are used to support the cost estimation are the documentation of assumptions, the documentation of constraints; and the documentation of risks.
Project documents' updates	The main documents that may be updated during this process are the assumption log, lessons learned register, and risk register.

3.4.3 The "Determine Budget" process

The "Determine Budget" process aggregates the estimated costs of activities and establishes a cost baseline. A budget includes all the funds required to execute the project. In reference to Figures 3.48 and 3.49, the main inputs, methods, and outputs of this process are detailed in Tables 3.64, 3.65, and 3.66, respectively.

The "Determine Budget" process		
Inputs	**Methods**	**Outputs**
1 Project management plan (e.g., cost management plan, resource management plan, and scope baseline)	1 Expert judgment	1 Cost baseline
2 Project documents (e.g., basis of estimates, cost estimates, project schedule, and risk register)	2 Cost aggregation	2 Project funding requirements
3 Business documents (e.g., business case and benefits management plan)	3 Data analysis (e.g., reserve analysis)	3 Project documents updates (e.g., cost estimates, project schedule, and risk register)
4 Agreements	4 Historical information review	
5 Enterprise environmental factors	5 Funding limit reconciliation	
6 Organizational process assets	6 Financing	

Figure 3.48: The "Determine Budget" process.

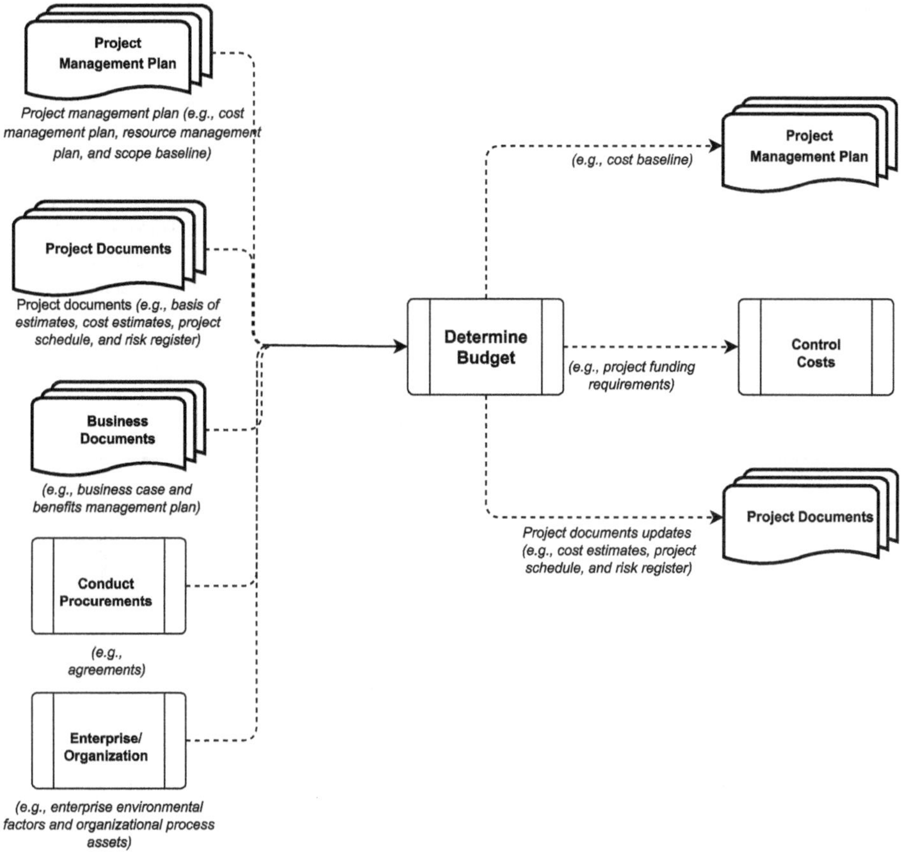

Figure 3.49: The "Determine Budget" process data flow diagram.

Table 3.64: Inputs to the "Determine Budget" process.

Input	Details
Project charter	As described earlier in Table 3.3.
Project management plan	The main components of the project management plan used as inputs to this process are the cost management plan, resource management plan, and scope baseline.
Project documents	The main documents used as inputs to this process are the basis of estimates, cost estimates, project schedule, and risk register.
Business documents	This includes the business case and benefits management plan.
Agreements	As described earlier in Table 3.1.

Table 3.64 (continued)

Input	Details
Enterprise environmental factors (EEFs)	The main EEFs used as inputs to this process are the marketplace conditions, published commercial information, and exchange rates and inflation.
Organizational process asset factors (OPAFs)	The main organizational process factors used as inputs to this process are the cost budgeting-related policies, procedures, and guidelines; lessons learned repository; available cost budgeting tools; and used reporting methods.

Table 3.65: Methods used during the "Determine Budget" process.

Method	Details
Expert judgment	This refers to all judgments and decisions made while determining the project budget. Such judgments require the project manager and his/her team to have adequate level of the required relevant skills and experience.
Cost aggregation	Cost estimates are aggregated by work tasks and activities according to the WBS.
Data analysis	The main data analysis method used during this process is the reserve analysis.
Historical information review	This includes project parameters to develop mathematical models to predict the total project costs.
Functional limit reconciliation	The spendeture should be reconciled with the project funding limits. In case of having a variance between the spendeture and the funding limits, the work schedule and probably the resource allocations should be revisited and rescheduled such that variance is eliminated.
Financing	This refers to soliciting financial resources to fund projects.

Table 3.66: Outputs of the "Determine Budget" process.

Output	Details
Cost baseline	This is the approved version of the time-phased project budget. It is used for comparison with actual results. It is developed as a summation of the approved budget items for the different scheduled activities.
Project funding requirements	This includes all what is needed to facilitate funding the project internally, or from an external funding authority.

Table 3.66 (continued)

Output	Details
Cost estimates	This includes quantitative assessments of the costs required to complete the project work, risk contingency fund, and management reserve to cover unplanned work.
Project documents' updates	The main documents that may be updated during this process are the cost estimates, project schedule, and risk register.

3.4.4 The "Control Costs" process

The "Control Costs" process is used to monitor the project status and update the project costs and managing changes to the cost baseline. Project cost control involves:

- Managing all CHANGE-REQUESTS in timely manners, ensuring that cost expenditures do not exceed the authorized funding by period, WBS component, and/or activity
- Monitoring cost performance to isolate and understand variances from the approved cost baseline
- Monitoring the work performance against funds spent
- Preventing unapproved CHANGE-REQUESTS from being included in the reported cost or resource usage

In reference to Figures 3.50 and 3.51, the main inputs, methods, and outputs of this process are detailed in Tables 3.67, 3.68, and 3.69, respectively.

The "Control Costs" process		
Inputs	**Methods**	**Outputs**
1 Project management plan (e.g., cost management plan, cost baseline, and performance measurement baseline) 2 Project documents (e.g., lessons learned register) 3 Project funding requirements 4 Work performance data 5 Organizational process assets	1 Expert judgment 2 Data analysis (e.g., earned value analysis, variance analysis, trend analysis, and reserve analysis) 3 To-complete performance index 4 Project management information system	1 Work performance information 2 Cost forecasts 3 Change requests 4 Project management plan updates (e.g., cost management plan, cost baseline, and performance measurement baseline) 5 Project documents updates (e.g., assumption log, basis of estimates, cost estimates, lessons learned register, and risk register)

Figure 3.50: The "Control Costs" process.

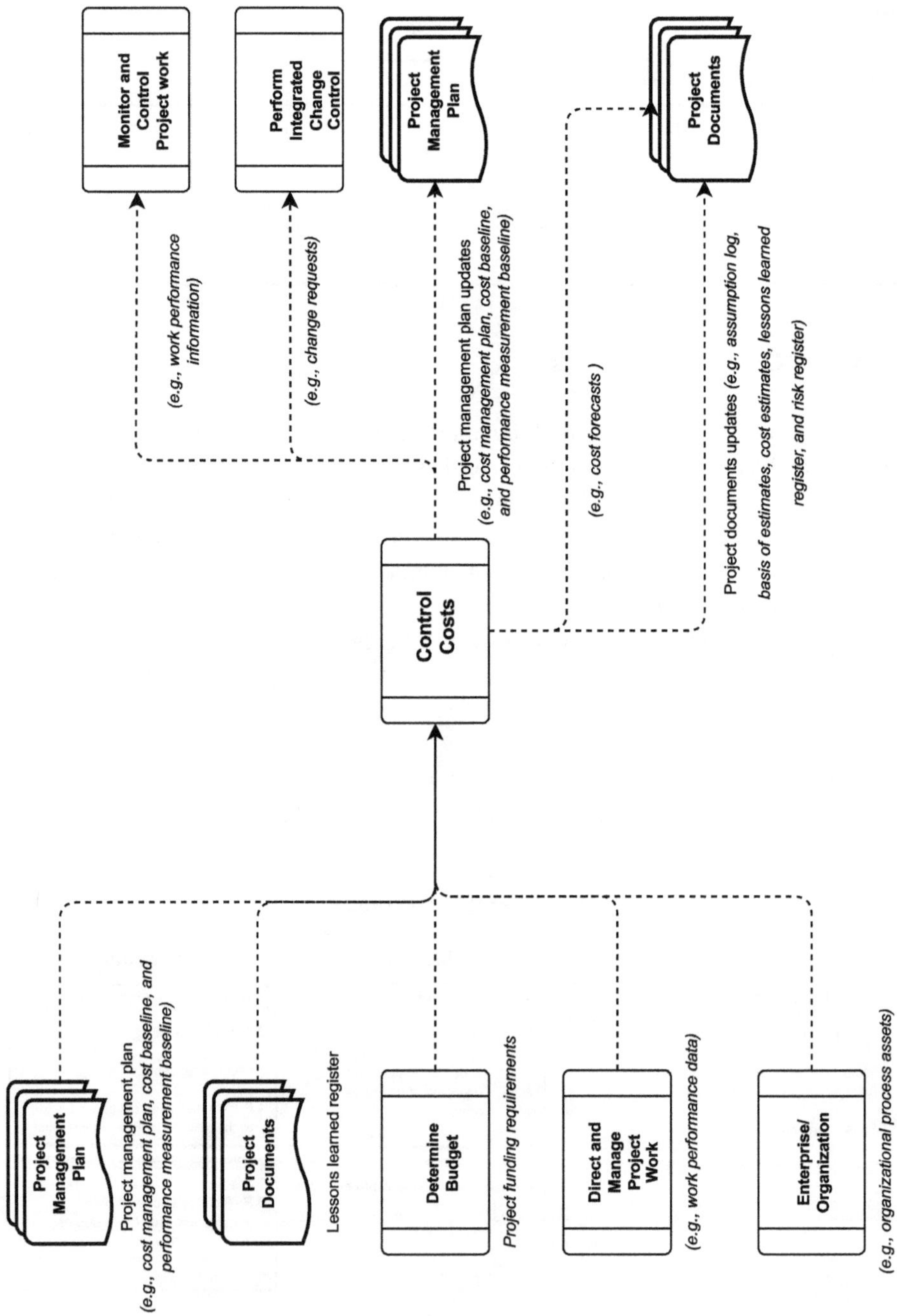

Figure 3.51: The "Control Costs" process data flow diagram.

Table 3.67: Inputs to the "Control Costs" process.

Input	Details
Project management plan	The main components of the project management plan used as inputs to this process are the cost management plan, cost baseline, and performance measurement baseline.
Project documents	The project's main document used as inputs to this process is the lessons learned register.
Project funding requirements	This includes the project projected spendeture along with any anticipated liabilities.
Agreements	As described earlier in Table 3.1.
Project work performance data	As described earlier in Table 3.9.
Organizational process asset factors (OPAFs)	The main organizational process factors used as inputs to this process are the existing cost control-related policies, procedures, and guidelines; cost control tools; and monitoring and control methods.

Table 3.68: Methods used during the "Control Costs" process.

Method	Details
Expert judgment	This refers to all judgments and decisions made while controlling the project costs and spendeture. Such judgments require the project manager and his/her team to have adequate level of the required relevant skills and experience.
Data analysis	The main data analysis methods used during this process are the earned-value analysis including the planned-for value, earned value, and actual costs; variance analysis including the schedule variance, cost variance, schedule performance index, and cost performance index; trend analysis including charts and forecasting; and reserve analysis.
To-complete performance index (TCPI)	The TCPI is the measure of the performance cost to be achieved with the remaining resources.
Project management information system (PMIS)	As described earlier in Table 3.8.

Table 3.69: Outputs of the "Control Costs" process.

Output	Details
Project work performance data	As described earlier in Table 3.9.
Cost forecasts	This includes the necessary cost forecasts of the anticipated costs of the project's various activities and tasks.
CHANGE-REQUESTS	As described earlier in Table 3.9.
Project management plan updates	The main components of the project management plan used as inputs to this process are the cost management plan, cost baseline, and performance measurement baseline.
Project documents' updates	The main documents that may be updated during this process are the assumption log, basis of estimates, cost estimates, lessons learned register, and risk register.

3.5 Project quality management

Project quality management includes the processes that are necessary to incorporate the organization's quality policy that is used to plan, manage, and control the project and the product quality requirements.

In reference to Figure 3.52, the project quality management processes are the Plan Quality Management, Manage Quality, and Control Quality, whereas Figure 3.53 depicts the relations of project quality management processes.

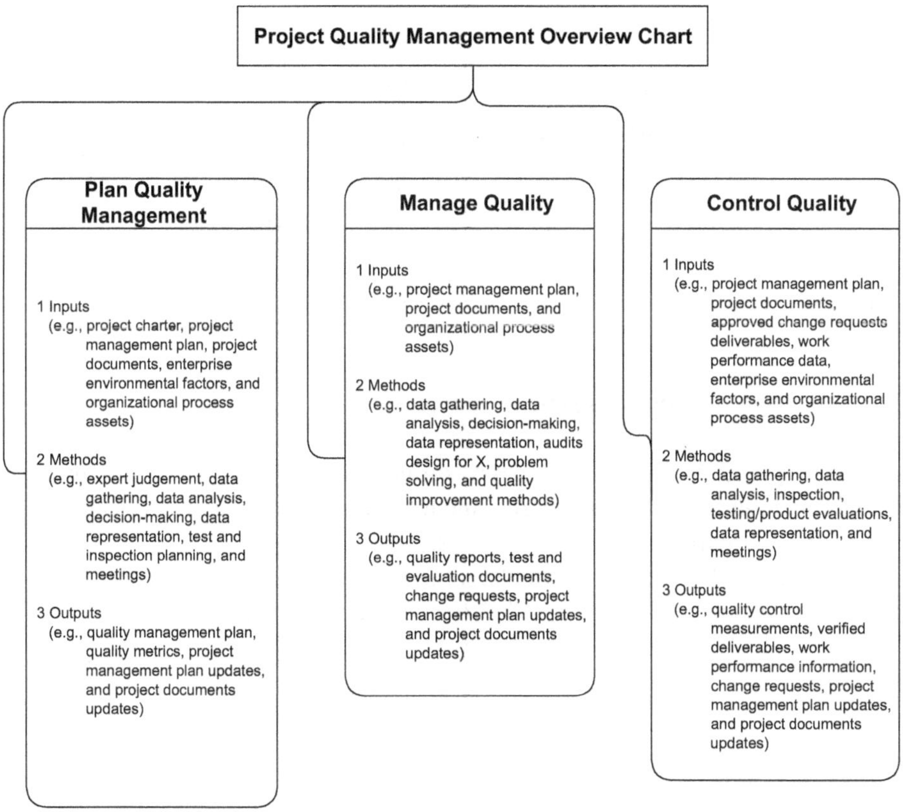

Figure 3.52: Project quality management overview chart.

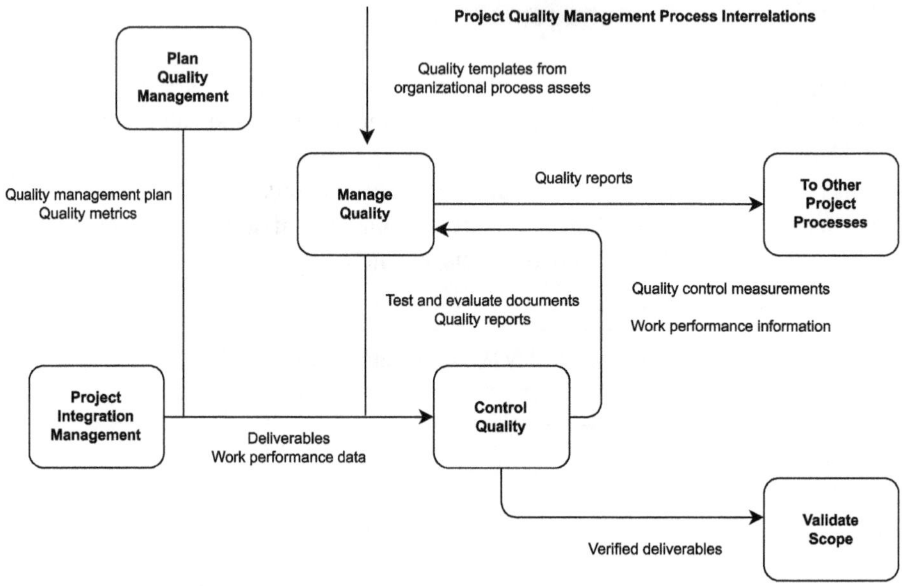

Figure 3.53: The project quality management processes inter-relations.

3.5.1 The "Plan Quality Management" process

The "Plan Quality Management" process is used to identify the project and product quality requirements and/or standards, and to document the project compliance with these quality requirements and/or standards.

In reference to Figures 3.54 and 3.55, the main inputs, methods, and outputs of this process are detailed in Tables 3.70, 3.71, and 3.72, respectively.

The "Plan Quality Management" Process		
Inputs	**Methods**	**Outputs**
1 Project charter	1 Expert judgment	1 Quality management plan
2 Project management plan (e.g., requirements management plan, risk management plan, stakeholder engagement plan, and scope baseline)	2 Data gathering (e.g., benchmarking, brainstorming, and interviews)	2 Quality metrics
3 Project documents (e.g., assumption log, requirements documentation, requirements traceability matrix, risk register, and stakeholder register)	3 Data analysis (e.g., cost-benefit analysis and cost of quality)	3 Project management plan updates (e.g., risk management plan and scope baseline)
4 Enterprise environmental factors	4 Decision-making (e.g., multicriteria decision analysis)	4 Project documents updates (e.g., lessons learned register, requirements traceability matrix, risk register, and stakeholder register)
5 Organizational process assets	5 Data representation (e.g., flowcharts, logical data model, matrix diagrams, and mind mapping)	
	6 Test and inspection planning	
	7 Meetings	

Figure 3.54: The "Plan Quality Management" process.

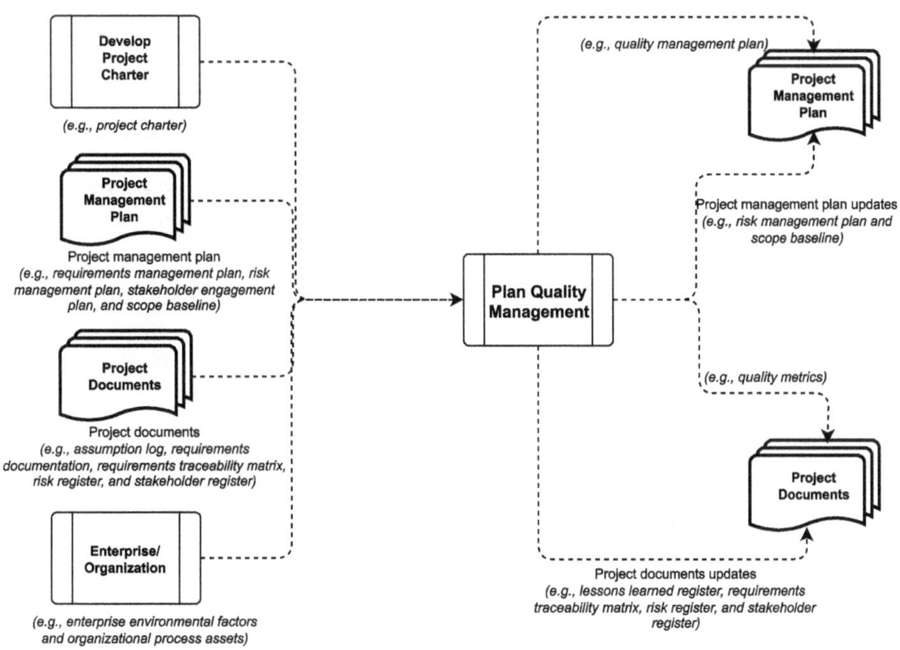

Figure 3.55: The "Plan Quality Management" process data flow diagram.

Table 3.70: Inputs to the "Plan Quality Management" process.

Input	Details
Project charter	As described earlier in Table 3.3.
Project management plan	The main components of the project management plan used as inputs to this process are the requirements management plan, risk management plan, stakeholder management plan, and scope baseline.
Project documents	The main documents used as inputs to this process are the assumption log, requirements documentation, requirements traceability matrix, risk register, and stakeholder register.
Enterprise environmental factors (EEFs)	The main EEFs used as inputs to this process are the relevant governmental regulations, rules, standards, and guidelines; geographical distribution of facilities; organizational structure and culture; and marketplace conditions.
Organizational process asset factors (OPAFs)	The main OPAFs used as inputs to this process are the organizational quality management policies, procedures, and guidelines; quality templates; and lessons learned register.

Table 3.71: Methods used during the "Plan Quality Management" process.

Method	Details
Expert judgment	This refers to all judgments and decisions made while planning for the project quality management. Such judgments require the project manager and his/her team to have adequate level of the required relevant skills and experience.
Data gathering	The main data gathering methods used during this Process are the benchmarking, brainstorming, and interviews.
Data analysis	The main methods used for data analysis are the cost-benefit analysis and cost of quality (e.g., *prevention cost*, *appraisal costs*, and *failure costs*).
Decision-making	This includes the multi-CRITERIA decision analysis that is used to identify the key issues and suitable alternatives to be prioritized as a set of implementation decisions.
Data representation	The main data representation methods used during this process are the flowcharts, logical data model, matrix diagrams, and mind mapping.
Test and inspection planning	As described earlier in Table 3.35.
Meetings	This refers to all meetings that take place while planning for the project quality management.

Table 3.72: Outputs of the "Plan Quality Management" process.

Output	Details
Quality management plan	The main quality components that are defined in this plan are the project quality standards and its deliverables, quality roles and responsibilities in the project, and quality control and management activities to be carried on during the project life cycle.
Quality metrics	This will be used to verify the project and the product's compliance to the quality requirements and standards.
Project management plan updates	The main components of the project management plan that may be updated during this process are the risk management plan and scope baseline.
Project documents' updates	The main documents that may be updated during this process are the lessons learned register, requirements traceability matrix, risk register, and stakeholder register.

3.5.2 The "Manage Quality" process

The "Manage Quality" process translates the quality management plan into a set of quality activities to incorporate the organization's quality policies into the project.

In reference to Figures 3.56 and 3.57, the main inputs, methods, and outputs of this process are detailed in Tables 3.73, 3.74, and 3.75, respectively.

The "Manage Quality" Process		
Inputs	**Methods**	**Outputs**
1 Project management plan (e.g., quality management plan)	1 Data gathering (e.g., checklists)	1 Quality reports
2 Project documents (e.g., lessons learned register, quality control measurements, quality metrics, and risk report)	2 Data analysis (e.g., alternatives analysis, document analysis, process analysis, and root cause analysis)	2 Test and evaluation documents
3 Organizational process assets	3 Decision-making (e.g., multicriteria decision analysis)	3 Change requests
	4 Data representation (e.g., affinity diagrams, cause-and-effect diagrams, flowcharts, histograms, matrix diagrams, and scatter diagrams)	4 Project management plan updates (e.g., quality management plan, scope baseline, schedule baseline, and cost baseline)
	5 Audits	5 Project documents updates (e.g., issue log, lessons learned register, and risk register)
	6 Design for X	
	7 Problem solving	
	8 Quality improvement methods	

Figure 3.56: The "Manage Quality" process.

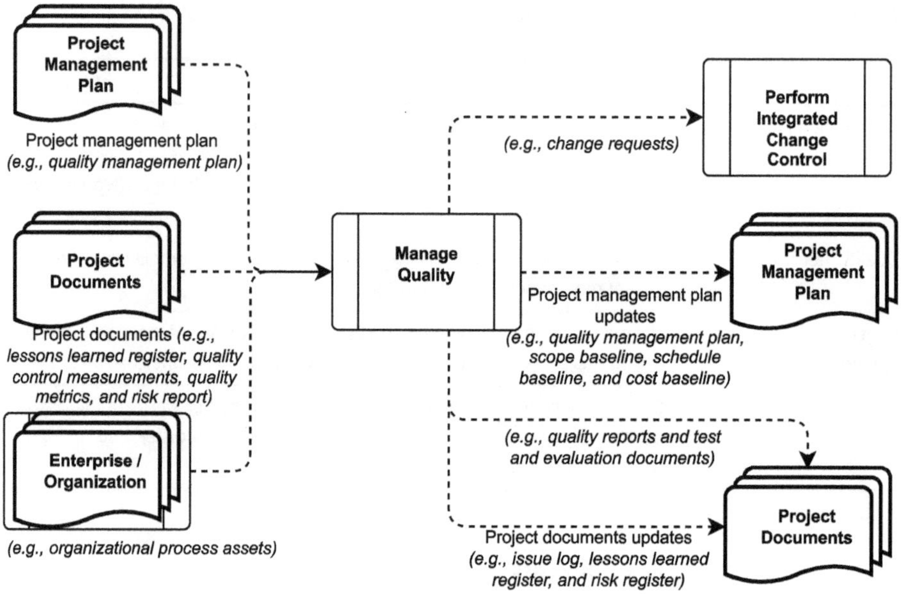

Figure 3.57: The "Manage Quality" process data flow diagram.

Table 3.73: Inputs to the "Manage Quality" process.

Input	Details
Project management plan	The main component of the project management plan used as inputs to this process is the quality management plan.
Project documents	The project's main documents used as inputs to this process are the lessons learned register, quality control measurements, quality metrics, and risk register.
Organizational process asset factors (OPAFs)	The main organizational process factors used as inputs to this process are the organizational quality management policies, procedures, and guidelines; quality templates; quality results from similar projects and lessons learned repository.

Table 3.74: Methods used during the "Manage Quality" process.

Method	Details
Data gathering	The main data gathering method used during this process is the checklist method.
Data analysis	The main methods used for data analysis are the alternatives' analysis, documents analysis, process analysis, and root-cause analysis (RCA).
Decision-making	This includes the multi-CRITERIA decision analysis that is used to identify the key issues and suitable alternatives to be prioritized as a set of decisions for managing the project quality.
Data representation	The main data representation methods used during this process are the affinity diagrams, cause-and-effect diagrams, flowcharts, histograms, matrix diagrams, and scatter diagrams.
Audits	The main quality audit objectives are to identify the BEST-PRACTICES; identify any nonconformity, gaps, and shortcomings; share BEST-PRACTICES; offer assistance to help improve team productivity; and highlight audits' contributions in the lessons learned repository.
Problem solving	The main problem-solving methods used during this process are meant to define the problem and to identify its root cause; and decide on the best solution for the problem, implement that solution, and then verify it.
Quality improvement methods	Quality improvements take place typically based on the quality control process findings and recommendations.

Table 3.75: Outputs of the "Manage Quality" process.

Output	Details
Quality reports	These reports all of the project and the product quality monitoring and control conclusions.
Test and evaluation documents	These documents present all of the project's test and evaluation results.
CHANGE-REQUESTS	As described earlier in Table 3.9.
Project management plan updates	The main components of the project management plan that may be updated during this process are the quality management plan, scope baseline, schedule baseline, and cost baseline.
Project documents' updates	The main documents that may be updated during this process are the issue log, lessons learned register, and risk register.

3.5.3 The "Control Quality" process

The "Control Quality" process is used to monitor and record the outcomes of the quality management activities to assess the project outcomes' performance. It determines if the project outcomes comply with the requirements, regulations, specifications, and standards.

In reference to Figures 3.58 and 3.59, the main inputs, methods, and outputs of this process are detailed in Tables 3.76, 3.77, and 3.78, respectively.

The "Control Quality" process		
Inputs	**Methods**	**Outputs**
1 Project management plan (e.g., quality management plan)	1 Data gathering (e.g., checklists, check sheets, statistical sampling, and questionnaires and surveys)	1 Quality control measurements
2 Project documents (e.g., lessons learned register, quality metrics, and test and evaluation documents)	2 Data analysis (e.g., performance reviews and root cause analysis)	2 Verified deliverables
3 Approved change requests	3 Inspection	3 Work performance information
4 Deliverables	4 Testing/product evaluations	4 Change requests
5 Work performance data	5 Data representation (e.g., cause-and-effect diagrams, control charts, histogram, and scatter diagrams)	5 Project management plan updates (e.g., quality management plan)
6 Enterprise environmental factors	6 Meetings	6 Project documents updates (e.g., issue log, lessons learned register, risk register, and test and evaluation documents)
7 Organizational process assets		

Figure 3.58: The "Control Quality" process.

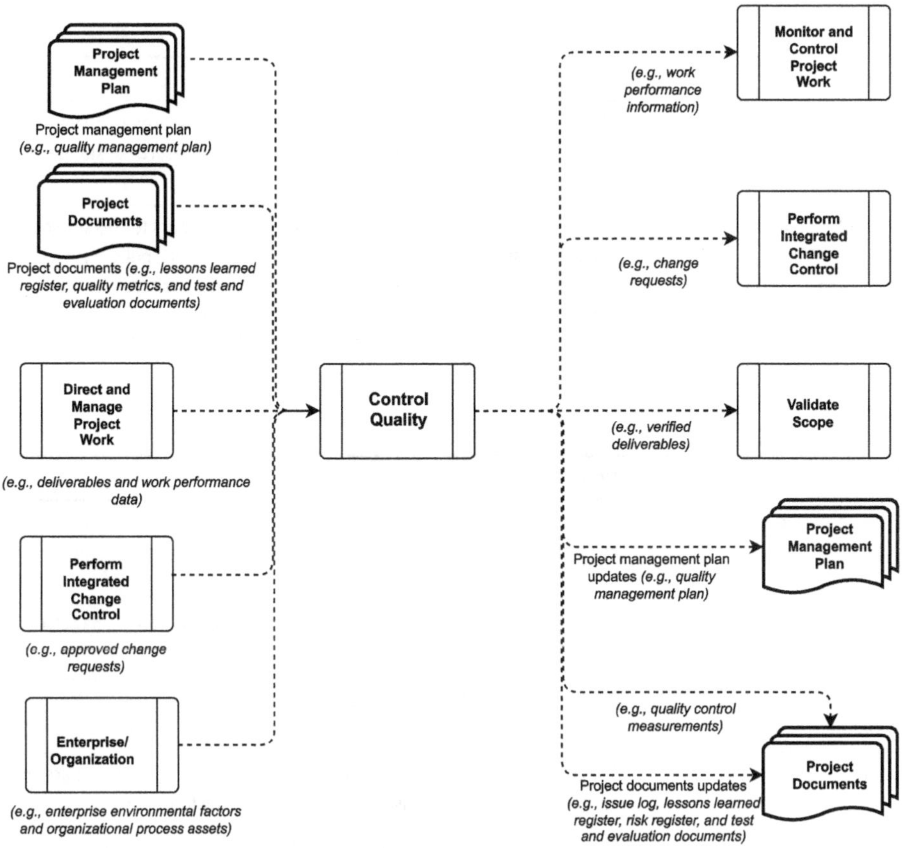

Figure 3.59: The “Control Quality” process data flow diagram.

Table 3.76: Inputs to the “Control Quality” process.

Input	Details
Project management plan	The main component of the project management plan used as inputs to this process is the quality management plan.
Project documents	The project’s main documents used as inputs to this process are the lessons learned register, quality control measurements, quality metrics, and test and evaluation documents.
Approved CHANGE-REQUESTS	As described earlier in Table 3.7.
Deliverables	As described earlier in Table 3.9.
Project work performance data	As described earlier in Table 3.9.

Table 3.76 (continued)

Input	Details
Enterprise environmental factors (EEFs)	The main EEFs used as inputs to this process are the project quality management systems; and governmental regulations, rules, standards, and guidelines.
Organizational process asset factors (OPAFs)	The main organizational process factors used as inputs to this process are the organizational quality management standards, policies, procedures, and guidelines; quality templates; quality results from similar projects, and defect reporting procedures.

Table 3.77: Methods used during the "Control Quality" process.

Method	Details
Data gathering	The main data gathering methods used during this process are the checklists, check sheets, statistical sampling, questionnaires, and surveys.
Data analysis	The main methods used for data analysis are the performance reviews and root-cause analysis.
Inspection and testing evaluations	As described earlier in Table 3.35.
Data representation	The main data representation methods used during this process are the cause-and-effect diagrams, control charts, and histograms.
Meetings	This refers to the meetings that take place in relevance to controlling the project quality.

Table 3.78: Outputs of the "Control Quality" process.

Output	Details
Quality control measurements	This refers to the documented results of the control-quality activities.
Verified deliverables	As described earlier in Table 3.34.
Project work performance data	As described earlier in Table 3.9.
CHANGE-REQUESTS	As described earlier in Table 3.9.
Project management plan updates	The main component of the project management plan that may be updated during this process is the quality management plan.
Project documents' updates	The main documents that may be updated during this process are the issue log, lessons learned register, risk register, and test and evaluation documents.

3.6 Project resource management

Project resource management includes the processes that are necessary to identify, acquire, and manage resources. These processes ensure that the right resources will be available at the right time and place.

In reference to Figure 3.60, the project resource management processes are the Plan Resource Management, Estimate Activity Resources, Acquire Resources, Develop Team, Manage Team, and Control Resources.

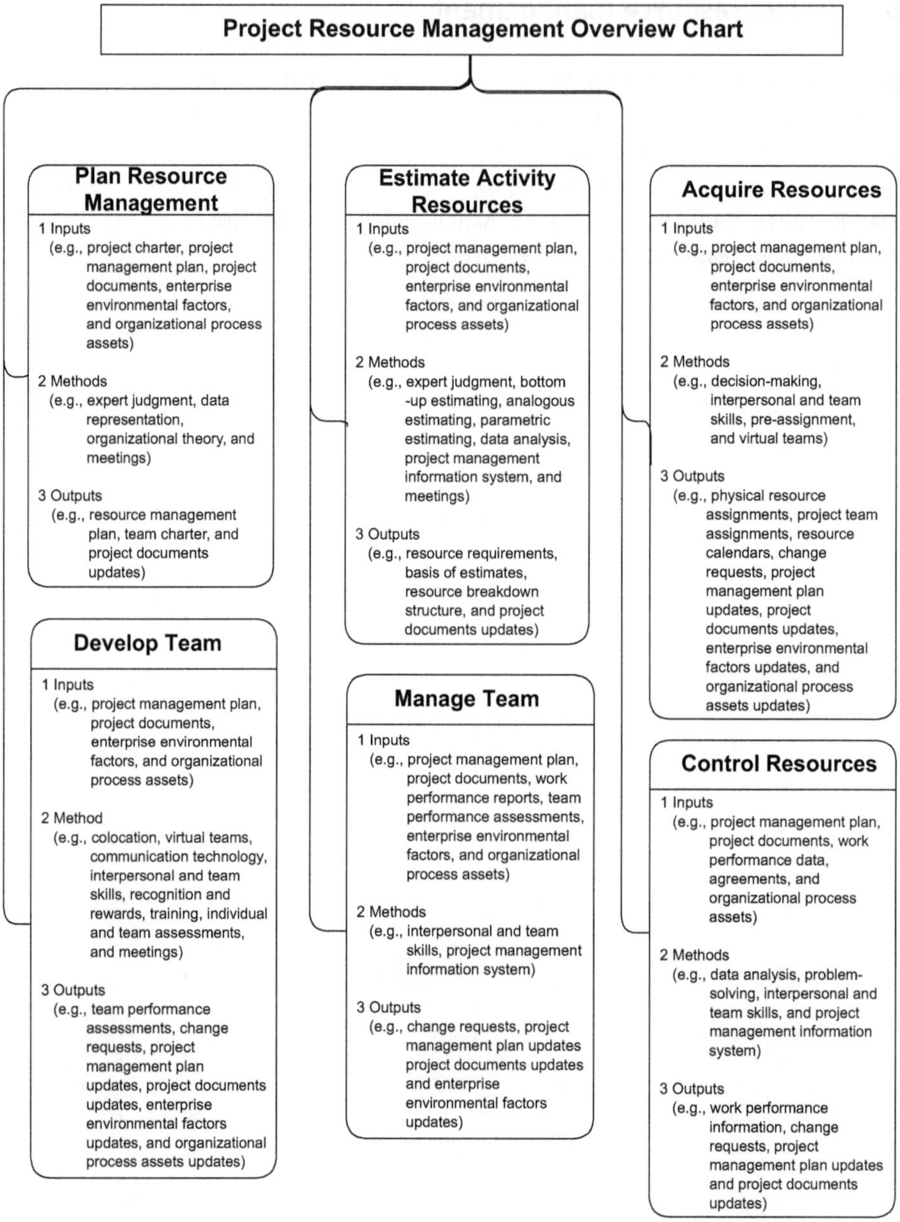

Figure 3.60: Project Resource Management overview chart.

3.6.1 The "Plan Resource Management" process

The "Plan Resource Management" process can help the project manager in estimating, acquiring, managing, and utilizing the project human and physical resources.

In reference to Figures 3.61 and 3.62, the main inputs, methods, and outputs of this process are detailed in Tables 3.79, 3.80, and 3.81, respectively.

The "Plan Resource Management" Process		
Inputs	**Methods**	**Outputs**
1 Project charter	1 Expert judgment	1 Resource management plan
2 Project management plan (e.g., quality management plan and scope baseline)	2 Data representation (e.g., hierarchical charts, responsibility assignment matrix, and text-oriented formats)	2 Team charter
3 Project documents (e.g., project schedule, requirements documentation, risk register, and stakeholder register)	3 Organizational theory	3 Project documents updates (e.g., assumption log and risk register)
4 Enterprise environmental factors	4 Meetings	
5 Organizational process assets		

Figure 3.61: The "Plan Resource Management" process.

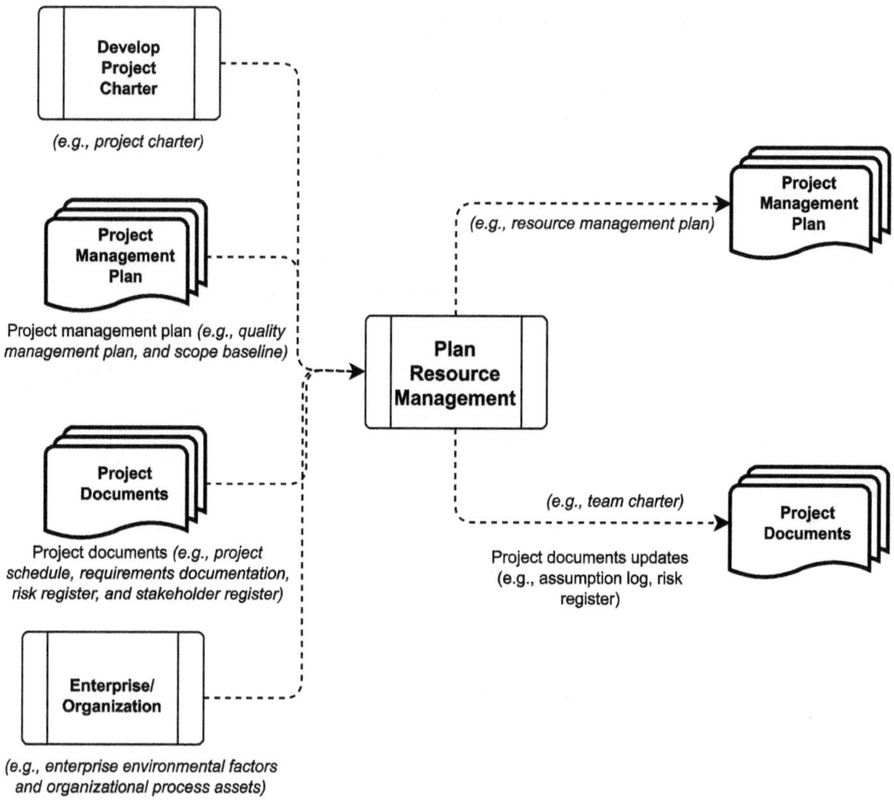

Figure 3.62: The "Plan Resource Management" process data flow diagram.

Table 3.79: Inputs to the "Plan Resource Management" process.

Input	Details
Project charter	As described earlier in Table 3.3.
Project management plan	The main components of the project management plan used as inputs to this process are the quality management plan and scope baseline.
Project documents	The main documents used as inputs to this process are the project schedule, requirements documentation, risk register, and stakeholder register.
Enterprise environmental factors (EEFs)	The main EEFs used as inputs to this process are the organizational culture and structure, resources and facilities' geographic distribution, resources' competencies, and marketplace conditions.
Organizational process asset factors (OPAFs)	The main OPAFs used as inputs to this process are the human resource policies and procedures, salary policies, safety and security policies, and information of similar projects.

Table 3.80: Methods used during the "Plan Resource Management" process.

Method	Details
Expert judgment	This refers to all judgments and decisions made while planning for the project resource management. Such judgments require the project manager and his/her team to have adequate level of the required relevant skills and experience.
Organizational theory	This refers to how teams and organizational units behave. It is used mainly to shorten the project time, reduce its cost, and complete it with less work.
Data representation	The main data representation methods used during this process are the hierarchical charts including the work breakdown structure, organizational breakdown structure , and resource breakdown structure; assignment matrix; and text-oriented formats.
Meetings	This refers to the meetings that take place in relevance to planning for the project resource management.

Table 3.81: Outputs of the "Plan Resource Management" process.

Output	Details

Resource management plan	The main components of the resource management plans are the:
	– Methods for deriving the project organization charts
	– Methods for identifying and acquiring resources
	– Methods for building the project team
	– Methods for defining the project team's roles and responsibilities:
	  – A *role* is the position such as programmer
	  – A *responsibility* is the duties assigned to a given role
	  – A *competency* is the set of skills that enable a person to take over his/her assigned role
	  – An *authority* is the rights to use resources, make decisions, sign approvals, and so forth
	– Methods for training, developing, controlling, recognizing, and managing the project team.

Team charter	This includes the project team values; team communications guidelines, decision-making process, conflict resolution process, meeting guidelines, and project team agreements.
Project documents' updates	The main documents that may be changed during this process are the assumption log and risk register.

3.6.2 The "Estimate Activity Resources" process

The "Estimate Activity Resources" process is used to estimate the project team resources, and the type and quantities of materials, equipment, and supplies necessary to carry on the project work.

In reference to Figures 3.63 and 3.64, the main inputs, methods, and outputs of this process are detailed in Tables 3.82, 3.83, and 3.84, respectively.

The "Estimate Activity Resources" Process		
Inputs	**Methods**	**Outputs**
1 Project management plan (e.g., resource management plan, scope baseline)	1 Expert judgment	1 Resource requirements
2 Project documents (e.g., activity attributes, activity list, assumption log, cost estimates, resource calendars, and risk register)	2 Bottom-up estimating	2 Basis of estimates
3 Enterprise environmental factors	3 Analogous estimating	3 Resource breakdown structure
4 Organizational process assets	4 Parametric estimating	4 Project documents updates (e.g., activity attributes, assumption log, and lessons learned register)
	5 Data analysis (e.g., alternatives analysis)	
	6 Project management information system	
	7 Meetings	

Figure 3.63: The "Estimate Activity Resources" process.

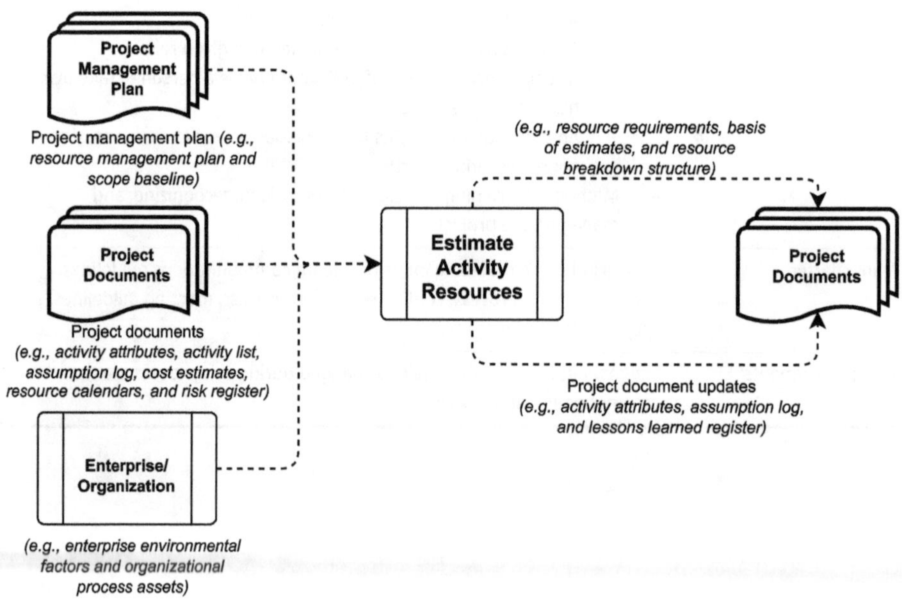

Figure 3.64: The "Estimate Activity Resources" process data flow diagram.

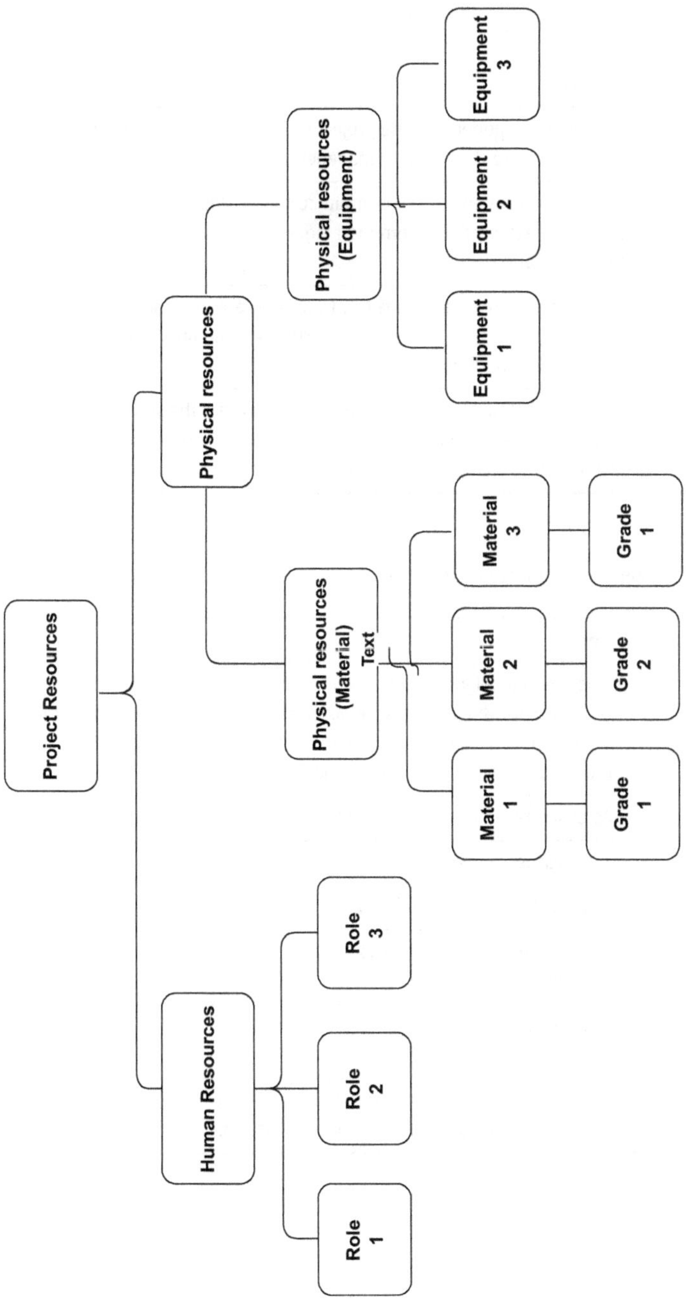

Figure 3.65: A sample resource breakdown structure.

Table 3.82: Inputs to the "Estimate Activity Resources" process.

Input	Details
Project management plan	The main components of the project management plan used as inputs to this process are the resource management plan and scope baseline.
Project documents	The main documents used as inputs to this process are the activity attributes, activity list, assumption log, cost estimates, resource calendars, and risk register.
Enterprise environmental factors (EEFs)	The main EEFs used as inputs to this process are the resource location, availability, and skills; organizational culture; estimated data; and marketplace conditions.
Organizational process asset factors (OPAFs)	The main OPAFs used as inputs to this process are the staffing policies and procedures, handling suppliers' policies and procedures, and historical information of similar projects.

Table 3.83: Methods used during the "Estimate Activity Resources" process.

Method	Details
Expert judgment	This refers to all judgments and decisions made while estimating resources for the project work activities. Such judgments require the project manager and his/her team to have adequate level of the required relevant skills and experience.
Bottom-up estimating	As described earlier in Table 3.50.
Analogous estimating	As described earlier in Table 3.50.
Parametric estimating	As described earlier in Table 3.50.
Data analysis	The main data analysis method used during this process is the alternatives' analysis.
Project management information system (PMIS)	As described earlier in Table 3.8.
Meetings	This refers to the meetings that take place in relevance to planning for the project work activities resource estimation.

Table 3.84: Outputs of the "Estimate Activity Resources" process.

Output	Details
Resource requirements	This depicts the types and quantities of the project's required resources.
Basis of estimates	This includes the methods for estimates development, assumptions made for estimates, constraints, estimates' ranges, and so forth.
Resource breakdown structure (RBS)	In reference to Figure 3.65, an RBS presents a hierarchical representation of resources by category and type.
Project documents' updates	The main documents that may be updated during this process are the activity attributes, assumption log, and the lessons learned register.

3.6.3 The "Acquire Resources" process

The "Acquire Resources" process is used to obtain the project team, facilities, equipment, materials, and supplies that are necessary to complete the project work.

In reference to Figures 3.66 and 3.67, the main inputs, methods, and outputs of this process are detailed in Tables 3.85, 3.86, and 3.87, respectively.

The "Acquire Resources" Process		
Inputs	**Methods**	**Outputs**
1 Project management plan (e.g., resource management plan and procurement management plan, and cost baseline)	1 Decision-making (e.g., multicriteria decision analysis)	1 Physical resource assignments
2 Project documents (e.g., project schedule, resource calendars, resource requirements, and stakeholder register)	2 Interpersonal and team skills (e.g., negotiation)	2 Project team assignments
3 Enterprise environmental factors	3 Pre-assignment	3 Resource calendars
4 Organizational process assets	4 Virtual teams	4 Change requests
		5 Project management plan updates (e.g., resource management plan, cost baseline)
		6 Project documents updates (e.g., lessons learned register project schedule, resource breakdown structure, resource requirements, risk register stakeholder register)
		7 Enterprise environmental factors updates
		8 Organizational process assets updates

Figure 3.66: The "Acquire Resources" process.

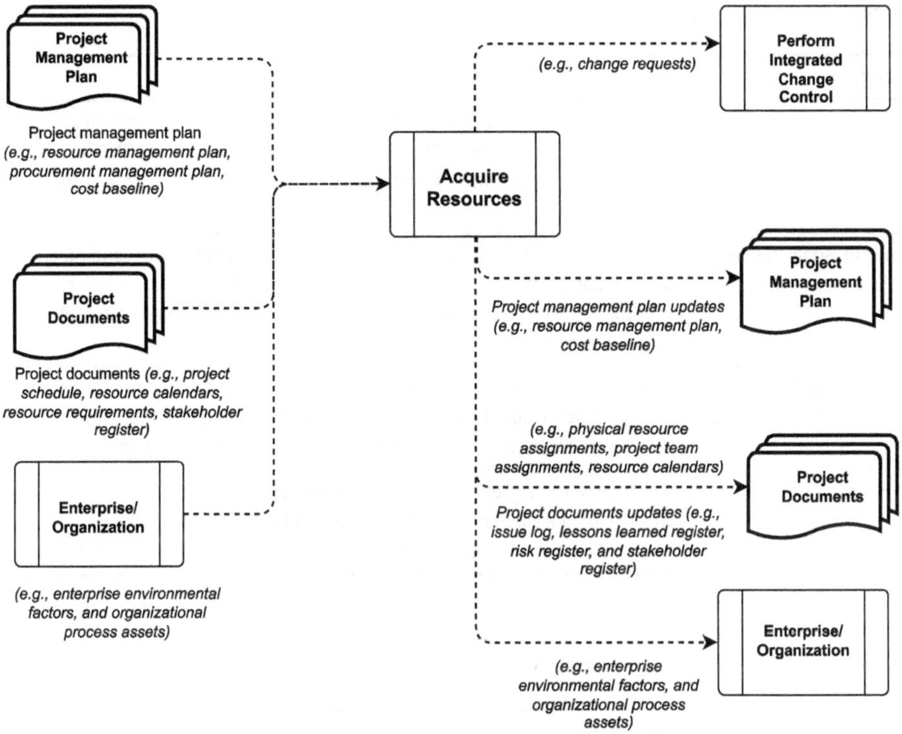

Figure 3.67: The "Acquire Resources" process data flow diagram.

Table 3.85: Inputs to the "Acquire Resources" process.

Input	Details
Project management plan	The main components of the project management plan used as inputs to this process are the resource management plan, procurement management plan, and cost estimates.
Project documents	The main documents used as inputs to this process are the project schedule, resource calendar, resource requirements, and stakeholder register.
Enterprise environmental factors (EEFs)	The main EEFs used as inputs to this process are the existing information on resources such as their availability, location, and skills and experience levels; organizational structure; and marketplace conditions.
Organizational process asset factors (OPAFs)	The main OPAFs used as inputs to this process are the staffing policies and procedures and information of similar projects.

Table 3.86: Methods used during the "Acquire Resources" process.

Method	Details
Decision-making	The main decision-making method used during this process is the multi-CRITERIA decision analysis. The main elements of the used SELECTION-CRITERIA are the resource availability, cost, ability, experience, knowledge, skills, and attitude.
Interpersonal manager	Negotiation is the main skill that is required in project managers and their project teams. The project team members may need to negotiate their matters with their functional managers, project managers, and probably external organizations and suppliers.
Pre-assignment	This refers to the resources who were already assigned to the project. Pre-assignments include the team members who have already been assigned in earlier process before completing the resource management plan.
Virtual teams	A virtual team is a group of people with a shared goal who fulfill their roles with little or no time spent meeting face to face. Virtual team makes it possible to establish a team without having its members to be physically in the same place.

Table 3.87: Outputs of the "Acquire Resources" process.

Output	Details
Physical resource assignments	This refers to the material, equipment, supplies, locations, and other physical resources that will be used by each resource during the project.
Project team assignments	This refers to the team members and their roles and responsibilities for the project.
Resource calendars	A resource calendar depicts the working days, shifts, business hours, weekends, public holidays, and so forth.
CHANGE-REQUESTS	As described earlier in Table 3.9.
Project management plan updates	The main components of the project management plan that may be changed during this process are the resource management plan and cost baseline.
Project documents' updates	The main documents that may be updated during this process are the lessons learned register, project schedule, resource breakdown structure, resource requirements, risk register, and stakeholder register.
Enterprise environmental factor (EEF) updates	This includes the resource availability within the organization, and the amount of consumable resources that have been used throughout the project life cycle.
Organizational process asset factor (OPAF) updates	This includes the documentation of resource acquirements, assignments, and allocations.

3.6.4 The "Develop Team" process

The "Develop Team" process is used to improve the project team members' competencies, interactions, and their work environment. This process improves teamwork, enhances interpersonal skills and competencies, motivates employees, and improves the project performance.

In reference to Figures 3.68 and 3.69, the main inputs, methods, and outputs of this process are detailed in Tables 3.88, 3.89, and 3.90, respectively.

The "Develop Team" Process		
Inputs	**Methods**	**Outputs**
1 Project management plan (e.g., Resource management plan)	1 Colocation	1 Team performance assessments
2 Project documents (e.g., Lessons learned register, project schedule, project team assignments, resource calendars, team charter)	2 Virtual teams	2 Change requests
3 Enterprise environmental factors	3 Communication technology	3 Project management plan Updates (e.g., Resource management plan)
4 Organizational process assets	4 Interpersonal and team skills (e.g., Conflict management, influencing, motivation, negotiation, team building)	4 Project documents updates (e.g., Lessons learned register, project schedule, project team assignments, resource calendars, team charter)
	5 Recognition and rewards	5 Enterprise environmental factors updates
	6 Training	6 Organizational process assets updates
	7 Individual and team assessments	
	8 Meetings	

Figure 3.68: The "Develop Team" process.

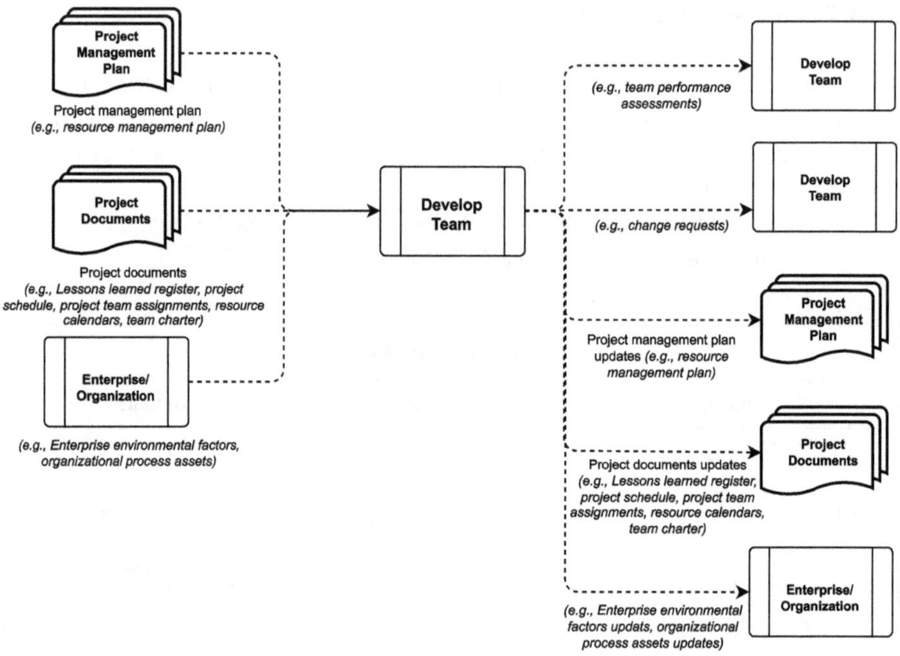

Figure 3.69: The "Develop Team" process data flow diagram.

Table 3.88: Inputs to the "Develop Team" process.

Input	Details
Project management plan	The main component of the project management plan used as an input to this process is the resource management plan.
Project documents	The main documents used as inputs to this process are the lessons learned register, project schedule, project team assignments, resource calendars, and team charter.
Enterprise environmental factors (EEFs)	The main EEFs used as inputs to this process are the human resources system policies, team members' skills and competencies, and team members' geographical distribution.
Organizational process asset factors (OPAFs)	The main OPAFs used as inputs to this process are the relevant historical data, and lessons learned repository.

Table 3.89: Methods used during the "Develop Team" process.

Method	Details
Co-location	This involves placing most of the project team members in the same physical location to enhance their ability to perform as a collaborative team.
Virtual teams	As described earlier in Table 3.86.
Communication Technology	The main methods used to transfer information to the project stakeholders include conversations, meetings, databases, documents, websites, and social media. The communication technology SELECTION-CRITERIA depends on many factors such the urgency of the need for the information, availability and reliability of technology, ease of use technology, project environment, and sensitivity and confidentiality of information.
Interpersonal and team skills	This includes the conflict management skills, ability to influence others, ability to motivate others, negotiation skills, and team building skills.
Recognition and rewards	This includes all what it takes to recognize and reward well-performing members of the project team.
Training	This includes all training activities meant to leverage the skills and competencies of the project team members.
Team assessments	This includes any tool that is used to assess the performance of the project team members.
Meetings	This refers to the meetings that take place in relevance to the project team development activities.

Table 3.90: Outputs of the "Develop Team" process.

Output	Details
Team performance assessments	This presents the performance assessments of the project team members throughout the project life cycle.
CHANGE-REQUESTS	As described earlier in Table 3.9.
Project management plan updates	The main component of the project management plan that may be changed during this process is the resource management plan.
Project documents' updates	The main documents that may be updated during this process are the lessons learned register, project schedule, project team assignments, resource calendars, and team charter.
Enterprise environmental factor (EEF) updates	The main EEFs that may be updated during this process are the employee development plan records and skill assessments.
Organizational process asset factor (OPAF) updates	The main OPAFs that may be updated during this process are the training requirements and personnel assessment.

3.6.5 The "Manage Team" process

The "Manage Team" process tracks the performance of the team members, provides them with feedback, resolves their issues, and manages their changes.

In reference to Figures 3.70 and 3.71, the main inputs, methods, and outputs of this process are detailed in Tables 3.91, 3.92, and 3.93, respectively.

The "Manage Team" Process		
Inputs	**Methods**	**Outputs**
1 Project management plan (e.g., Resource management plan) 2 Project documents (e.g., issue log, lessons learned register, project team assignments, team charter) 3 Work performance reports 4 Team performance assessments 5 Enterprise environmental factors 6 Organizational process assets	1 Interpersonal and team skills (e.g., conflict management, decision-making, emotional intelligence, influencing, leadership) 2 Project management information system	1 Change requests 2 Project management plan updates (e.g., resource management plan, schedule baseline, cost baseline) 3 Project documents updates (e.g., issue log, lessons learned register, project team assignments) 4 Enterprise environmental factors updates

Figure 3.70: The "Manage Team" process.

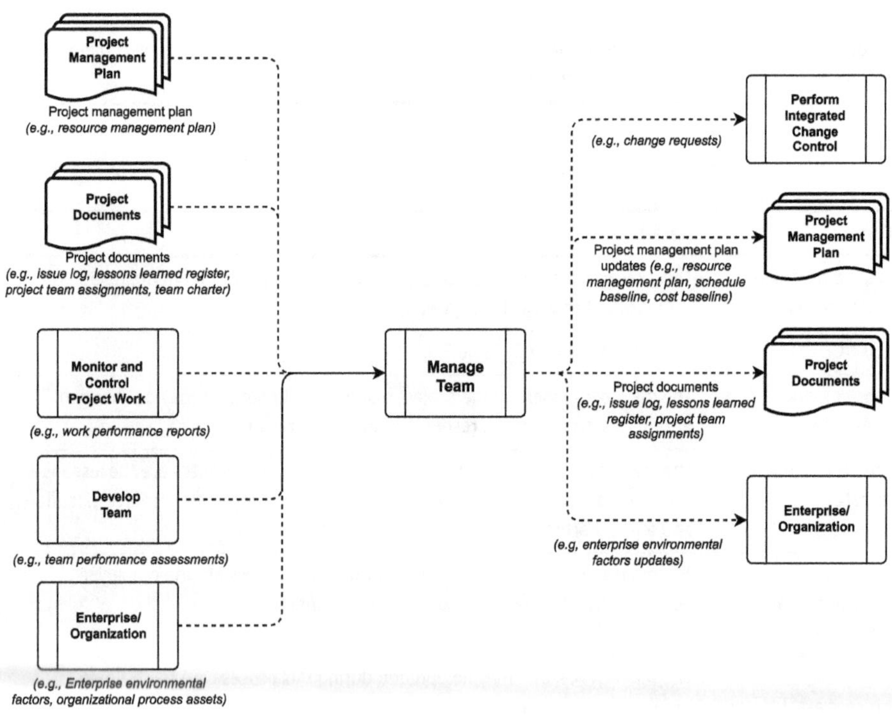

Figure 3.71: The "Manage Team" process data flow diagram.

Table 3.91: Inputs to the "Manage Team" process.

Input	Details
Project management plan	The main component of the project management plan used as an input to this process is the resource management plan.
Project documents	The main documents used as inputs to this process are the issue log, lessons learned register, project team assignments, and team charter.
Project work performance reports	As described earlier in Table 3.15.
Team performance assessments	This refers to the ongoing assessments of the project team's performance.
Enterprise environmental factors (EEFs)	The main enterprise environmental factor used as an input to this process is the human resource management policies.
Organizational process asset factors (OPAFs)	The main organizational process asset used as an input to this process is the certificates of appreciation.

Table 3.92: Methods used during the "Manage Team" process.

Method	Details
Interpersonal and team skills	This includes the conflict management methods such as withdraw or avoid, smooth or accommodate, compromise or reconcile, force or direct, and collaborate or solve problems; decision-making; emotional intelligence; ability to influence others; and ability to lead.
Project management information system (PMIS)	As described earlier in Table 3.8.

Table 3.93: Outputs of the "Manage Team" process.

Output	Details
CHANGE-REQUESTS	As described earlier in Table 3.9.
Project management plan updates	The main components of the project management plan that may be changed during this process are the resource management plan, schedule baseline, and cost baseline.
Project documents' updates	The main documents that may be updated during this process are the issue log, lessons learned register, and project team assignments.
Enterprise environmental factor (EEF) updates	The main EEFs that may be updated during this process are the inputs to the organizational performance appraisal systems and personnel skills.

3.6.6 The "Control Resources" process

The "Control Resources" process ensures that the project's physical resources are available as planned for. Also, it monitors the planned for versus actual utilization of resources and take corrective action as necessary.

In reference to Figures 3.72 and 3.73, the main inputs, methods, and outputs of this process are detailed in Tables 3.94, 3.95, and 3.96, respectively.

The "Control Resources" Process		
Inputs	**Methods**	**Outputs**
1 Project management plan (e.g.,resource management plan)	1 Data analysis (e.g., alternatives analysis, cost-benefit analysis, performance reviews, and trend analysis)	1 Work performance information
2 Project documents (e.g., issue log, lessons learned register, physical resource assignments, project schedule, resource breakdown structure, resource requirements, risk register)	2 Problem solving	2 Change requests
3 Work performance data	3 Interpersonal and team skills (e.g., negotiation and influencing)	3 Project management plan updates (e.g., resource management plan, schedule baseline, and cost baseline)
4 Agreements	4 Project management information system	4 Project documents updates (e.g., assumption log, issue log, lessons learned register, physical resource assignments, resource breakdown structure, and risk register)
5 Organizational process assets		

Figure 3.72: The "Control Resources" process.

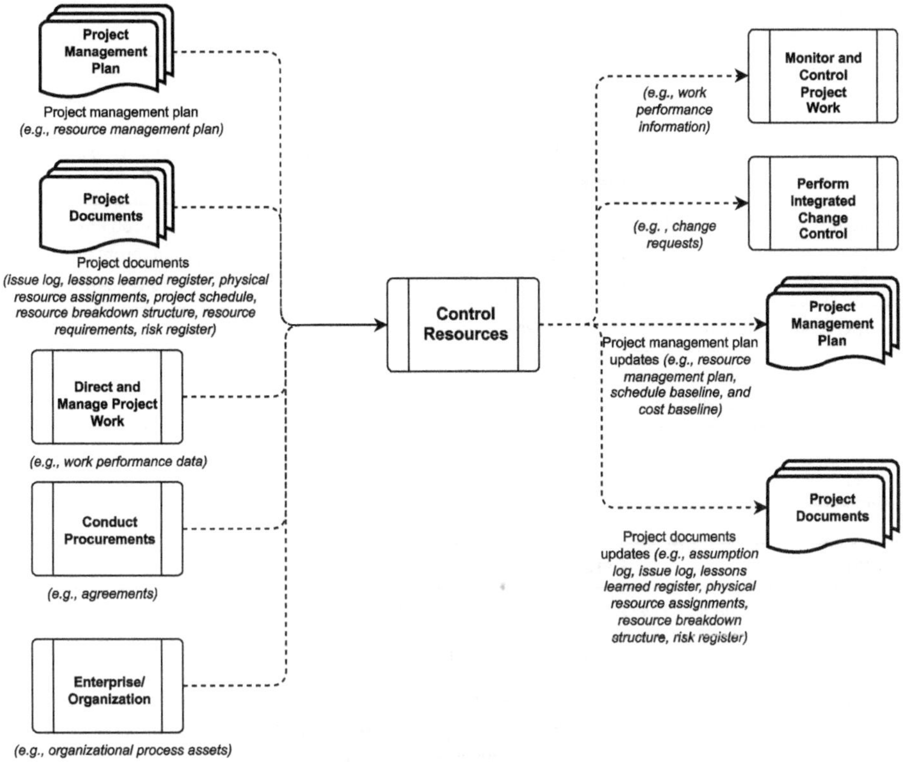

Figure 3.73: The "Control Resources" process data flow diagram.

Table 3.94: Inputs to the "Control Resources" process.

Input	Details
Project management plan	The main component of the project management plan used as an input to this process is the resource management plan.
Project documents	The main documents used as inputs to this process are the issue log, lessons learned register, physical resource equipment, project schedule, resource breakdown structure, resource requirements, and risk register.
Project work performance data	As described earlier in Table 3.9.
Agreements	As described earlier in Table 3.1.
Organizational process asset factors (OPAFs)	The main OPAFs used as inputs to this process are the resource control and assignment policies, issues' escalation procedure, and lessons learned from similar projects.

Table 3.95: Methods used during the "Control Resources" process.

Method	Details
Data analysis	The main data analysis methods used during this process are the alternatives' analysis, cost-benefit analysis, performance reviews, and trend analysis.
Problem solving	This method goes through six steps to solve a problem: problem identification, problem definition, problem investigation, problem analyzation, problem solving, and solution checking and validating.
Interpersonal and team skills	This includes the negotiation and influencing skills.
Project management information system (PMIS)	As described earlier in Table 3.8.

Table 3.96: Outputs of the "Control Resources" process.

Output	Details
Project work performance data	As described earlier in Table 3.9.
CHANGE-REQUESTS	As described earlier in Table 3.9.
Project management plan updates	The main components of the project management plan that may be changed during this process are the resource management plan, schedule baseline, and cost baseline.
Project documents' updates	The main documents that may be updated during this process are the assumption log, issue log, lessons learned register, physical resource assignments, and risk register.

3.7 Project communications management

Project communications management includes the processes that are necessary to ensure that the project informational needs are satisfied through a set of activities designed to achieve effective information exchange. This is the exchange of information including ideas, instructions, and/or emotions. This can be in written or spoken, formal or informal, through gestures or media.

Misunderstanding can be reduced by using the 5Cs method that must be supported by communication skills including listening, awareness of cultural and personal differences, and using persuasive language. The 5Cs method stands for **Correct** grammar and spelling, **Concise** expression and elimination of excess words, **Clear** purpose and expression directed to the needs of the reader, **Coherent** logical flow of ideas, and **Controlling** flow of words and ideas.

In reference to Figure 3.74, the project communications management processes are the plan communications management, manage communications, and monitor communications.

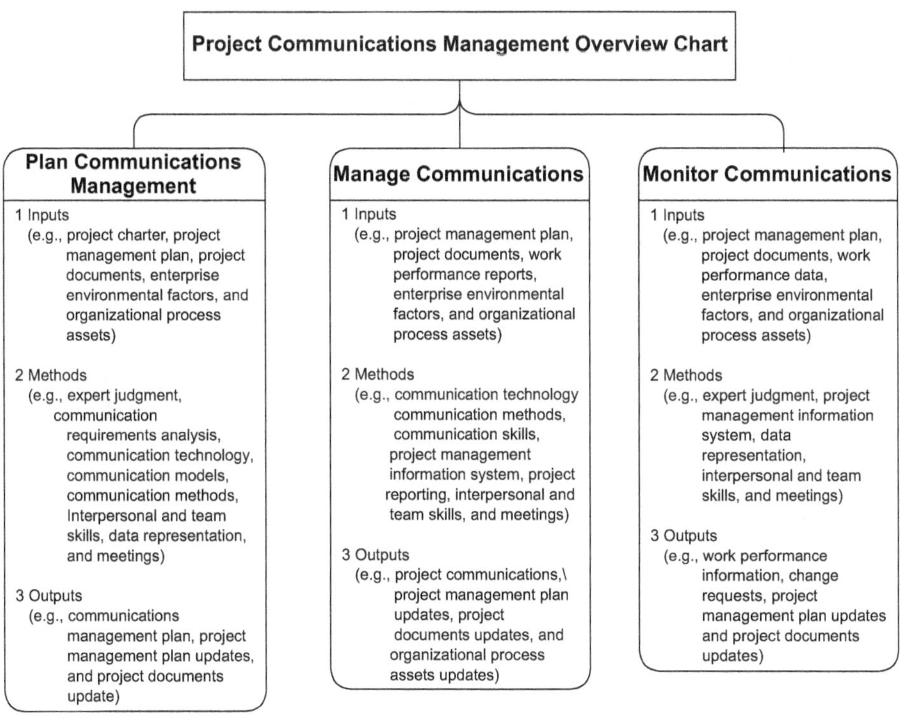

Figure 3.74: Project communications management overview chart.

3.7.1 The "Plan Communications Management" process

The "Plan Communications Management" process is used to plan for the project communications.

In reference to Figures 3.75 and 3.76, the main inputs, methods, and outputs of this process are detailed in Tables 3.97, 3.98, and 3.99, respectively.

The "Plan Communications Management" Process		
Inputs	**Methods**	**Outputs**
1 Project charter	1 Expert judgment	1 Communications management plan
2 Project management plan (e.g., resource management plan, stakeholder engagement plan)	2 Communication requirements analysis	2 Project management plan updates (e.g., stakeholder engagement plan)
3 Project documents (e.g., requirements documentation, stakeholder register)	3 Communication technology	3 Project documents updates (e.g., project schedule, stakeholder register)
4 Enterprise environmental factors	4 Communication models	
5 Organizational process assets	5 Communication methods	
	6 Interpersonal and team skills (e.g., communication styles assessment, political awareness, cultural awareness)	
	7 Data representation (e.g., stakeholder engagement assessment matrix)	
	8 Meetings	

Figure 3.75: The "Plan Communications Management" process.

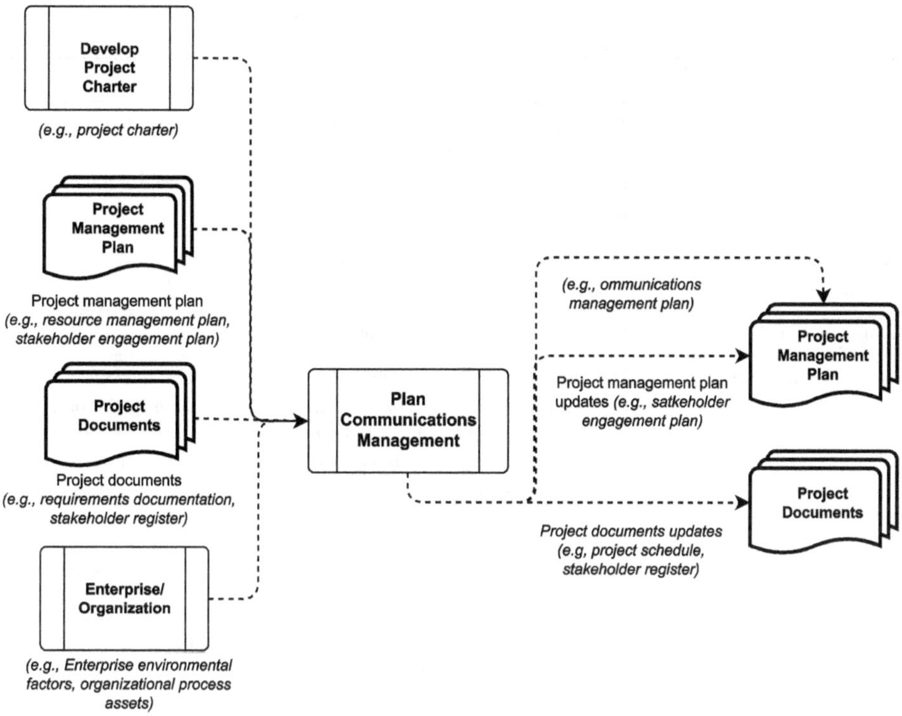

Figure 3.76: The "Plan Communications Management" process data flow diagram.

Table 3.97: Inputs to the "Plan Communications Management" process.

Input	Details
Project charter	As described earlier in Table 3.3.
Project management plan	The main components of the project management plan used as inputs to this process are the resource management plan, and stakeholders' engagement plan.
Project documents	The main documents used as inputs to this process are the requirements documentation and stakeholder register.
Enterprise environmental factors (EEFs)	The main EEFs used as inputs to this process are the organizational culture, organizational politics, organizational governance, human resources policies, stakeholder risk thresholds, and geographical distribution of resources.
Organizational process asset factors (OPAFs)	The main OPAFs used as inputs to this process are the organizational policies and procedures for social media, ethics, security, risks, CHANGE-REQUESTS, and data management; organizational communications requirements; and communication data from similar projects.

Table 3.98: Methods used during the "Plan Communications Management" process.

Method	Details
Expert judgment	This refers to all judgments and decisions made while planning for the project communications management. Such judgments require the project manager and his/her team to have adequate level of the required relevant skills and experience.
Communication requirements analysis	This analysis determines the project and stakeholders' informational needs.
Communication technology	As described earlier in Table 3.89.
Interpersonal and team skills	This includes the communication style assessment used to assess possible communication styles and choose the right one for the project; political awareness that helps the project manager to plan for the project communications based on the project environment; and cultural awareness that helps understand the differences between individuals, groups, and organizations.
Data representation	The stakeholder engagement matrix is the main data representation used during this process.
Meetings	This refers to the meetings that take place in relevance to the planning for the project communications management.
Communication models and methods	The main communication models used during this process are the basic sender/receiver communication model. With this model, information is first encoded and then transmitted, and once arrived at the receiver side, information gets decoded and used; and interactive communication model. With this model, once information is received, the receiver acknowledges the reception of information, and then decoded. The main communication methods used during this process are the interactive communication, push communication, pull communication, interpersonal communication, small group communication, public communication, mass communication, and networks and social computing communication.

Table 3.99: Outputs of the "Plan Communications Management" process.

Output	Details
Communication management plan	The communication management plan is an integral component of the overall project management plan. It shows how project communication is planned for, structured, implemented, and monitored.
Project management plan updates	The main component of the project management plan that may be changed during this process is the stakeholders' engagement plan.
Project documents' updates	The main documents that may be updated during this process are the project schedule and stakeholder register.

3.7.2 The "Manage Communications" process

The "Manage Communications" process is used to ensure effective and timely collection, creation, distribution, storage, retrieval, management, monitoring, and disposition of the project information.

In reference to Figures 3.77 and 3.78, the main inputs, methods, and outputs of this process are detailed in Tables 3.100, 3.101, and 3.102, respectively.

The "Manage Communications" Process		
Inputs	**Methods**	**Outputs**
1 Project management plan (e.g., resource management plan, communications management plan, stakeholder engagement plan)	1 Communication technology	1 Project communications
2 Project documents (e.g., change log, issue log, lessons learned register, quality report, risk report, stakeholder register)	2 Communication methods	2 Project management plan updates (e.g., communications management plan, stakeholder engagement plan)
3 Work performance reports	3 Communication skills (e.g., communication competence, feedback, nonverbal, presentations)	3 Project documents updates (e.g., issue log, lessons learned register, project schedule, risk register, stakeholder register)
4 Enterprise environmental factors	4 Project management information system	4 Organizational process assets updates
5 Organizational process assets	5 Project reporting	
	6 Interpersonal and team skills (e.g., active listening, conflict management, cultural awareness, meeting management, networking, political awareness)	
	7 Meetings	

Figure 3.77: The "Manage Communications" process.

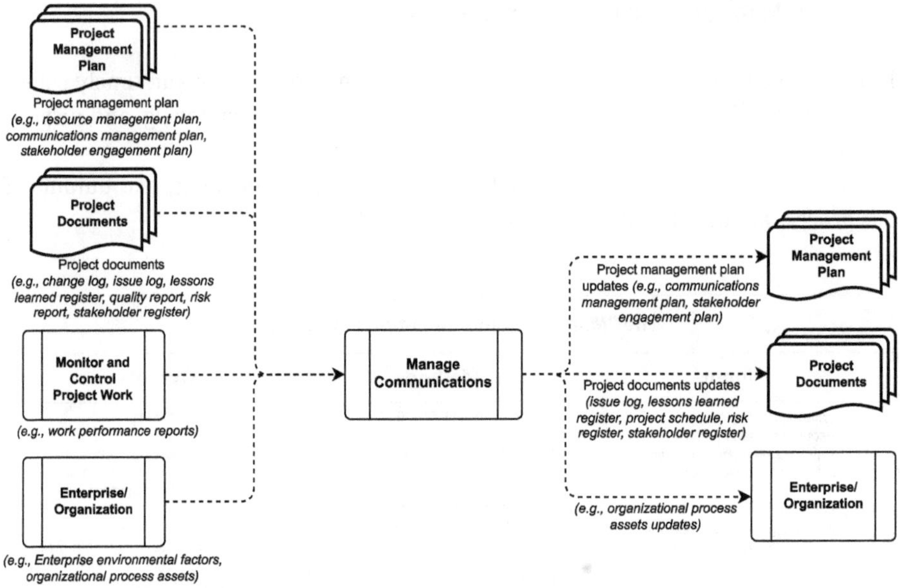

Figure 3.78: The "Manage Communications" process data flow diagram.

Table 3.100: Inputs to the "Manage Communications" process.

Input	Details
Project management plan	The main components of this plan used as inputs to this process are the resource management plan, communications management plan, and stakeholders' engagement plan.
Project documents	The main documents used as inputs to this process are the change log, issue log, lessons learned register, quality report, risk report, and stakeholder register.
Project work performance reports	As described earlier in Table 3.15.
Enterprise environmental factors (EEFs)	The main EEFs used as inputs to this process are the organizational culture, organizational political climate, organizational governance framework, human resources policies, stakeholder risk thresholds, communication channels, and so forth.
Organizational process asset factors (OPAFs)	The main OPAFs used as inputs to this process are the corporate policies and procedures for social media, ethics, security, risks, CHANGE-REQUESTS, and data management; organizational communications requirements; guidelines for information development, exchange, storage, and retrieval; data from similar projects; and lessons learned repository.

Table 3.101: Methods used during the "Manage Communications" process.

Method	Details
Communication technology	As described earlier in Table 3.89.
Communication methods	As described earlier in Table 3.98.
Communication skills	The main communication skills used during this process are the ability to communicate clearly and competently, always providing feedback after the communication takes place, ability to communicate nonverbally, ability to use presentations, and so forth.
Project management information system (PMIS)	As described earlier in Table 3.8.
Project reporting	This refers to the collection and distribution of the project information.
Interpersonal and team skills	The main interpersonal and team skills considered during this process are listen actively, ability to manage conflicts, being able to deal with cultural and political matters, being able to effectively manage meetings and networking.
Meetings	This refers to the meetings that take place in relevance to the project communications management.

Table 3.102: Outputs of the "Manage Communications" process.

Output	Details
Project communications	This includes all communications that take place throughout the project life cycle such as performance reports, deliverable status, schedule progress, cost incurred, and presentations.
Project management plan updates	The main components of the project management plan used as inputs to this process are the communications management plan and stakeholders' engagement plan.
Project documents' updates	The main documents that may be updated during this process are the issue log, lessons learned register, project schedule, risk register, and stakeholder register.
Organizational process asset factor (OPAF) updates	This includes the updates that take place on the project records, and project reports and presentations.

3.7.3 The "Monitor Communications" process

The "Monitor Communications" process is used to ensure that the project and its stakeholders' informational needs are satisfied.

In reference to Figures 3.79 and 3.80, the main inputs, methods, and outputs of this process are detailed in Tables 3.103, 3.104, and 3.105, respectively.

The "Monitor Communications" Process		
Inputs	**Methods**	**Outputs**
1 Project management plan (e.g., resource management plan, communications management plan, stakeholder engagement plan)	1 Expert judgment	1 Work performance information
2 Project documents (e.g., issue log, lessons learned register, project communications)	2 Project management information sys	2 Change requests
3 Work performance data	3 Data analysis (e.g., stakeholder engagement assessment matrix)	3 Project management plan updates (e.g., communications management plan, and stakeholder engagement plan)
4 Enterprise environmental factors	4 Interpersonal and team skills (e.g., observation/conversation)	4 Project documents updates (e.g., issue log, lessons learned register, stakeholder register)
5 Organizational process assets	5 Meetings	

Figure 3.79: The "Monitor Communications" process.

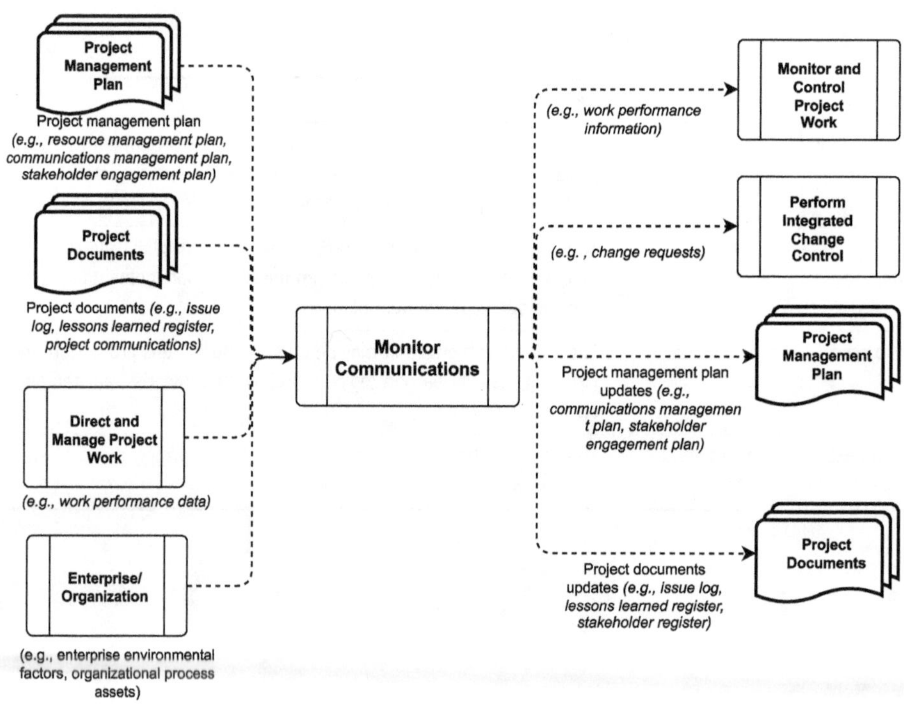

Figure 3.80: The "Monitor Communications" process data flow diagram.

Table 3.103: Inputs to the "Monitor Communications" process.

Input	Details
Project management plan	The main components of the project management plan used as inputs to this process are the resource management plan, communications management plan, and stakeholders' engagement plan.
Project documents	The main documents used as inputs to this process are the issue log, lessons learned register, and project communications.
Project work performance data	As described earlier in Table 3.9.
Enterprise environmental factors (EEFs)	The main EEFs used as inputs to this process are the organizational culture, organizational political climate, organizational governance framework, human resources policies, stakeholder risk thresholds, communication channels, and so forth.
Organizational process asset factors (OPAFs)	The main OPAFs used as inputs to this process are the corporate policies and procedures for social media, ethics, security, risks, change, and data management; organizational communications requirements; guidelines for information development, exchange, storage, and retrieval; data from similar projects; and lessons learned repository.

Table 3.104: Methods used during the "Monitor Communications" process.

Method	Details
Expert judgment	This refers to all judgments and decisions made while monitoring the project communications. Such judgments require the project manager and his/her team have adequate level of the required relevant skills and experience.
Project management information system (PMIS)	As described earlier in Table 3.8.
Data representation	The main data representation method that is used during this process is the stakeholder engagement assessment matrix, which depicts the communications effectiveness.
Interpersonal and team skills	The main interpersonal and team skills considered during this process are the observation and conversation.
Meetings	This refers to the meetings that take place in relevance to monitoring the project communications.

Table 3.105: Outputs of the "Monitor Communications" process.

Output	Details
Project work performance data	As described earlier in Table 3.9.
CHANGE-REQUESTS.	As described earlier in Table 3.9.
Project management plan updates	The main components of the project management plan used as inputs to this process are the communications management plan and stakeholders' engagement plan.
Project documents' updates	The main documents that may be updated during this process are the issue log, lessons learned register, and stakeholder register.

3.8 Project risk management

Project risk management includes the processes that are necessary to conduct project risk management planning; identify and analyze these risks; and plan, implement, and monitor responses to these risks. Project risk management ensures that all types of risks are considered, understood, and responded to in timely manners.

In reference to Figure 3.81, the project risk management processes are the plan risk management, identify risks, perform qualitative risk analysis, perform quantitative risk analysis, plan risk responses, implement risk responses, and monitor risks.

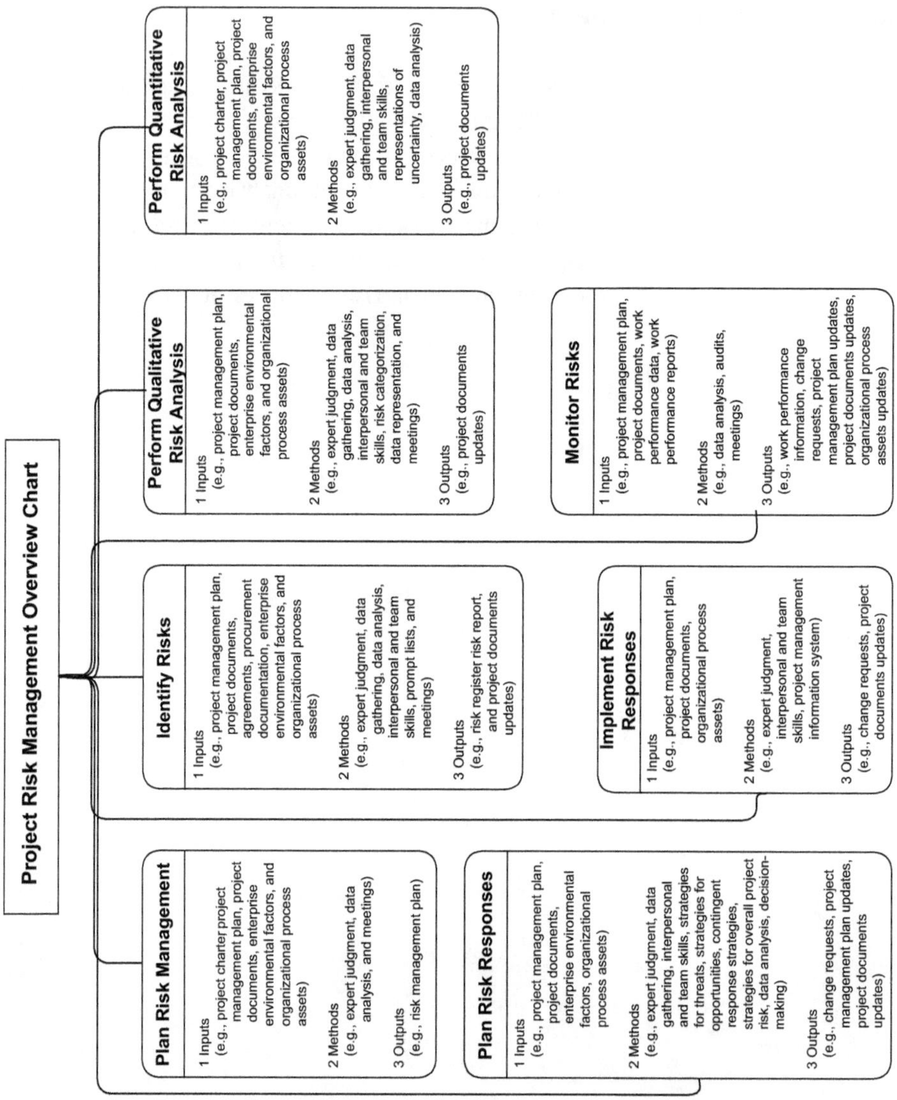

Figure 3.81: Project risk management overview chart.

3.8.1 The "Plan Risk Management" process

The "Plan Risk Management" process is used to define how to conduct the project risk management activities. It should begin when a project is started and should end early in the project.

In reference to Figures 3.82 and 3.83, the main inputs, methods, and outputs of this process are detailed in Tables 3.106, 3.107, and 3.108, respectively.

The "Plan Risk Management" Process		
Inputs	**Methods**	**Outputs**
1 Project charter	1 Expert judgment	1 Risk management plan
2 Project management plan (e.g., all components)	2 Data analysis (e.g., stakeholder analysis)	
3 Project documents (e.g., stakeholder register)	3 Meetings	
4 Enterprise environmental factors		
5 Organizational process assets		

Figure 3.82: The "Plan Risk Management" process.

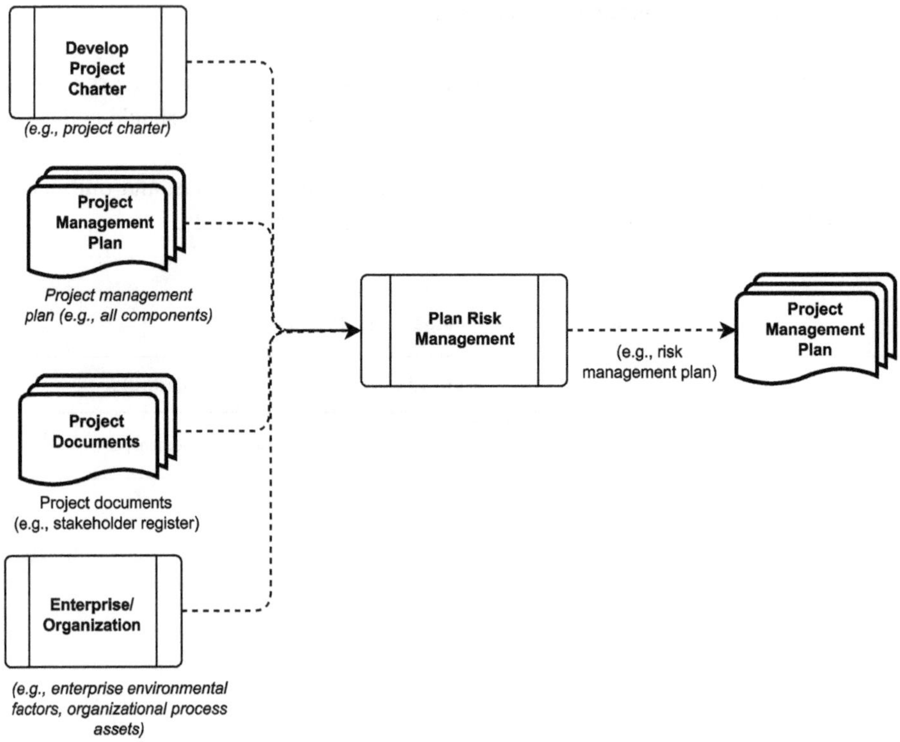

Figure 3.83: The "Plan Risk Management" process data flow diagram.

Table 3.106: Inputs to the "Plan Risk Management" process.

Input	Details
Project charter	As described earlier in Table 3.3.
Project management plan	Planning for risk management requires that all approved subsidiary plans are taken into consideration in order to make the risk management plan consistent with them.
Project documents	The project's main document used as inputs to this process is the stakeholder register.
Enterprise environmental factors (EEFs)	The main EEFs used as an input to this process is the risk thresholds.
Organizational process asset factors (OPAFs)	The main OPAFs used as inputs to this process are the organizational risk policy, risk categories, risk statement formats, risk management plan template, and so forth.

Table 3.107: Methods used during the "Plan Risk Management" process.

Method	Details
Expert judgment	This refers to all judgments and decisions made while planning for the project risk management. Such judgments require the project manager and his/her team to have adequate level of the required relevant skills and experience.
Data analysis	The main data representation method that is used during this process is the stakeholders analysis that is used to determine risks that are associated with the project stakeholders.
Meetings	This refers to the meetings that take place in relevance to risk management planning.

Table 3.108: Outputs of the "Plan Risk Management" process.

Output	Details
Project documents' updates	The project's main document that may be updated during this process is the risk management plan, which includes several elements: – Risk management strategy. This element describes the general approach to managing project risks. – Risk management method. This element defines the approaches, tools, and data sources used to perform the project risk management. – Roles and responsibilities. This element defines the risk management team members and clarifies their responsibilities. – Timing. This element schedules the project risk management processes throughout the project life cycle. – Risk categories and risk breakdown structure. – Definitions of risk probability and impacts. – Probability and impact matrix. – Reporting formats.

3.8.2 The "Identify Risks" process

The "Identify Risks" process is used to identify and document the project risks and their characteristics.

In reference to Figures 3.84 and 3.85, the main inputs, methods, and outputs of this process are detailed in Tables 3.109, 3.110, and 3.111, respectively.

The "Identify Risks" Process		
Inputs	**Methods**	**Outputs**
1 Project management plan (e.g., requirements management plan, schedule management plan, cost management plan, quality management plan, resource management plan, risk management plan, scope baseline, schedule baseline, cost baseline)	1 Expert judgment	1 Risk register
2 Project documents (e.g., assumption log, cost estimates, duration estimates, issue log, lessons learned register, requirements documentation, resource requirements, stakeholder register)	2 Data gathering (e.g., brainstorming, checklists, interviews)	2 Risk report
3 Agreements	3 Data analysis (e.g., root cause analysis, assumption and constraint analysis, SWOT analysis, document analysis)	3 Project documents updates (e.g., assumption log, issue log, lessons learned register)
4 Procurement documentation	4 Interpersonal and team skills (e.g., facilitation)	
5 Enterprise environmental factors	5 Prompt lists	
6 Organizational process assets	6 Meetings	

Figure 3.84: The "Identify Risks" process.

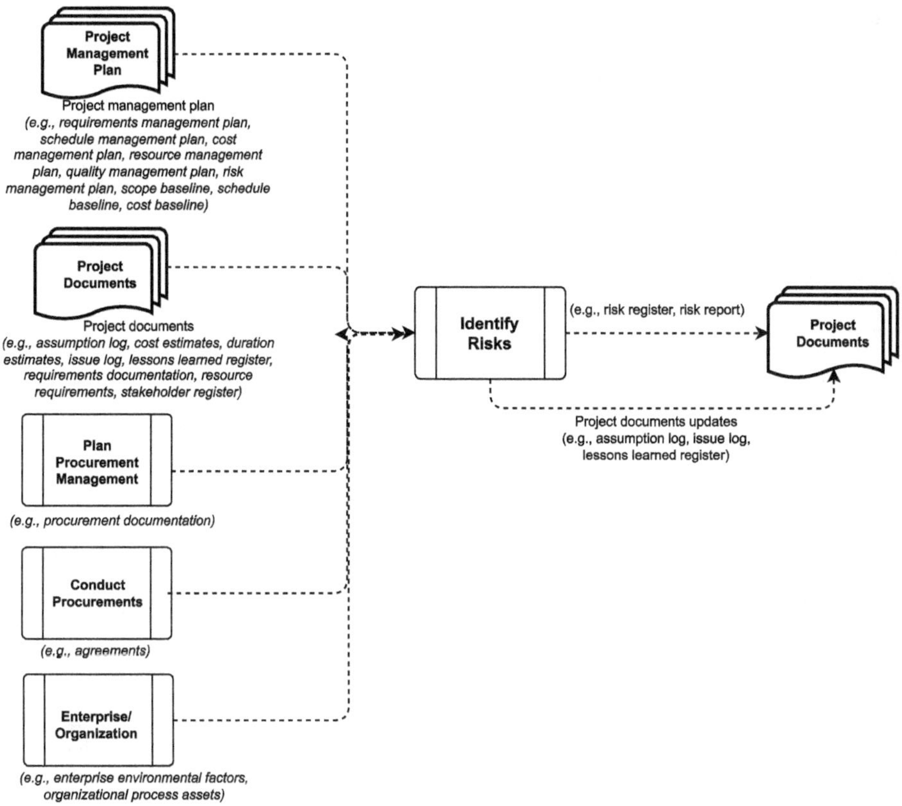

Figure 3.85: The "Identify Risks" process data flow diagram.

Table 3.109: Inputs to the "Identify Risks" process.

Input	Details
Project management plan	The main components of the project management plan used as inputs to this process are the requirements management plan, schedule management plan, cost management plan, quality management plan, resource management plan, risk management plan, scope baseline, schedule baseline, and cost baseline.
Project documents	The main documents used as inputs to this process are the assumption log, cost estimates, duration estimates, issue log, lessons learned register, requirements documentation, resource requirements, and stakeholder register.

Table 3.109 (continued)

Input	Details
Agreements	As described earlier in Table 3.1.
Procurement documentation	Procurement documentation is updated throughout the project life cycle and can be reviewed for risks.
Enterprise environmental factors (EEFs)	The main EEFs used as inputs to this process are the commercial risk databases, academic studies, benchmarking results, and information from current and similar projects.
Organizational process asset factors (OPAFs)	The main OPAFs used as inputs to this process are the project data, organizational and project process control system, and risk statement formats.

Table 3.110: Methods used during the "Identify Risks" process.

Method	Details
Expert judgment	This refers to all judgments and decisions made while identifying the project risks. Such judgments require the project manager and his/her team to have adequate level of the required relevant skills and experience.
Data gathering	The main data gathering methods used during this process are the brainstorming, checklists, and interviews.
Data analysis	The main data analysis methods used during this process are the root-cause analysis, assumption-and-constraint analysis, SWOT analysis, and documents analysis.
Interpersonal and team skills	Facilitation is the main interpersonal and team skill used during this process.
Prompt lists	This is a predetermined list of risk categories that may rise to the project.
Meetings	This refers to the meetings that take place in relevance to risk management planning.

Table 3.111: Outputs of the "Identify Risks" process.

Output	Details
Risk register	This contains the results of the following processes: perform qualitative risk analysis, plan risk responses, implement risk responses, and monitor risks. At the end of the identify risk process, the risk register can contain things like the list of identified risks; risk owner; and list of risk responses.
Risk report	This contains all risks identified during this process.
Project documents' updates	The main documents that may be updated during this process are the assumption log, issue log, and lessons learned register.

3.8.3 The "Perform Qualitative Risk Analysis" process

The "Perform Qualitative Risk Analysis" process is used to prioritize the project risks by assessing mainly their probability and impact.

In reference to Figures 3.86 and 3.87, the main inputs, methods, and outputs of this process are detailed in Tables 3.112, 3.113, and 3.114, respectively.

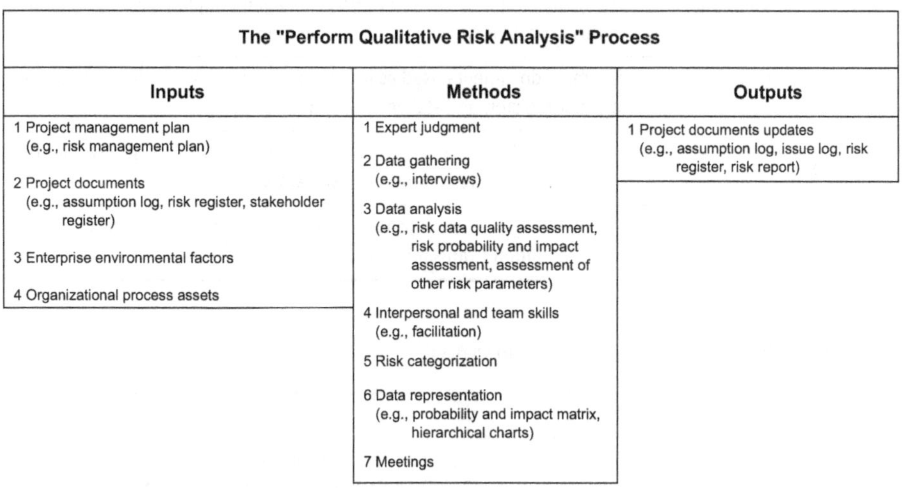

Figure 3.86: The "Perform Qualitative Risk Analysis" process.

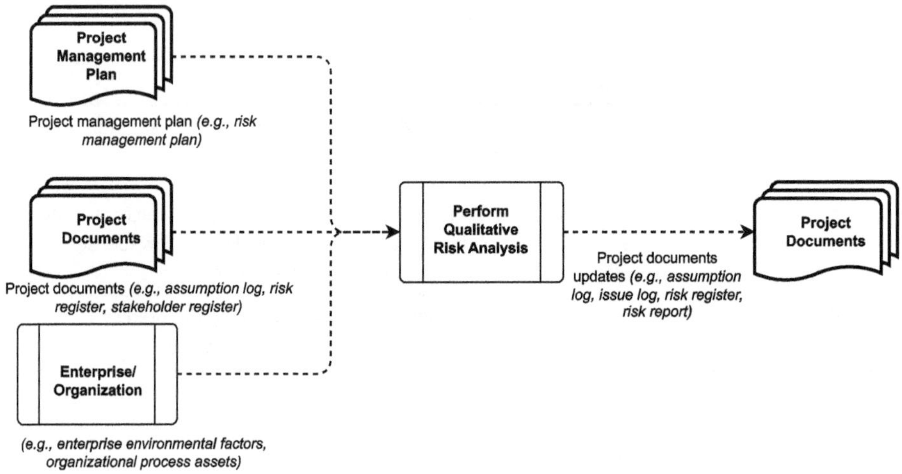

Figure 3.87: The "Perform Qualitative Risk Analysis" process data flow diagram.

Table 3.112: Inputs to the "Perform Qualitative Risk Analysis" process.

Input	Details
Project management plan	The main component of the project management plan used as an input to this process is the risk management plan.
Project documents	The main documents used as inputs to this process are the assumption log, risk register, and stakeholder register.
Enterprise environmental factors (EEFs)	The main EEFs used as inputs to this process are the industry studies of similar projects and commercial risk databases.
Organizational process asset factors (OPAFs)	The main organizational process asset used as an input to this process is the information from similar projects.

Table 3.113: Methods used during the "Perform Qualitative Risk Analysis" process.

Method	Details
Expert judgment	This refers to all judgments and decisions made while performing the project qualitative risk analysis. Such judgments require the project manager and his/her team to have adequate level of the required relevant skills and experience.
Data gathering	Interview is the main data gathering method that is used during this process.

Table 3.113 (continued)

Method	Details
Data analysis	The main data analysis methods used during this process are the risk data quality assessment; risk probability and impact assessment; and assessment of risk urgency, manageability, controllability, detectability, connectivity, strategic impact, and propinquity.
Interpersonal and team skills	Facilitation is the main interpersonal and team skill used during this process.
Risk categorization	Project risks can be categorized by their sources.
Data representation	The main data representation methods used during this process are the probability and impact matrix and hierarchical charts.
Meetings	This refers to the meetings that take place in relevance to performing qualitative risk analysis.

Table 3.114: Outputs of the "Perform Qualitative Risk Analysis" process.

Output	Details
Project documents' updates	The main documents that may be updated during this process are the assumption log, issue log, risk register, and risk report.

3.8.4 The "Perform Quantitative Risk Analysis" process

The "Perform Quantitative Risk Analysis" process is used to numerically analyze the combined effect of the project risks.

In reference to Figures 3.88 and 3.89, the main inputs, methods, and outputs of this process are detailed in Tables 3.115, 3.116, and 3.117, respectively.

The "Perform Quantitative Risk Analysis" Process		
Inputs	**Methods**	**Outputs**
1 Project management plan (e.g., risk management plan, scope baseline, schedule baseline, cost baseline) 2 Project documents (e.g., assumption log, basis of estimates, cost estimates, cost forecasts, duration estimates, milestone list, resource requirements, risk register, risk report, schedule forecasts) 3 Enterprise environmental factors 4 Organizational process assets	1 Expert judgment 2 Data gathering (e.g., interviews) 3 Interpersonal and team skills (e.g., facilitation) 4 Representations of uncertainty 5 Data analysis (e.g., simulations, sensitivity analysis, decision tree analysis, influence diagrams)	1 Project documents updates (e.g., risk report)

Figure 3.88: The "Perform Quantitative Risk Analysis" process.

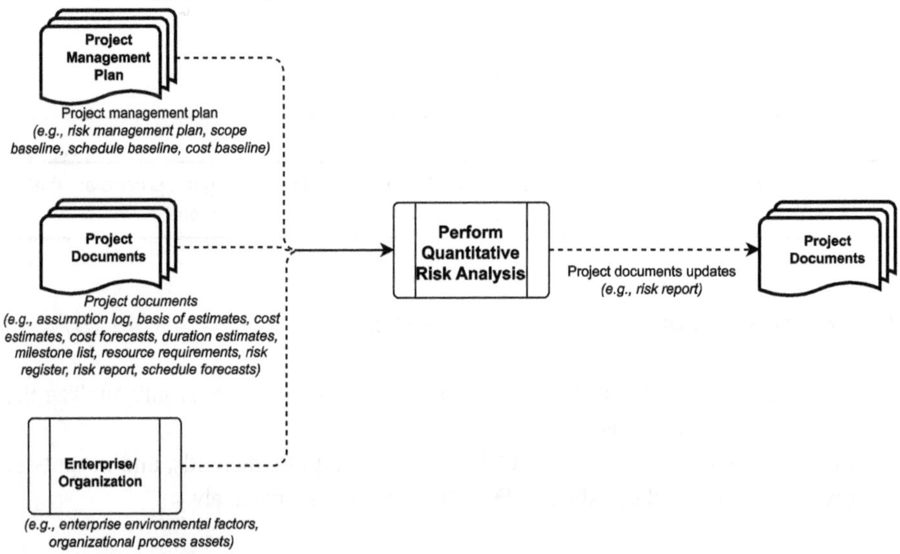

Figure 3.89: The "Perform Quantitative Risk Analysis" process data flow diagram.

Table 3.115: Inputs to the "Perform Quantitative Risk Analysis" process.

Input	Details
Project management plan	The main components of the project management plan used as inputs to this process are the risk management plan, scope baseline, schedule baseline, and cost baseline.
Project documents	The main documents used as inputs to this process are the assumption log, basis of estimates, cost estimates, cost forecasts, duration estimates, milestone list, resource requirements, risk register, and schedule forecasts.
Enterprise environmental factors (EEFs)	The main EEFs used as inputs to this process are the industry studies of similar projects and commercial risk databases.
Organizational process asset factors (OPAFs)	The main organizational process assets used as an input to this process is the information from similar projects.

Table 3.116: Methods used during the "Perform Quantitative Risk Analysis" process.

Method	Details
Expert judgment	This refers to all judgments and decisions made while performing the project quantitative risk analysis. Such judgments require the project manager and his/her team to have adequate level of the required relevant skills and experience.
Data gathering	Interviews is the main data gathering method that is used during this process.
Interpersonal and team skills	Facilitation is the main interpersonal and team skill used during this process.
Representation of uncertainty	A quantitative risk analysis model requires inputs that reflect individual project risks and uncertainties of the duration, cost, and resource requirements for the planned activities.
Data analysis	The main data analysis methods used during this process are the simulation method, sensitivity analysis, and decision tree analysis.

Table 3.117: Outputs of the "Perform Quantitative Risk Analysis" process.

Output	Details
Project documents' updates	The main documents that may be updated during this process are the assessment of overall project risk exposure, detailed probabilistic analysis of the project, trends in quantitative risk analysis, and recommended risk responses.

3.8.5 The "Plan Risk Responses" process

The "Plan Risk Responses" process is used to develop options, select strategies, and agree on actions to address and treat the project risks.

In reference to Figures 3.90 and 3.91, the main inputs, methods, and outputs of this process are detailed in Tables 3.118, 3.119, and 3.120, respectively.

The "Plan Risk Responses" Process		
Inputs	**Methods**	**Outputs**
1 Project management plan (e.g., resource management plan, risk management plan, and cost baseline)	1 Expert judgment	1 Change requests
2 Project documents (e.g., lessons learned register, project schedule, project team assignments, resource calendars, risk register, risk report, and stakeholder register)	2 Data gathering (e.g., interviews)	2 Project management plan updates (e.g., schedule management plan, cost management plan, quality management plan, resource management plan, procurement management plan, scope baseline, schedule baseline, and cost baseline)
3 Enterprise environmental factors	3 Interpersonal and team skills (e.g., facilitation)	3 Project documents updates (e.g., assumption log, cost forecasts, lessons learned registeproject schedule, project team assignments, risk register, and risk report)
4 Organizational process assets	4 Strategies for threats	
	5 Strategies for opportunities	
	6 Contingent response strategies	
	7 Strategies for overall project risk	
	8 Data analysis (e.g., alternatives analysis, cost-benefit analysis)	
	9 Decision-making (e.g., multicriteria decision analysis)	

Figure 3.90: The "Plan Risk Responses" process.

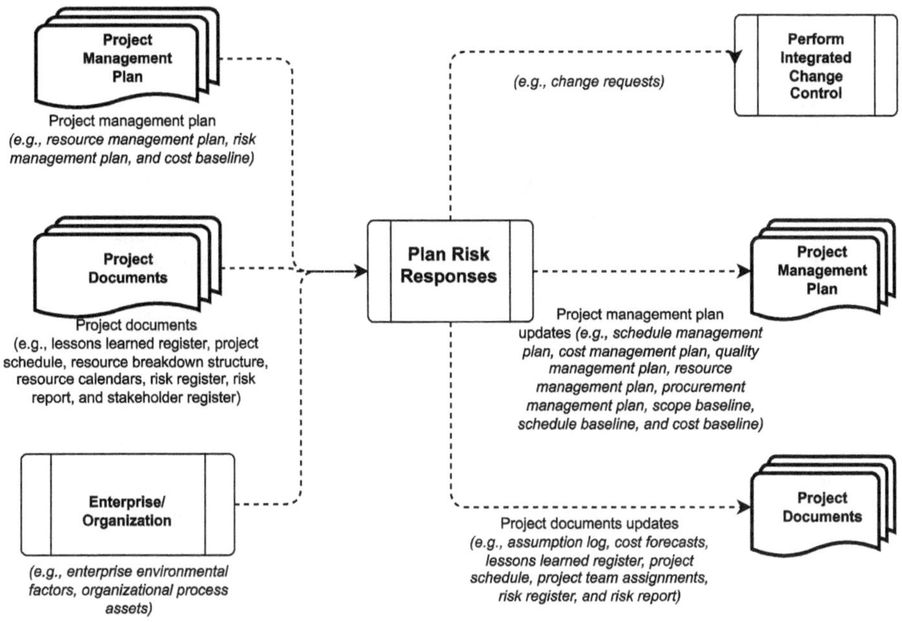

Figure 3.91: The "Plan Risk Responses" process data flow diagram.

Table 3.118: Inputs to the "Plan Risk Responses" process.

Input	Details
Project management plan	The main components of the project management plan used as inputs to this process are the resource management plan, risk management plan, and cost baseline.
Project documents	The main documents used as inputs to this process are the lessons learned register, project schedule, project team assignments, resource calendars, risk register, risk report, and stakeholder register.
Enterprise environmental factors (EEFs)	The main EEFs used as inputs to this process are the stakeholders' risk thresholds.
Organizational process asset factors (OPAFs)	The main OPAFs used as inputs to this process are the risk management plan templates; risk register; risk report; and the lessons learned from similar projects.

Table 3.119: Methods used during the "Plan Risk Responses" process.

Method	Details
Expert judgment	This refers to all judgments and decisions made while planning for the responses to the project risks. Such judgments require the project manager and his/her team to have adequate level of the required relevant skills and experience.
Data gathering	Interviews is the main data gathering method that is used during this process.
Interpersonal and team skills	Facilitation is the main interpersonal and team skill used during this process.
Strategies for threats	The main methods used during this process to deal with threats are escalation, avoiding, transferring, mitigation, and accepting threats.
Strategies for opportunities	The main methods used during this process to deal with opportunities are escalation, exploitation, sharing, enhancing, and accepting opportunities.
Contingent response strategies	Responses are designed to be used when certain events occur. However, it is preferred to create a detailed response plan that will be executed under certain events such as the event of missing intermediate milestones.
Strategies for overall project risk	The main methods used during this process to deal with the overall project risks are avoiding, exploiting, transferring and sharing, mitigating and enhancing, and accepting risks.
Data analysis	The main data analysis method used during this process is the multi-CRITERIA decision analysis.
Decision-making	The main decision-making method used during this process is the multi-CRITERIA decision analysis.

Table 3.120: Outputs of the "Plan Risk Responses" process.

Output	Details
Project management plan updates	The main components of the project management plan that may be updated during this process are the schedule management plan, cost management plan, quality management plan, resource management plan, procurement management plan, scope baseline, schedule baseline, and cost baseline.
Project documents' updates	The main documents that may be updated during this process are the assumption log, cost forecasts, lessons learned register, project schedule, project team assignments, risk register, and risk report.

3.8.6 The "Implement Risk Responses" process

The "Implement Risk Responses" process is used to implement a set of predetermined risk responses.

In reference to Figures 3.92 and 3.93, the main inputs, methods, and outputs of this process are detailed in Tables 3.121, 3.122, and 3.123, respectively.

The "Implement Risk Responses" Process		
Inputs	**Methods**	**Outputs**
1 Project management plan (e.g., risk management plan) 2 Project documents (e.g., lessons learned register, risk register, and risk report) 3 Organizational process assets	1 Expert judgment 2 Interpersonal and team skills (e.g., influencing) 3 Project management information system	1 Change requests 2 Project documents updates (e.g., issue log, lessons learned register, project team assignments, risk register, and risk report)

Figure 3.92: The "Implement Risk Responses" process.

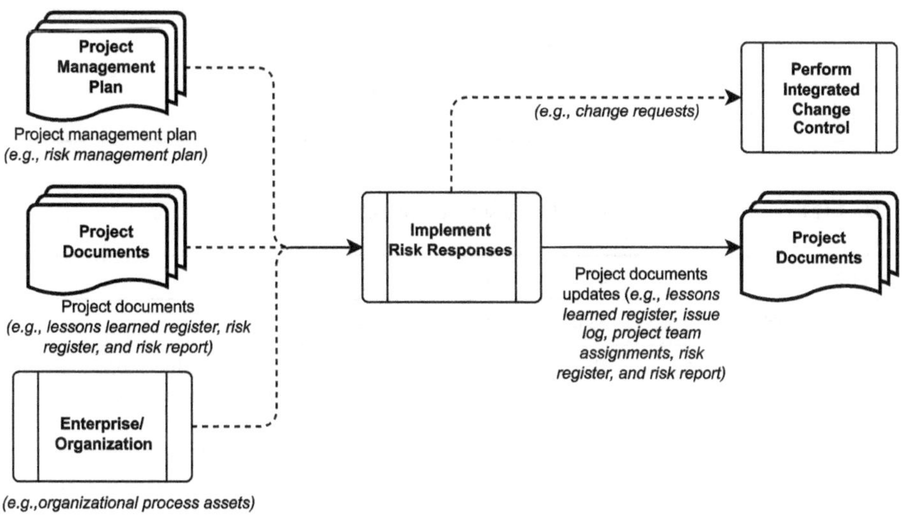

Figure 3.93: The "Implement Risk Responses" process data flow diagram.

Table 3.121: Inputs to the "Implement Risk Responses" process.

Input	Details
Project management plan	The main component of the project management plan used as an input to this process is the risk management plan.
Project documents	The main documents used as inputs to this process are the lessons learned register, risk register, and risk report.
Organizational process asset factors (OPAFs)	The main organizational process asset used as an input to this process is the lessons learned repository.

Table 3.122: Methods used during the "Implement Risk Responses" process.

Method	Details
Expert judgment	This refers to all judgments and decisions made while implementing the responses to the project risks. Such judgments require the project manager and his/her team to have adequate level of the required relevant skills and experience.
Interpersonal and team skills	Influencing is the main interpersonal and team skill that is used during this process.
Project management information system (PMIS)	As described earlier in Table 3.8.

Table 3.123: Outputs of the "Implement Risk Responses" process.

Output	Details
CHANGE-REQUESTS	As described earlier in Table 3.9.
Project documents' updates	The main documents that may be updated during this process are the issue log, lessons learned register, project team assignments, risk register, and risk report.

3.8.7 The "Monitor Risks" process

The "Monitor Risks" process is used to monitor the implementation of the predefined risk responses and track them, identify them, and analyze new risks.

In reference to Figures 3.94 and 3.95, the main inputs, methods, and outputs of this process are detailed in Tables 3.124, 3.125, and 3.126, respectively.

The "Monitor Risks" Process		
Inputs	**Methods**	**Outputs**
1 Project management plan (e.g., risk management plan) 2 Project documents (e.g., issue log, lessons learned register, risk register, and risk report) 3 Work performance data 4 Work performance reports	1 Data analysis (e.g, technical performance analysis, and reserve analysis) 2 Audits 3 Meetings	1 Work performance information 2 Change requests 3 Project management plan updates (e.g., any component) 4 Project documents updates (e.g., assumption log, issue log, lessons learned register, risk register, and risk report) 5 Organizational process assets updates

Figure 3.94: The "Monitor Risks" process.

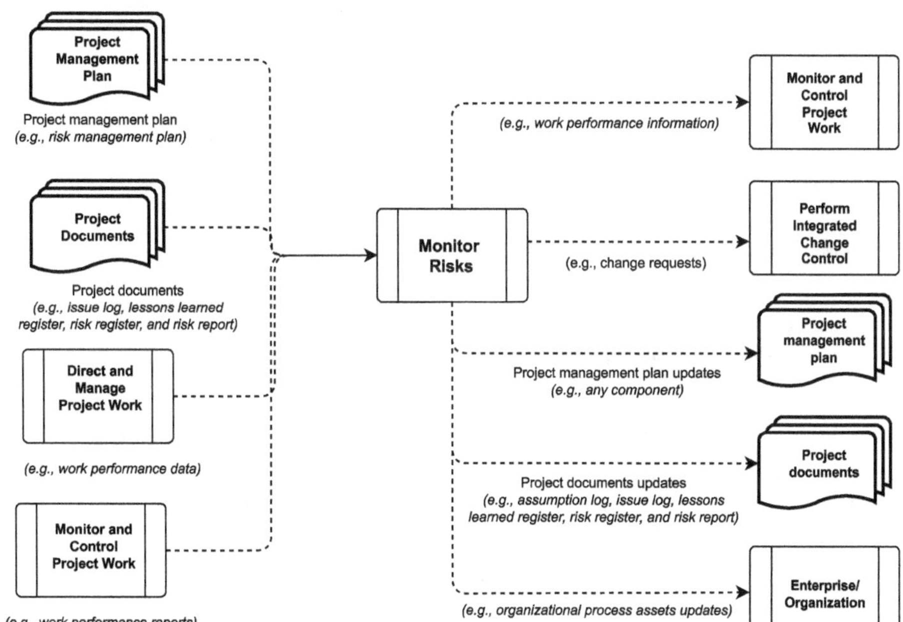

Figure 3.95: The "Monitor Risks" process data flow diagram.

Table 3.124: Inputs to the "Monitor Risks" process.

Input	Details
Project management plan	The main component of the project management plan used as an input to this process is the risk management plan.
Project documents	The main documents used as inputs to this process are the issue log, lessons learned register, risk register, and risk report.
Project work performance data	As described earlier in Table 3.9.
Project work performance reports	As described earlier in Table 3.15.

Table 3.125: Methods used during the "Monitor Risks" process.

Method	Details
Data analysis	The main data analysis methods used during this process are the technical performance analysis and reserve analysis.
Audits	Risk audits may be used to assess the effectiveness of the risk management process.
Meetings	This refers to the meetings that take place while monitoring the project risks.

Table 3.126: Outputs of the "Monitor Risks" process.

Output	Details
CHANGE-REQUESTS	As described earlier in Table 3.9.
Project work performance data	As described earlier in Table 3.9.
Project management plan	All components of the project management plan are subject to be updated during this process.
Project documents' updates	The main documents that may be updated during this process are the assumption log, issue log, lessons learned register, risk register, and risk report.
Organizational process asset factor (OPAF) updates	The main OPAFs that may be updated during this process are the risk breakdown structure; and the templates for risk management plan, risk register, and risk report.

3.9 Project procurement management

Project procurement management includes the processes that are necessary to purchase products, solutions, and/or services needed from outside the project. It includes the processes required to develop and manage agreements such as contracts and purchase orders.

In reference to Figure 3.96, the project procurement management processes are the Plan Procurement Management, Conduct Procurements, and Control Procurements.

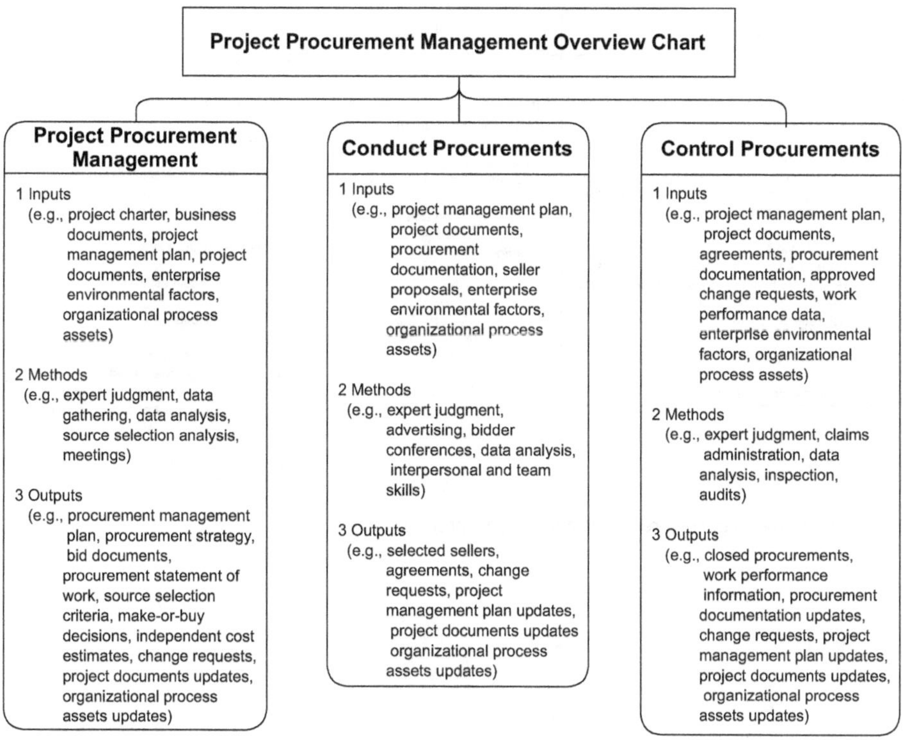

Figure 3.96: Project Procurement Management overview chart.

3.9.1 The "Plan Procurement Management" process

The "Plan Procurement Management" process is used to document the procurement decisions and identify possible sellers.

In reference to Figures 3.97 and 3.98, the main inputs, methods, and outputs of this process are detailed in Tables 3.127, 3.128, and 3.129, respectively.

The "Plan Procurement Management" Process

Inputs	Methods	Outputs
1 Project charter	1 Expert judgment	1 Procurement management plan
2 Business documents (e.g., business case, benefits management plan)	2 Data gathering (e.g., market research)	2 Procurement strategy
3 Project management plan (e.g., scope management plan, quality management plan, resource management plan, scope baseline)	3 Data analysis (e.g., make-or-buy analysis)	3 Bid documents
4 Project documents (e.g., milestone list, project team assignments, requirements documentation, requirements traceability matrix, resource requirements, risk register, stakeholder register)	4 Source selection analysis	4 Procurement statement of work
5 Enterprise environmental factors	5 Meetings	5 Source selection criteria
6 Organizational process assets		6 Make-or-buy decisions
		7 Independent cost estimates
		8 Change requests
		9 Project documents updates (e.g., lessons learned register, milestone list, requirements documentation, requirements traceability matrix, risk register, stakeholder register)
		10 Organizational process assets updates

Figure 3.97: The "Plan Procurement Management" process.

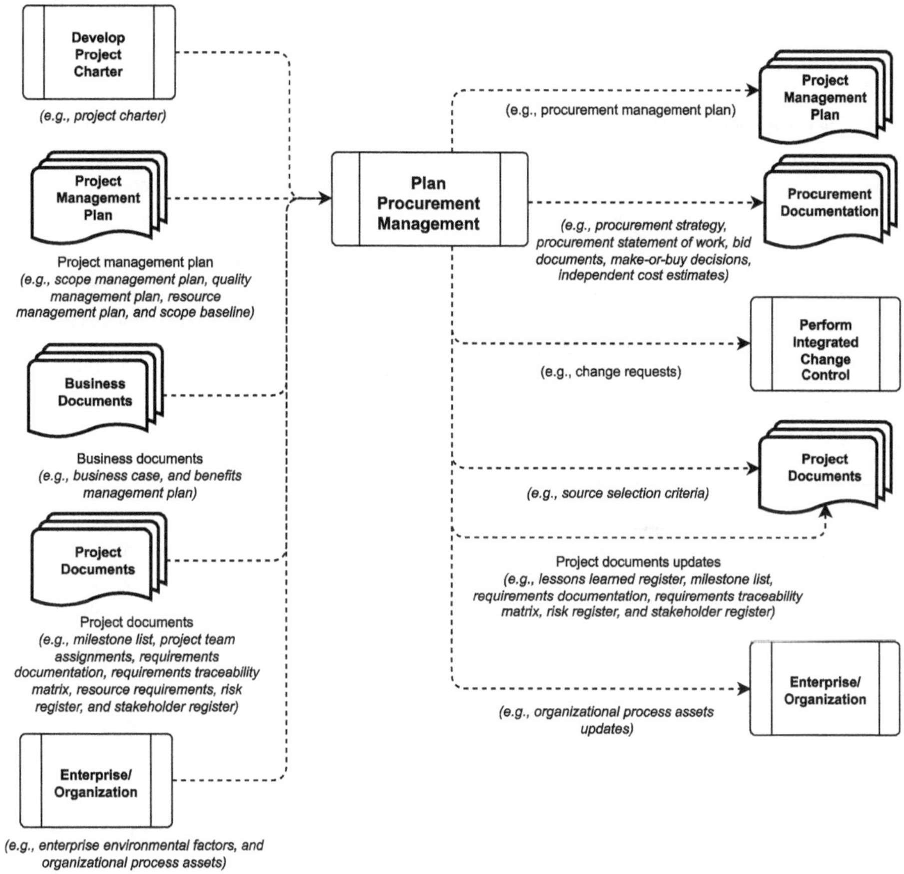

Figure 3.98: The "Plan Procurement Management" process data flow diagram.

Table 3.127: Inputs to the "Plan Procurement Management" process.

Input	Details
Project charter	As described earlier in Table 3.3.
Project management plan	The main components of the project management plan used as inputs to this process are the scope management plan, quality management plan, resource management plan, and scope baseline.
Project documents	The main documents used as inputs to this process are the milestone list, project team assignments, requirements documentation, requirements traceability matrix, resource requirements, risk register, and stakeholder register.
Business documents	This includes the business case and benefits management plan.

Table 3.127 (continued)

Input	Details
Enterprise environmental factors (EEFs)	The main EEFs used as inputs to this process are the marketplace conditions, potential suppliers, products and services availability in the market, contract management systems, and financial accounting payment systems.
Organizational process asset factors (OPAFs)	The main OPAFs used as inputs to this process are the preapproved suppliers' lists; formal procurement policies, procedures, and guidelines; fixed-price contract; cost-reimbursement contracts; time and material contracts; and performance-based contracts.

Table 3.128: Methods used during the "Plan Procurement Management" process.

Method	Details
Expert judgment	This refers to all judgments and decisions made while planning for the procurement management. Such judgments require the project manager and his/her team to have adequate level of the required relevant skills and experience.
Data gathering	The market research method is the main data gathering method used during this process.
Data analysis	The make-or-buy method is the main data analysis method used during this process.
Source selection analysis	The main source selection analysis methods used during this process are the least-cost analysis, qualifications-only analysis, quality-based analysis, cost-based analysis, sole-source analysis, and fixed budget analysis.
Meetings	This refers to the meetings that take place while planning for the project procurement management.

Table 3.129: Outputs of the "Plan Procurement Management" process.

Output	Details
Procurement management plan	The procurement management plan defines how to coordinate procurements with the schedule development and control processes, timetabling the procurement activities, procurement metrics for managing contracts, stakeholder roles and responsibilities, and constraints and assumptions that affect procurements.

Table 3.129 (continued)

Output	Details
Procurement strategy	This is to determine the project delivery method, the type of legally binding agreement(s), and how the procurement will advance through the procurement phases.
Bid documents	The main bidding documents come out of this process are a request for information, a request for quotation, and a request for proposal.
Procurement statement of work	This includes a description of the required services. It can be revised as required throughout the procurement process.
Source SELECTION-CRITERIA	This includes many factors such as the capability and capacity, product cost, delivery dates, and management experience.
Make-or-buy decisions	A make-or-buy analysis leads to a decision to whether it is better to carry on a particular part of the project work by the project team or outsource it to outsider resources.
Independent cost estimates	For large procurements, the procuring organization may elect to prepare its own estimate or have a cost estimate prepared by an outside professional estimator to serve as a benchmark on the proposed responses.
CHANGE-REQUESTS	As described earlier in Table 3.9.
Project documents' updates	The project documents that may be updated during this process are the lessons learned register, milestone list, requirements documentation, requirements traceability matrix, risk register, and stakeholder register.
Organizational process asset factor (OPAF) updates	The main OPAFs that may be updated during this process are the information on the qualified suppliers and service providers.

3.9.2 The "Conduct Procurements" process

The "Conduct Procurements" process is used to solicit and obtain proposals from service providers. Then select one of them and negotiate and sign a contract with the selected service provider.

In reference to Figures 3.99 and 3.100, the main inputs, methods, and outputs of this process are detailed in Tables 3.130, 3.131, and 3.132, respectively.

The "Conduct Procurements" Process		
Inputs	**Methods**	**Outputs**
1 Project management plan (e.g., scope management plan, requirements management plan, communications management plan, risk management plan, procurement management plan, configuration management plan, cost baseline)	1 Expert judgment	1 Selected sellers
2 Project documents (e.g., lessons learned register, project schedule, requirements documentation, risk register, stakeholder register)	2 Advertising	2 Agreements
3 Procurement documentation	3 Bidder conferences	3 Change requests
4 Seller proposals	4 Data analysis (e.g., proposal evaluation)	4 Project management plan updates (e.g., requirements management plan, quality management plan, communications management plan, risk management plan, procurement management plan, scope baseline, schedule baseline, cost baseline)
5 Enterprise environmental factors	5 Interpersonal and team skills (e.g., negotiation)	5 Project documents updates (e.g., lessons learned register, requirements documentation, requirements traceability matrix, resource calendars, risk register, stakeholder register)
6 Organizational process assets		6 Organizational process assets updates

Figure 3.99: The “Conduct Procurements” process.

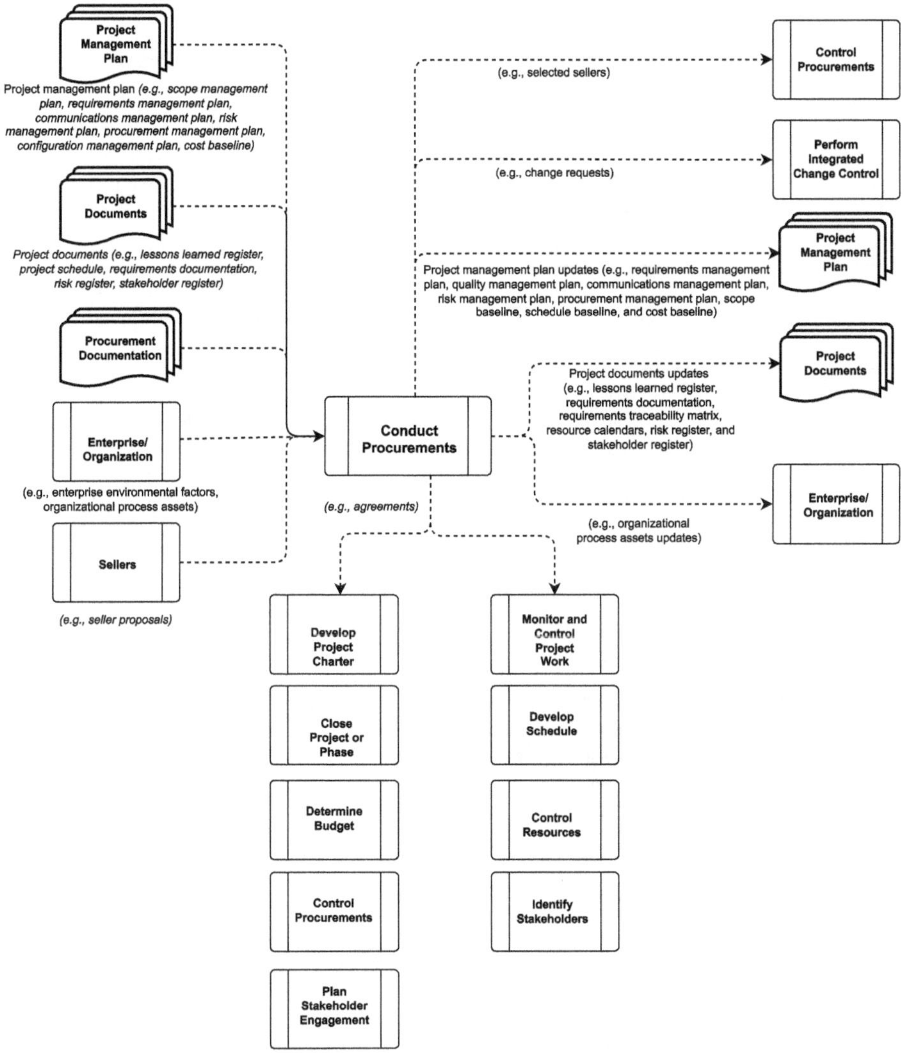

Figure 3.100: The “Conduct Procurements” process data flow diagram.

Table 3.130: Inputs to the "Conduct Procurements" process.

Input	Details
Project management plan	The main components of the project management plan used as inputs to this process are the scope management plan, requirements management plan, communications management plan, risk management plan, procurement management plan, configuration management plan, and cost baseline.
Project documents	The main documents used as inputs to this process are the lessons learned register, project schedule, requirements documentation, risk register, and stakeholder register.
Procurement documentation	The main procurement documents used as inputs to this process are the bid documents, procurement statement of work, independent cost estimates, and source SELECTION-CRITERIA.
Service providers' proposals	Proposals are typically prepared in response to a procurement request for proposal document. The evaluation team reviews and evaluates each submitted proposal and then selects the best supplier/service provider.
Enterprise environmental factors (EEFs)	The main EEFs used as inputs to this process are the local procurement laws and regulations, marketplace conditions, and contract management systems.
Organizational process asset factors (OPAFs)	The main OPAFs used as inputs to this process are the list of prequalified and selected service providers, organizational policies that affect service providers, and invoicing and payment processes' financial policies.

Table 3.131: Methods used during the "Conduct Procurements" process.

Method	Details
Expert judgment	This refers to all judgments and decisions made while conducting the project procurements. Such judgments require the project manager and his/her team to have adequate level of the required relevant skills and experience.
Advertising	Advertising is using available and affordable communication medium to share with potential suppliers/service providers the intention to initiate a project to produce a particular product or solution or to solicit a particular service.
Bidder conferences	Bidder conferences are held to conduct meetings between the customer and the prospective service provider prior to their proposals' submission.
Data analysis	The proposals' evaluation method is the main data analysis method used during this process.
Interpersonal and team skills	The negotiation method is the main interpersonal and team skill used during this process.

Table 3.132: Outputs of the "Conduct Procurements" process.

Output	Details
Shortlisted service providers	This includes the initially selected sellers/bidders/service providers based on the evaluation of their proposals.
Agreements and contracts	As described earlier in Table 3.1.
CHANGE-REQUESTS	As described earlier in Table 3.9.
Project management plan updates	The main components of the project management plan that may be updated during this process are the requirements management plan, quality management plan, communications management plan, risk management plan, scope baseline, schedule baseline, and cost baseline.
Project documents' updates	The main documents that may be updated during this process are the lessons learned register, requirements documentation, requirements traceability matrix, resource calendars, risk register, and stakeholder register.
Organizational process asset factor (OPAF) updates	The main OPAFs that may be updated during this process are the list of the prospect suppliers/service providers, list of the prequalified suppliers/service providers, and information on their relevant experience.

3.9.3 The "Control Procurements" process

The "Control Procurements" process is used to manage the procurement relationships, monitor the contract performance and make changes and corrections as appropriate, and close out contracts.

In reference to Figures 3.101 and 3.102, the main inputs, methods, and outputs of this process are detailed in Tables 3.133, 3.134, and 3.135, respectively.

The "Control Procurements" Process		
Inputs	**Methods**	**Outputs**
1 Project management plan (e.g., requirements management plan, risk management plan, procurement management plan, change manageme plan, and schedule baseline)	1 Expert judgment	1 Closed procurements
2 Project documents (e.g., assumption log, lessons learned regist milestone list, quality reports, requirem documentation, requirements traceabili matrix, risk register, and stakeholder register)	2 Claims administration	2 Work performance information
3 Agreements	3 Data analysis (e.g., performance reviews, earned value analysis, and trend analysis)	3 Procurement documentation updates
4 Procurement documentation	4 Inspection	4 Change requests
5 Approved change requests	5 Audits	5 Project management plan updates (e.g., risk management plan, procurement management plan, schedule baseline, and cost baseline)
6 Work performance data		6 Project documents updates (e.g., lessons learned register, resource requirements, requirements traceability matrix, risk register, and stakeholder register)
7 Enterprise environmental factors		7 Organizational process assets updates
8 Organizational process assets		

Figure 3.101: The “Control Procurements” process.

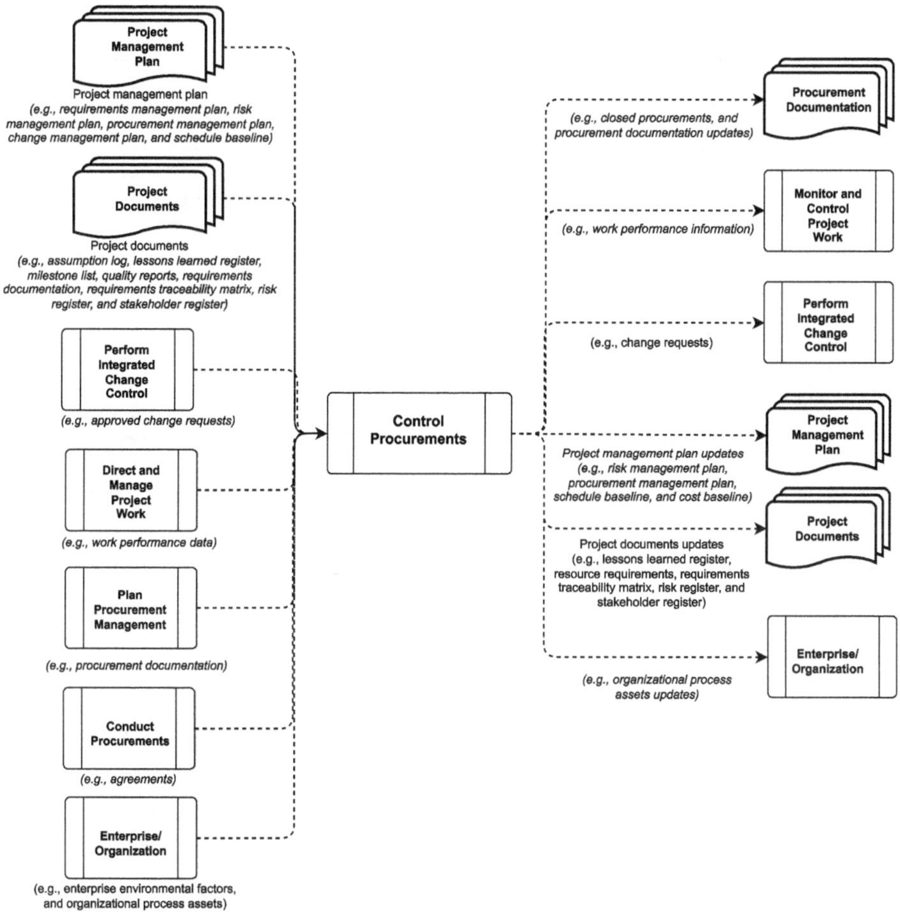

Figure 3.102: The "Control Procurements" process data flow diagram.

Table 3.133: Inputs to the "Control Procurements" process.

Input	Details
Project management plan	The main components of the project management plan used as inputs to this process are the requirements management plan, risk management plan, procurement management plan, change management plan, and schedule baseline.
Project documents	The main documents used as inputs to this process are the assumption log, lessons learned register, milestone list, quality reports, requirements documentation, requirements traceability matrix, risk register, and stakeholder register.

Table 3.133 (continued)

Input	Details
Agreements	As described earlier in Table 3.1.
Procurement documentation	This includes the statement of work, payment information, contractor work performance information, plans, drawings, and correspondence details.
Approved CHANGE-REQUESTS	As described earlier in Table 3.7.
Project work performance data	As described earlier in Table 3.9.
Enterprise environmental factors (EEFs)	This includes the contract change control system, financial and account payable system, and the organizational buy code of ethics.
Organizational process asset factors (OPAFs)	The procurement policies are the main organizational process asset used during this process.

Table 3.134: Methods used during the "Control Procurements" process.

Method	Details
Expert judgment	This refers to all judgments and decisions made while controlling the project procurements. Such judgments require the project manager and his/her team to have adequate level of the required relevant skills and experience.
Claims administration	Constructive changes are requested when the customer and service provider cannot agree on compensation for the change, and that is called claims. Claims become disputes and then appeals if they cannot be resolved. Claims are documented, processed, monitored, and managed throughout the contract life cycle according to its terms and conditions.
Data analysis	This includes the performance reviews, earned-value analysis, and trend analysis.
Inspection	As described earlier in Table 3.35.
Audits	Audits are structured reviews of the procurement processes. Observations should be brought to the attention of the project managers at both sides, customer and service provider.

Table 3.135: Outputs of the "Control Procurements" process.

Output	Details
Closed procurements	Procurement closure requirements are defined in the contract's terms and conditions, and included in the procurement management plan. All deliverables should have been shipped on time and meet the technical and quality requirements. There should be no outstanding claims or invoices and all payments should have been made.
Project work performance data	As described earlier in Table 3.9.
Procurement documentation updates	The main procurement documentation that may be updated during this process are the contract, schedules, unapproved/approved CHANGE-REQUESTS; technical documentation, work performance information; deliverables; performance reports; invoices and payment records; and inspection results.
CHANGE-REQUESTS	As described earlier in Table 3.9.
Project management plan updates	The main components of the project management plan that may be updated during this process are the risk and procurement management plans, schedule baseline, and cost baseline.
Project document updates	The main documents that may be updated during this process are the lessons learned register, resource requirements, requirements traceability matrix, risk register, and stakeholder register.
Organizational process asset factors (OPAF) updates	The main OPAFs that may be updated during this process are the payment schedules and requests, performance evaluation documentation, and so forth.

3.10 Project stakeholder management

Project stakeholder management includes the processes that are necessary to identify the people who impact on or be impacted by the project, to analyze stakeholder expectations and their impact on the project, and to develop appropriate management strategies for effectively engaging stakeholders in project decisions and execution.

In reference to Figure 3.103, the project stakeholder management processes are the Identify Stakeholders, Plan Stakeholder Engagement, Manage Stakeholder Engagement, and Monitor Stakeholder Engagement.

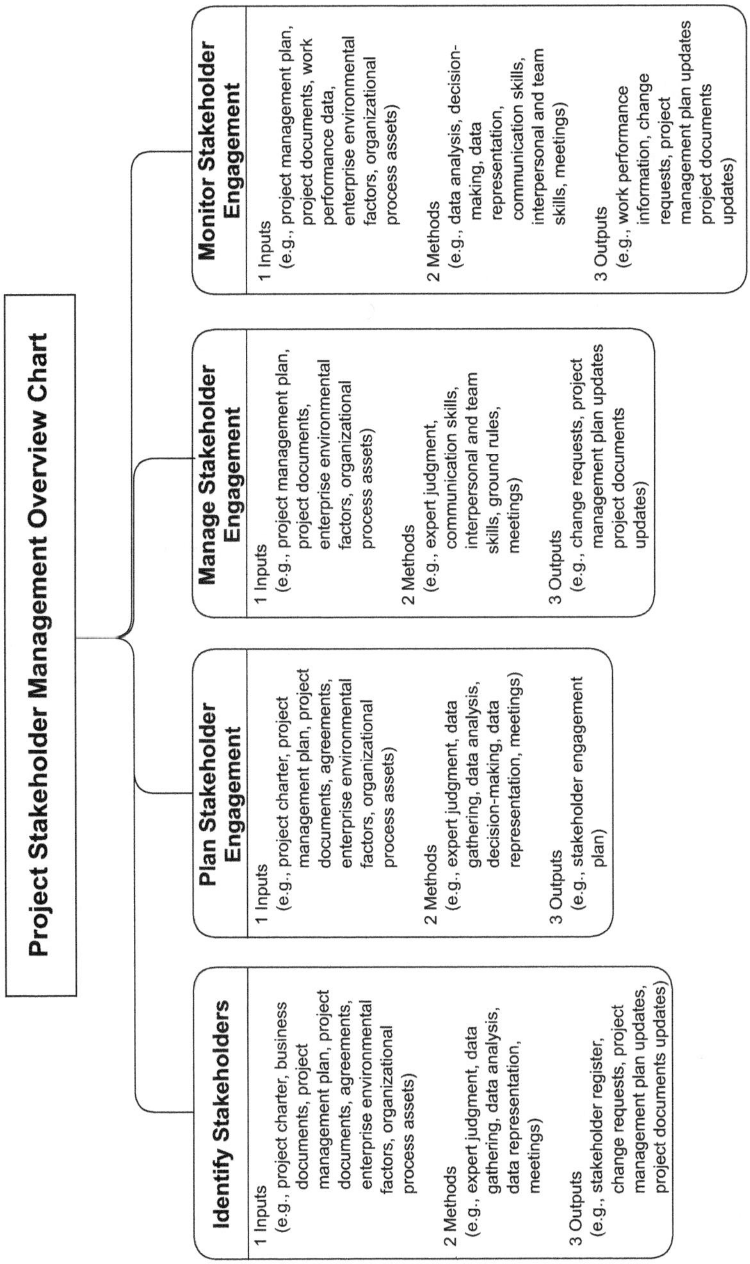

Figure 3.103: Project stakeholder management overview chart.

3.10.1 The "Identify Stakeholders" process

The "Identify Stakeholders" process is used to identify the project stakeholders and their interests, involvement, inter-dependencies, influence, and so forth.

In reference to Figures 3.104 and 3.105, the main inputs, methods, and outputs of this process are detailed in Tables 3.136, 3.137, and 3.138, respectively.

The "Identify Stakeholders"Process		
Inputs	**Methods**	**Outputs**
1 Project charter	1 Expert judgment	1 Stakeholder register
2 Business documents (e.g., business case, benefits management plan)	2 Data gathering (e.g., questionnaires and surveys, brainstorming)	2 Change requests
3 Project management plan (e.g., communications management plan, stakeholder engagement plan)	3 Data analysis (e.g., stakeholder analysis, document analysis)	3 Project management plan updates (e.g., requirements management plan, communications management plan, risk management plan, stakeholder engagement plan)
4 Project documents (e.g., change log, issue log, requirements documentation)	4 Data representation (e.g., stakeholder mapping/ representation)	4 Project documents updates (e.g., assumption log, issue log, risk register)
5 Agreements	5 Meetings	
6 Enterprise environmental factors		
7 Organizational process assets		

Figure 3.104: The "Identify Stakeholders" process.

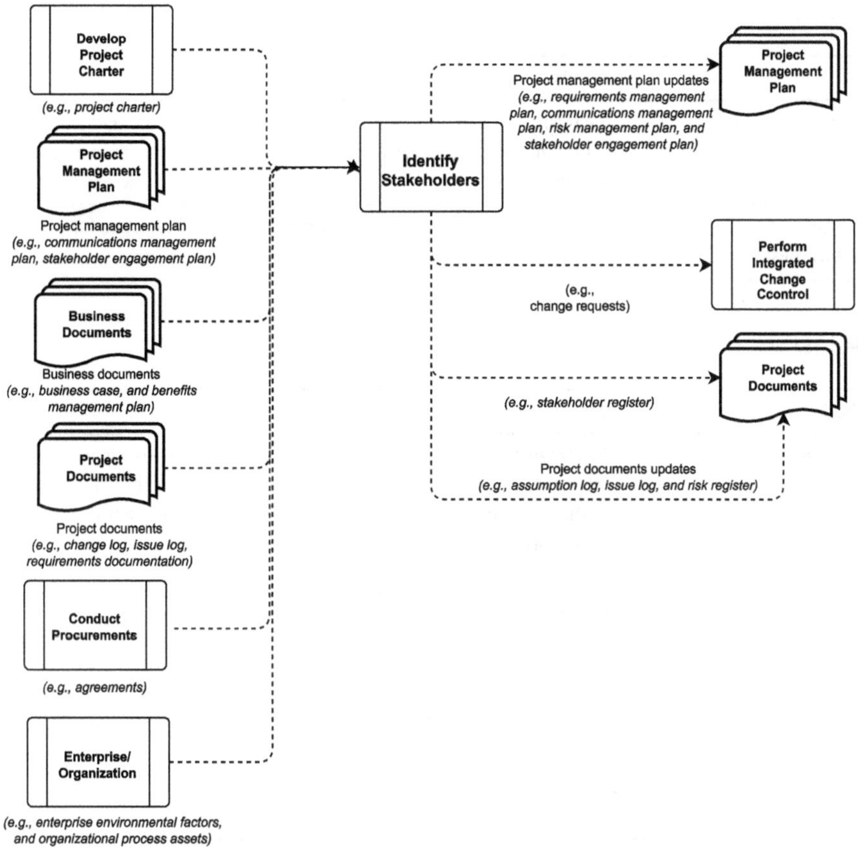

Figure 3.105: The "Identify Stakeholders" process data flow diagram.

Table 3.136: Inputs to the "Identify Stakeholders" process.

Input	Details
Project charter	As described earlier in Table 3.3.
Business documents	This includes the business case and benefits management plan.
Project management plan	The main components of the project management plan used as inputs to this process are the communications management plan and stakeholder engagement plan.
Project documents	The main documents used as inputs to this process are the change log, issue log, and the requirements documentation.
Agreements	As described earlier in Table 3.1.

Table 3.136 (continued)

Input	Details
Enterprise environmental factors (EEFs)	The main EEFs used as inputs to this process are the organizational culture, politics, and governance framework; governance standards; and so forth.
Organizational process asset factors (OPAFs)	This includes the stakeholder register templates and registers from similar projects.

Table 3.137: Methods used during the "Identify Stakeholders" process.

Method	Details
Expert judgment	This refers to all judgments and decisions made while identifying the project stakeholders. Such judgments require the project manager and his/her team to have adequate level of the required relevant skills and experience.
Data gathering	The main data gathering methods used during this process are the questionnaires, surveys, and brainstorming.
Data analysis	The main data analysis methods used during this process are the stakeholder analysis and documents analysis.

Table 3.138: Outputs of the "Identify Stakeholders" process.

Output	Details
Stakeholder register	This contains information about the project stakeholders such as their identification information; assessment information; and stakeholder classification.
CHANGE-REQUESTS	As described earlier in Table 3.9.
Project management plan updates	The main components of the project management plans are the requirements management plan, communications management plan, risk management plan, and stakeholders' engagement plan.
Project documents' updates	The main documents that may be updated during this process are the assumption log, issue log, and risk register.

3.10.2 The "Plan Stakeholder Engagement" process

The "Plan Stakeholder Engagement" process is used to involve the project stakeholders based on their needs, expectations, interests, and potential impact on the project.

In reference to Figures 3.106 and 3.107, the main inputs, methods, and outputs of this process are detailed in Tables 3.139, 3.140, and 3.141, respectively.

The "Plan Stakeholder Engagement" Process		
Inputs	**Methods**	**Outputs**
1 Project charter	1 Expert judgment	1 Stakeholder engagement plan
2 Project management plan (e.g., resource management plan, communications management plan, and risk management plan)	2 Data gathering (e.g., benchmarking)	
3 Project documents (e.g., assumption log, change log, issue log, project schedule, risk register, and stakeholder register)	3 Data analysis (e.g., assumption and constraint analysis, and root cause analysis)	
4 Agreements	4 Decision-making (e.g., prioritization/ranking)	
5 Enterprise environmental factors	5 Data representation (e.g., mind mapping, and stakeholder engagement assessment matrix)	
6 Organizational process assets	6 Meetings	

Figure 3.106: The "Plan Stakeholder Engagement" process.

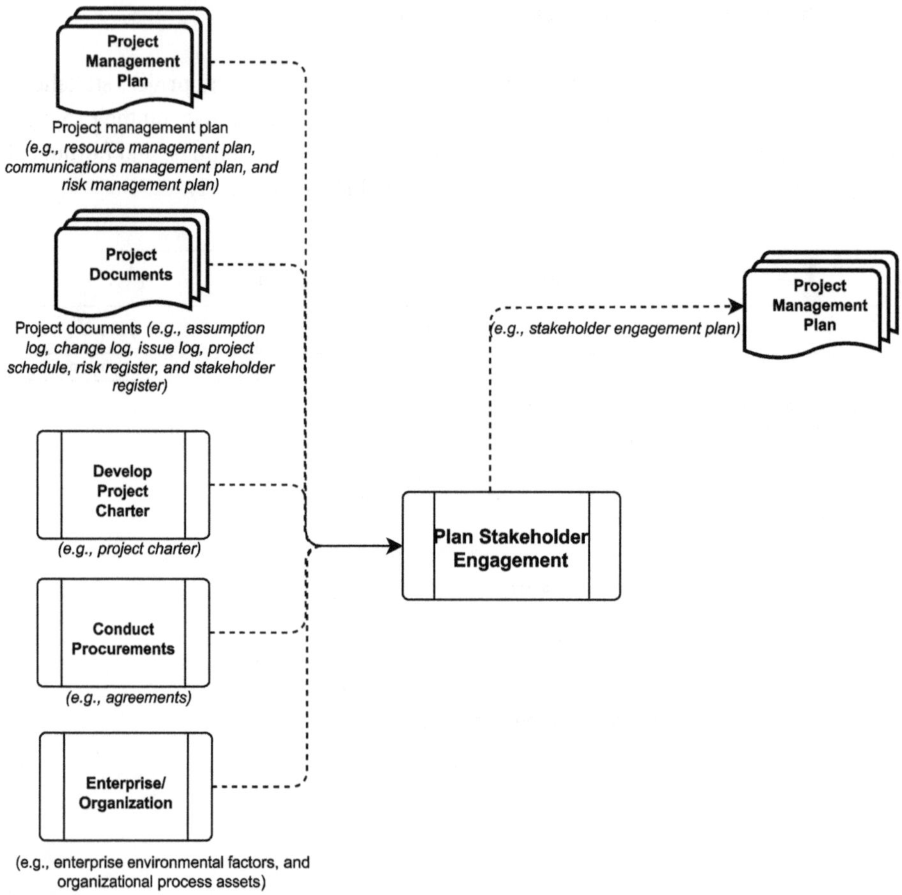

Figure 3.107: The "Plan Stakeholder Engagement" process data flow diagram.

Table 3.139: Inputs to the "Plan Stakeholder Engagement" process.

Input	Details
Project charter	As described earlier in Table 3.3.
Project management plan	The main components of the project management plan used as inputs to this process are the resource management plan, communications management plan, and risk management plan.
Project documents	The main documents used as inputs to this process are the assumption log, change log, issue log, project schedule, risk register, and stakeholder register.
Agreements	As described earlier in Table 3.1.

Table 3.139 (continued)

Input	Details
Enterprise environmental factors (EEFs)	The main EEFs used as inputs to this process are the organizational culture and politics, personnel policies, and communication channels.
Organizational process asset factors (OPAFs)	The main OPAFs used as inputs to this process are the inside and outside organizational environment and culture, communication channels and strategies, and organizational policies.

Table 3.140: Methods used during the "Plan Stakeholder Engagement" process.

Method	Details
Expert judgment	This refers to all judgments and decisions made while planning for the project stakeholder management. Such judgments require the project manager and his/her team to have adequate level of the required relevant skills and experience.
Data gathering	Benchmarking is the main data gathering method used during this process.
Data analysis	The main data analysis methods used during this process are the constraint analysis and root-cause analysis.
Decision-making	Prioritization and ranking is the main method used during this process in support to making decisions.
Data representation	The main data representation methods used during this process are the mid-mapping, and stakeholder engagement assessment matrix.
Meetings	This refers to the meetings that take place while planning for the project stakeholders' engagements.

Table 3.141: Outputs of the "Plan Stakeholder Engagement" process.

Output	Details
Stakeholders' engagement plan	This is an integral component of the project management plan. It identifies the strategies and actions required to promote productive involvement of stakeholders in the decision-making process.

3.10.3 The "Manage Stakeholder Engagement" process

The "Manage Stakeholder Engagement" process is used to communicate with stakeholders to meet their expectations, address their issues, and foster appropriate stakeholder involvement.

In reference to Figures 3.108 and 3.109, the main inputs, methods, and outputs of this process are detailed in Tables 3.142, 3.143, and 3.144, respectively.

The "Manage Stakeholder Engagement" Process		
Inputs	**Methods**	**Outputs**
1 Project management plan (e.g., communications management plan, risk management plan, stakeholder engagement plan, and change management plan)	1 Expert judgment	1 Change requests
2 Project documents (e.g., change log, issue log, lessons learned register, and stakeholder register)	2 Communication skills (e.g., feedback)	2 Project management plan updates (e.g., communications management plan and stakeholder engagement plan)
3 Enterprise environmental factors	3 Interpersonal and team skills (e.g., conflict management, cultural awareness, negotiation, observation/conversation, and political awareness)	3 Project documents updates (e.g., change log, issue log, lessons learned register, and stakeholder register)
4 Organizational process assets	4 Ground rules	
	5 Meetings	

Figure 3.108: The "Manage Stakeholder Engagement" process.

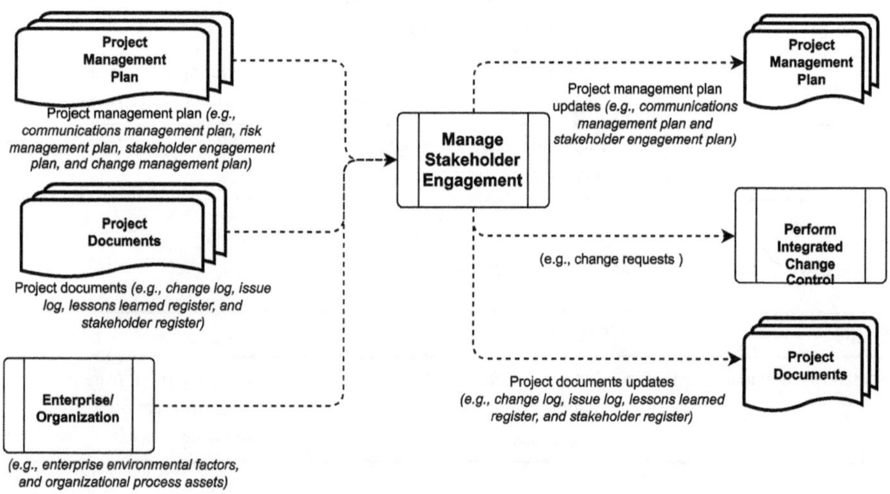

Figure 3.109: The "Manage Stakeholder Engagement" process data flow diagram.

Table 3.142: Inputs to the "Manage Stakeholder Engagement" process.

Input	Details
Project management plan	The main components of the project management plan used as inputs to this process are the communications management plan, risk management plan, stakeholders' engagement plan, and change management plan.
Project documents	The main documents used as inputs to this process are the change log, issue log, lessons learned register, and stakeholder register.
Enterprise environmental factors (EEFs)	The main EEFs used as inputs to this process are the organizational culture and politics, personnel policies, and so forth.
Organizational process asset factors (OPAFs)	The main OPAFs used as inputs to this process are the organizational policies and procedures and the organizational communications channel.

Table 3.143: Methods used during the "Manage Stakeholder Engagement" process.

Method	Details
Expert judgment	This refers to all judgments and decisions made while managing the project's stakeholder engagements. Such judgments require the project manager and his/her team to have adequate level of the required relevant skills and experience.
Communication skills	The main communication skills that are required during this process are the formal and informal conversation skills, discussing issues, holding and managing meetings in a professional manner, preparing progress reports, conducting surveys and questionnaires, and so forth.
Interpersonal and team skills	This includes conflict management skills, cultural awareness, negotiation skills, observation and conversation skills, and political awareness.
Meetings	This refers to the meetings that take place while managing the project's stakeholder engagements.

Table 3.144: Outputs of the "Manage Stakeholder Engagement" process.

Output	Details
CHANGE-REQUESTS	As described earlier in Table 3.9.
Project management plan updates	The main components of the project management plan that may be updated during this process are the communications management plan and stakeholders' engagement plan.
Project document updates	The main documents that may be updated during this process are the change log, issue log, lesson learned register, and stakeholder register.

3.10.4 The "Monitor Stakeholder Engagement" process

The "Monitor Stakeholder Engagement" process is used to monitor the stakeholders' relationships and tailoring strategies for engaging stakeholders through modification of engagement strategies and plans.

In reference to Figures 3.110 and 3.111, the main inputs, methods, and outputs of this process are detailed in Tables 3.145, 3.146, and 3.147, respectively.

The "Monitor Stakeholder Engagement" Process		
Inputs	**Methods**	**Outputs**
1 Project management plan (e.g., resource management plan, communications management plan, and stakeholder engagement plan)	1 Data analysis (e.g., alternatives analysis, root cause analysis, and stakeholder analysis)	1 Work performance information
2 Project documents (e.g., issue log, lessons learned register, project communications, risk register, and stakeholder register)	2 Decision-making (e.g., multicriteria decision analysis, voting, data representation, and stakeholder engagement assessment matrix)	2 Change requests
3 Work performance data	4 Communication skills (e.g., feedback and presentations)	3 Project management plan updates (e.g., Resource management plan, communications managemen plan, and stakeholder engagement plan)
4 Enterprise environmental factors	5 Interpersonal and team skills (e.g., active listening, cultural awareness, leadership, networking, and political awareness)	4 Project documents updates (e.g., Issue log, lessons learned register, risk register, and stakeholder register)
5 Organizational process assets	6 Meetings	

Figure 3.110: The "Monitor Stakeholder Engagement" process.

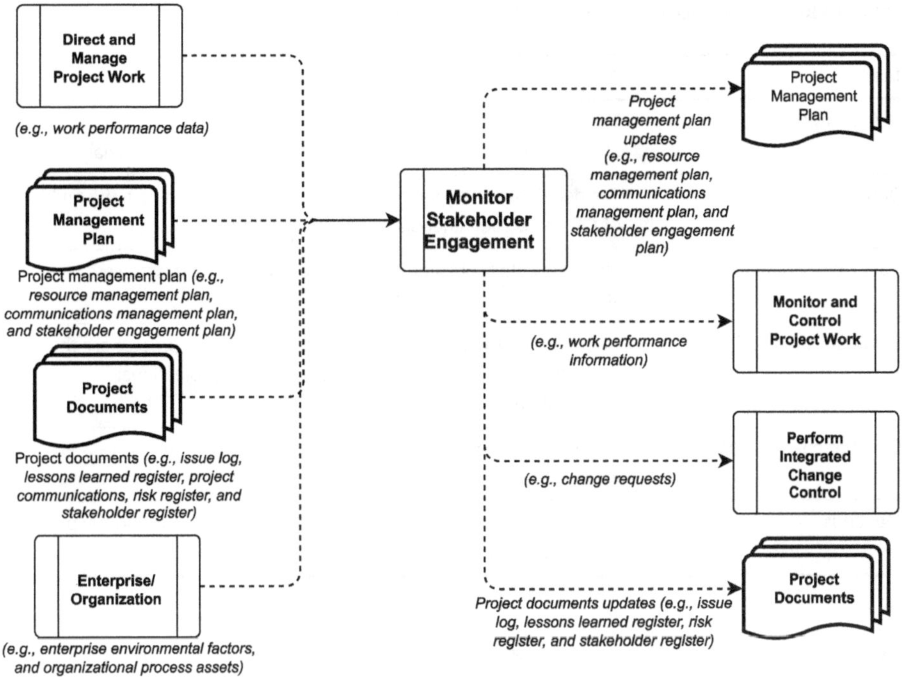

Figure 3.111: The "Monitor Stakeholder Engagement" process data flow diagram.

Table 3.145: Inputs to the "Monitor Stakeholder Engagement" process.

Input	Details
Project management plan	The main components of the project management plan used as inputs to this process are the resource management plan, communications management plan, and stakeholders' engagement plan.
Project documents	The main documents used as inputs to this process are the issue log, lessons learned register, project communications, risk register, and stakeholder register.
Project work performance data	As described earlier in Table 3.9.
Enterprise environmental factors (EEFs)	The main EEFs used as inputs to this process are the organizational culture and politics, personnel policies, and so forth.
Organizational process asset factors (OPAFs)	The main OPAFs used as inputs to this process are the organizational policies and procedures, and the organizational communication channel.

Table 3.146: Methods used during the "Monitor Stakeholder Engagement" process.

Method	Details
Data analysis	The main data analysis methods used during this process are the alternatives' analysis, root-cause analysis, and stakeholder analysis.
Decision-making	The decision-making methods used during this process are the multi-CRITERIA decision analysis and voting methods.
Data representation	The main data representation method used during this process is the stakeholder engagement matrix.
Communication skills	The main communication methods used during this process are the ability to provide articulated feedback and give professional presentations.
Interpersonal and team skills	The main interpersonal and team skills used during this process are the active listening, cultural awareness, leadership, and political awareness.
Meetings	This refers to the meetings that take place while monitoring the project stakeholder engagements.

Table 3.147: Outputs of the "Monitor Stakeholder Engagement" process.

Output	Details
Project work performance data	As described earlier in Table 3.9
CHANGE-REQUESTS	As described earlier in Table 3.9.
Project management plan updates	The main components of the project management plan that may be updated during this process are the resource management plan, communications management plan, and stakeholders' engagement plan.
Project document updates	The main documents that may be updated during this process are the issue Log, lesson learned register, risk register, and stakeholder register.

Chapter 4
Software engineering and software project management: the IEEE-CS perspective

This chapter provides an executive summary of the IEEE-Computer Society (IEEE-CS) perspective of software engineering and software project management (SPM) [3].

In the late 1990s and early 2000s, the IEEE-CS managed to formalize the discipline of software engineering and defined a set of 15 knowledge areas for that discipline. The IEEE-CS presented these knowledge areas in their Software Engineering Body of Knowledge (SWEBOK). The seventh of these knowledge areas is entitled "Software Engineering Management," which can be viewed as the backbone for SPM.

This chapter comes in four sections. Section 4.1 overviews the discipline of software engineering. Section 4.2 describes the 15 knowledge areas of software engineering in accordance with the IEEE-CS SWEBOK. These knowledge areas are the software requirements, software design, software construction, software testing, software maintenance, software configuration management, software engineering management, software engineering process, software engineering models and methods, software quality, software engineering economics, software engineering professional practice, computing foundation, mathematical foundation, and engineering foundation. Section 4.3 presents the international professional certifications that involve software engineering and SPM. Section 4.4 elaborates on the software engineering management knowledge area and compares it with the Project Management Institute (PMI) perspective of project management.

The context of this chapter serves the course learning outcomes CLO1 and CLO3 that were specified in the Preface (e.g., *refer to the "suggested course learning outcomes and their mapping to IET Competencies"*).

4.1 Software engineering

The following are among the information that would be acquired with respect to software engineering:

4.1.1 What is software?

Software is a computer program that comes along with specific documentation. Software solutions can be developed for particular customers or for markets.

https://doi.org/10.1515/9783111206868-004

4.1.2 What are the attributes of good software?

A good software provides the required functionalities and performance levels. A good software is expected to be maintainable, reliable, secured, efficient, and easy-to-use.

4.1.3 What is the software engineering discipline?

4.1.4 What are the fundamental software engineering activities?

The fundamental software engineering activities are the software specification, software development, software validation, and software evolution.

4.1.5 What are the differences between software engineering and computer science?

Computer science focuses on computer-related theories and fundamentals, while software engineering focuses on the practicalities of developing and delivering useful software solutions.

4.1.6 What are the differences between software engineering and systems engineering?

Software engineering is part of the systems engineering, which is concerned with all aspects of computer-software-based systems development including hardware, software, and process engineering.

4.1.7 What are the challenges facing software engineering?

The main challenges that software engineering is facing are the need to cope with the increasing diversity, the demands for reduced delivery times, and the need to develop trustworthy software solutions.

4.1.8 How much is the cost of software engineering?

About 60% of software costs are mainly development costs and the left 40% are testing costs.

4.1.9 What are the best software engineering techniques?

While all software projects have to be professionally managed and developed, different methods are good for different types of systems.

4.1.10 Software engineers versus engineers and computer scientists

The following is a comparison among the roles of software engineer, engineer, and computer scientist:

- **The computer scientist role**. A computer scientist proves theorems about algorithms, designs computer programming languages, and defines knowledge representation schemes. He/she can take his/her time to complete his/her tasks.
- **The engineer role**. An engineer develops solutions that solve specific problems for his/her clients. He/she uses computers and programming languages along with relevant tools and methods to complete his/her tasks and activities within predefined time boundaries.
- **The software engineer role**. A software engineer works in multi-application software and IT domains. He/she collects and analyzes the requirements of his/her client and design, implement, test, and deploy their targeted solutions. Similar to engineers, software engineers are restricted with certain time boundaries to complete their tasks.

4.2 Software Engineering Body of Knowledge (SWEBOK)

The IEEE-CS SWEBOK [3] presents 15 software engineering knowledge areas: software requirements, software design, software construction, software testing, software maintenance, software configuration management, software engineering management, software engineering process, software engineering models and methods, software quality, software engineering professional practice, software engineering economics, computing foundation, mathematical foundation, and engineering foundation. This section elaborates on each of these 15 knowledge areas.

Table 4.1 presents the main international standards that support software and systems engineering.

Table 4.1: The main international standards that support the discipline of software engineering.

Supporting standards – early versions	Supporting standards – updated versions
ISO/IEC TR 19759:2005 Software Engineering: a guide to the Software Engineering Body of Knowledge (SWEBOK) Link: https://www.iso.org/Standard/33897.html	ISO/IEC TR 19759:2015 Software Engineering: a guide to the Software Engineering Body of the Knowledge (SWEBOK) Link: https://www.iso.org/Standard/67604.html
ISO/IEC/IEEE 24765:2010 Systems and Software Engineering – Vocabulary Link: https://www.iso.org/Standard/50518.html	ISO/IEC/IEEE 24765:2017 Systems and Software Engineering – Vocabulary. Link: https://www.iso.org/Standard/71952.html

4.2.1 The software requirements knowledge area

A software requirement is an essential functional or nonfunctional feature that must be:
- identified, analyzed, and counted in during the software design and implementation; and
- tested, qualified, and brought into an operational mode as part of the overall software solution.

The software requirements knowledge area is concerned with elicitation, analysis, specification, verification, and validation of software requirements.

In general, the IEEE-CS SWEBOK classifies the software requirements into the following classes:
1. **Class#1: The Product and Process Requirements.** The product requirements refer to what is needed in the targeted product, including the constraints imposed on the product by the customer.
2. **Class#2: The Functional and Nonfunctional Requirements.** The functional requirements include the functionality of the targeted software solution, while the nonfunctional requirements include the quality factors that constrain the solution. These quality factors include performance, privacy, security, safety, scalability, maintainability, serviceability, supportability, availability, interoperability, scalability, and reliability. Each quality factor must have a set of predefined quality metrics that can be used to assess and evaluate the concerned solution.
3. **Class#3: The System Requirements and Software Requirements.** A system is a composition of components such as hardware, software, firmware, people, information, techniques, tools, facilities, and services.

4. **Class#4: The Emergent Requirements.** Emergent requirements are those that cannot be addressed by a single component of the system, but rather they depend on how the software components interoperate.
5. **Class#5: The Quantifiable Requirements.** This refers to the nonfunctional requirements that should be stated such that they are precise, verified, clear, unambiguous, qualitative, and quantitative.

Table 4.2 depicts the structure of the software requirements knowledge area and Table 4.3 presents the international standards that support the software requirements knowledge area.

Table 4.2: The topics structure of the "software requirements" knowledge area.

Software requirement fundamentals	
– Definition of a software requirement – Product and process requirements – Functional and nonfunctional requirements	– Emergent properties – Quantifiable requirement – System requirements and software requirements
Requirement process	
– Process models – Process actors	– Process support and management – Process quality and improvement
Requirement elicitation	
– Requirement sources	– Elicitation techniques.
Requirement analysis	
– Requirement classification – Conceptual modeling – Requirement negotiation	– Architectural design and requirement allocation – Formal analysis
Requirement specification	
– System definition document – System requirement specification	– Software requirement specification
Requirements' validation	
– Requirements' reviews – Prototyping	– Model validation – Acceptance tests
Practical considerations	
– Iterative nature of the requirements process – Change management	– Requirements' attributes – Requirements' tracing – Measuring requirements

Table 4.3: The main international standards that support the software requirements knowledge area.

Supporting standards – early versions	Supporting standards – updated versions
ISO/IEC 19761:2011 Software Engineering – COSMIC: A Functional Size Measurement Method Link: https://www.iso.org/Standard/54849.html	Last reviewed and confirmed in 2019
ISO/IEC 20926:2009 Software and Systems Engineering – Software Measurement – IFPUG Functional Size Measurement Method Link: https://www.iso.org/Standard/51717.html	Last reviewed and confirmed in 2019
ISO/IEC 20968:2002 Software Engineering – Mk II Function Point Analysis – Counting Practices Manual Link: https://www.iso.org/Standard/35603.html	Last reviewed and confirmed in 2020
ISO/IEC 24570:2005 Software Engineering – NESMA Functional Size Measurement Method Version 2.1 – Definitions and Counting Guidelines for the Application of Function Point Analysis Link: https://www.iso.org/Standard/37289.html	ISO/IE 24570:2018 Software Engineering – NESMA Functional Size Measurement Method –Definitions and Counting Guidelines for the Application of Function Point Analysis Link: https://www.iso.org/Standard/72505.html
ISO/IEC 19501:2005 Information Technology – Open Distributed Processing – Unified Modeling Language (UML) Version 1.4.2 Link: https://www.iso.org/Standard/32620.html	Last reviewed and confirmed in 2019
ISO/IEC 19505:2012 [two parts] Information Technology – Object Management Group Unified Modeling Language (OMG UML) Link-Part I: https://www.iso.org/Standard/32624.html Link-Part II: https://www.iso.org/Standard/52854.html	Last reviewed and confirmed in 2017

4.2.2 The software design knowledge area

Software Design is best defined as the process of defining the architecture, components, interfaces, and other characteristics of a software solution/program/system.

Table 4.4 presents the structure of the software design knowledge area and Table 4.5 presents the international standards that support this knowledge area.

Table 4.4: The structure of the "software design" knowledge area.

Software design fundamentals	
– General design concepts – Context of software design	– Software design process – Software design principles
Key issues in software design	
– Concurrency – Content and handling of events – Data persistence – Distribution of components	– Error and exception handling and fault tolerance – Interaction and presentation – Security
Software structure and architecture	
– Architectural structures/viewpoints – Architectural styles – Design patterns	– Architectural design decisions – Families of programs and frameworks
User interface design	
– General user interface principles – User interface design issues – The design of user interaction modalities	– The design of information presentation – User interface design process – Localization and internationalization – Metaphors and conceptual models
Software design quality analysis and evaluation	
– Quality attributes – Measures	– Quality analysis and evaluation techniques
Software design notations	
– Structural descriptions: a static view of the software structure – Behavioral description: a dynamic view of the of software structure	
Software design strengths and methods	
– General strategies – Function-oriented design: A structured design	– Object-oriented design – Data structure-centered design – Component-based design

Table 4.5: The main international standards that support the software design knowledge area.

Supporting standards – early versions	Supporting standards – updated versions
ISO/IEC/IEEE 42010:2011 Systems and Software Engineering – Architecture Description Link: https://www.iso.org/Standard/50508.html	ISO/IEC/IEEE 42010:2022 Software, Systems, and Enterprise – Architecture Description Link: https://www.iso.org/Standard/74393.html

Table 4.5 (continued)

Supporting standards – early versions	Supporting standards – updated versions
IEEE Std. 1016-2009 Standard for Information Technology – Systems Design – Software Design Descriptions Link: https://ieeexplore.ieee.org/document/5167255	No updates
IEEE Std. 26514-2010 Standard Adoption of ISO/IEC 26514:2008 Systems and Software Engineering – Requirements for Designers and Developers of User Documentation Link: https://Standards.ieee.org/ieee/26514/4990/	ISO/IEC/IEEE 26514:2022 Systems and Software Engineering – Design and Development of Information for Users Link: https://www.iso.org/Standard/77451.html

4.2.3 Software construction knowledge area

In general, software construction is the creation of working software solutions through a combination of coding, verification, validation, unit testing, integration testing, and debugging. The software construction knowledge area is strongly connected to all other software engineering knowledge areas, particularly the software design and the software testing knowledge areas.

Table 4.6 presents the structure of the software construction knowledge area and Table 4.7 presents the international standards that support this knowledge area.

Table 4.6: The structure of the "software construction" knowledge area.

Software construction fundamentals	
– Minimizing complexity – Anticipating change – Software reuse	– Constructing software for verification – Software construction standards
Managing construction	
– Construction in life cycle models	– Construction planning – Construction measurement
Practical considerations	
– Construction design – Construction languages – Coding – Construction testing	– Construction for reuse – Construction with reuse – Construction quality – Integration

Table 4.6 (continued)

Construction techniques	
– API design and use	– Executable models
– Object-oriented runtime issues	– Grammar-based input processing
– Parameterization and generics	– Concurrency primitives
– Assumptions, design by contract, and defensive programming	– Middleware
– Error handling/exception/handling/fault tolerance	– Construction methods for distributed software
– State-based and table-driven construction techniques	– Constructing heterogeneous systems
– Runtime configuration and internationalization	– Performance analysis and tuning
	– Platform standards
	– Test-first programming
Software construction tools	
– Development environments	– Unit testing tools
– GUI builders	– Profiling, performance analysis, and slicing tools

Table 4.7: The main international standards that support the software construction knowledge area.

Supporting standards – early versions	Supporting standards – updated versions
ISO/IEC TR 24772:2013 Information Technology – Programming Languages – Guidance to Avoiding Vulnerabilities in Programming Languages Through Language Selection and Use Link: https://www.iso.org/Standard/61457.html	Part I: ISO/IEC TR 24772-1:2019 Programming Languages – Guidance to Avoiding Vulnerabilities in Programming Languages – Part 1: Language-Independent Guidance Link-Part I: https://www.iso.org/Standard/71091.html Part II: ISO/IEC TR 24772-2:2020 Programming Languages – Guidance to Avoiding Vulnerabilities in Programming Languages – Part 2: Ada Link-Part II: https://www.iso.org/Standard/71092.html Part III: ISO/IEC TR 24772-3:2020 Programming Languages – Guidance to Avoiding Vulnerabilities in Programming Languages – Part 3: C Link-Part III: https://www.iso.org/Standard/71093.html
IEEE Std. 1008-1987 Standard for Software Unit Testing Link: https://ieeexplore.ieee.org/document/27763	No updates
ISO/IEC/IEEE 29119 [4 Parts] (Draft) Software and Systems Engineering – Software Testing Link: https://www.iso.org/Standard/60245.html	ISO/IEC/IEEE 29119-4:2021 Software and Systems Engineering – Software Testing- Part 4: Test Techniques Link: https://www.iso.org/Standard/79430.html

Table 4.7 (continued)

Supporting standards – early versions	Supporting standards – updated versions
ISO/IEC/IEEE 26515:2012 Systems and Software Engineering – Developing User Documentation in an Agile Environment Link: https://ieeexplore.ieee.org/document/6170923/versions#versions	ISO/IEC/IEEE 26515:2018 Systems and Software Engineering – Developing Information for Users in an Agile Environment Link: https://ieeexplore.ieee.org/document/8584455
IEEE Std. 1517-2010 Standard for Information Technology – System and Software Life Cycle Processes – Reuse Process Link: https://ieeexplore.ieee.org/document/5551093	No updates

4.2.4 The software testing knowledge area

Software testing is best defined as a dynamic verification of a software solution to ensure that it behaves as expected on a finite set of TEST-CASES:

- Dynamic testing implies that the software solution is tested using a preselected set of inputs.
- Finite testing implies that the software solution is tested using a subset of all possible TEST-CASES selected from a large set of TEST-CASES. The subset SELECTION-CRITERIA depends on the:
  - requirements that the software solution must be tested for;
  - validation and verification requirements; and
  - any other implicit requirements.

Table 4.8 presents the structure of the software construction knowledge area and Table 4.9 presents the international standards that support this knowledge area.

Table 4.8: The structure of the "software testing" knowledge area.

Software testing fundamentals	
– Testing-related terminology – Key issues	– Relationship of testing to other activities
Test levels	
– The testing target	– Objectives of testing

Table 4.8 (continued)

Test techniques	
– Based on the software engineer's intuition and experience – Input domain-based techniques – Code-based techniques – Fault-based techniques	– Usage-based techniques – Model-based techniques – Techniques based on the nature of the application – Selecting/combining techniques
Test-related measure	
– Programs-under-test evaluation	– Test evaluation
Test process	
– Practical consideration	– Test activities
Software testing tools	
– Testing tools support	– Categories of tools

Table 4.9: The main international standards that support the software testing knowledge area.

Supporting standards – early versions	**Supporting standards – updated versions**
IEEE Std. 829-2008 Standard for Software and System Test Documentation Link: https://ieeexplore.ieee.org/document/4578383	No updates
IEEE Std. 1008-1987 Standard for Software Unit Testing Link: https://ieeexplore.ieee.org/document/27763	No updates
ISO/IEC/IEEE 29119 [four parts] (Draft) Software and Systems Engineering – Software Testing Link: https://www.iso.org/Standard/60245.html	ISO/IEC/IEEE 29119-4:2021 Software and Systems Engineering – Software Testing – Part 4: Test Techniques Link: https://www.iso.org/Standard/79430.html
IEEE Std. 26513-2010 Standard Adoption of ISO/IEC 26513:2009 Systems and Software Engineering – Requirements for Testers and Reviewers of Documentation Link: https://ieeexplore.ieee.org/document/5712772	26513-2017 ISO/IEC/IEEE International Standard – Systems and Software Engineering – Requirements for Testers and Reviewers of Information for Users Link: https://ieeexplore.ieee.org/document/8085435
IEEE Std. 1012-2012 Standard for System and Software Verification and Validation Link: https://ieeexplore.ieee.org/document/6204026	IEEE Std. 1012-2016 Standard for System, Software and Hardware Verification and Validation Link: https://ieeexplore.ieee.org/document/8055462

Table 4.9 (continued)

Supporting standards – early versions	Supporting standards – updated versions
IEEE Std. 1044-2009 Standard for Classification for Software Anomalies Link: https://ieeexplore.ieee.org/document/5399061	No updates

4.2.5 The software maintenance knowledge area

Software maintenance is an integral part of a software solution life cycle. However, it is not given enough attention in comparison with the software development. But recently, things started to change as more attention is being paid to software maintenance.

Table 4.10 presents the structure of the software maintenance knowledge area and Table 4.11 presents the international standard that supports this knowledge area.

Table 4.10: The structure of the "software maintenance" knowledge area.

Software maintenance fundamentals	
– Definitions and terminology – Nature maintenance – Need for maintenance	– Majority of maintenance costs – Evolution of software – Categories of maintenance
Key issues in software maintenance	
– Technical issues – Management issues	– Maintenance cost estimation – Software maintenance measurement
Maintenance process	
– Maintenance processes	– Maintenance activities
Techniques for maintenance	
– Program comprehension – Reengineering – Reverse engineering	– Migration – Retirement

Table 4.11: The main standard that supports the software maintenance knowledge area.

Supporting standards – early versions	Supporting standards – updated versions
IEEE Std. 14764-2006 (a.k.a. ISO/IEC 14764:2006) Standard for Software Engineering – Software Life Cycle Processes – Maintenance Link: https://www.iso.org/Standard/39064.html	ISO/IEC/IEEE 14764:2022 Software Engineering – Software Life Cycle Processes – Maintenance Link: https://www.iso.org/Standard/80710.html

4.2.6 Software configuration management knowledge area

Software configuration management is a supporting software life cycle process that benefits the project management, development and maintenance activities, and quality assurance activities. This applies to all software components that need to be controlled.

Table 4.12 presents the structure of the software configuration management knowledge area and Table 4.13 presents the international standard that supports this knowledge area.

Table 4.12: The structure of the "software configuration management" knowledge area.

Management of the software configuration management (SCM) process	
– Organizational context for SCM – Constraints and guidance for SCM process	– Planning for SCM – SCM plan – Surveillance of SCM
Key issues in software maintenance	
– Identifying to-be-controlled items	– Software library
Software configuration control	
– Requesting, evaluating, and approving software changes	– Implementing software changes – Deviation and waivers
Software configuration status accounting	
– Software configuration status information	– Software configuration status reporting
Software configuration auditing	
– Software functional configuration audit – Software physical configuration audit	– In-process audits of a software baseline
Software release management and delivery	
– Software building	– Software release management

Table 4.13: The main standard that supports the software configuration management knowledge area.

Supporting standards – early versions	Supporting standards – updated versions
IEEE Std. 828-2012 Standard for Configuration Management in Systems and Software Engineering Link: https://ieeexplore.ieee.org/document/6170935	No updates

4.2.7 The software engineering management knowledge area

Software engineering management is used to present an SPM approach. It is associated with the other 14 software engineering knowledge areas.

This SPM approach comprises six phases: initiation and scope definition phase, software project planning phase, software project enactment phase, review and evaluation phase, closure phase, and software engineering measurement phase.

Table 4.14 presents the structure of the software engineering management knowledge area and Table 4.15 presents the international standards that support this knowledge area.

Table 4.14: The structure of the "software engineering management" knowledge area.

Initiation and scope definition phase	
– Determination and negotiation of requirements – Feasibility analysis	– Process for the review and revision of requirements
Software project planning phase	
– Process planning – Determine deliverables – Effort, schedule, and cost estimation	– Resource allocation – Risk management – Quality management
Software project enactment phase	
– Implementation of plans – Software acquisition and supplier contract management – Reporting	– Implementation of measurement process – Monitor process – Control process
Review and evaluation phase	
– Determining satisfaction of requirements	– Reviewing and evaluating performance
Closure phase	
– Determining closure	– Closure activities
Software engineering measurement phase	
– Establish and sustain measurement commitment – Plan measurement process	– Perform the measurement process – Evaluate measurement

Table 4.15: The main international standards that support the software engineering management knowledge area.

Supporting standards – early versions	Supporting standards – updated versions
ISO/IEC/IEEE 16326:2009 Systems and Software Engineering – Life Cycle Processes – Project Management Link: https://www.iso.org/Standard/41977.html	ISO/IEC/IEEE 16326:2019 Systems and Software Engineering – Life Cycle Processes – Project Management Link: https://www.iso.org/Standard/75276.html
IEEE Std. 16085-2006 (a.k.a. ISO/IEC 16085:2006) Standard for Systems and Software Engineering – Software Life Cycle Processes – Risk Management Link: https://www.iso.org/Standard/40723.html	ISO/IEC/IEEE 16085:2021 System and Software Engineering – Life Cycle Processes – Risk Management Link: https://www.iso.org/Standard/74371.html
IEEE Std. 15939-2008 Standard Adoption of ISO/ IEC 15939:2007 Systems and Software Engineering – Measurement Process Link: https://www.iso.org/Standard/44344.html	ISO/IEC/IEEE 15939:2017 System and Software Engineering – Measurement Process Link: https://www.iso.org/Standard/71197.html
IEEE Std. 1062-1998 Recommended Practice for Software Acquisition Link: https://ieeexplore.ieee.org/document/741938	No updates
ISO/IEC/IEEE 26512:2011 Systems and Software Engineering – Requirements for Acquirers and Suppliers of User Documentation Link: https://www.iso.org/Standard/43071.html	ISO/IEC/IEEE 26512:2018 Systems and Software Engineering – Requirements for Acquirers and Suppliers of Information for Users Link: https://www.iso.org/Standard/72088.html
IEEE Std. 1028-2008 Standard for Software Reviews and Audits. Link: https://ieeexplore.ieee.org/document/4601584	No updates
IEEE Std. 1061-1998 Standard for Software Quality Metrics Method Link: https://ieeexplore.ieee.org/document/749159	No updates
ISO/IEC/IEEE 26515:2012 Systems and Software Engineering – Developing User Documentation in an Agile Environment Link: https://ieeexplore.ieee.org/document/6170923/versions#versions	ISO/IEC/IEEE 26515:2018 Systems and Software Engineering – Developing Information for Users in an Agile Environment Link: https://ieeexplore.ieee.org/document/8584455

4.2.8 The software engineering process knowledge area

Software engineering processes are concerned with the project activities that are carried on by software engineers to develop, maintain, and operate software solutions.

These processes include the software requirement-specific processes, software design-specific processes, software construction-specific processes, software testing-specific processes, software maintenance-specific processes, software configuration-specific processes, software management-specific processes, and software quality assurance and control-specific processes.

Table 4.16 presents the structure of the software engineering process knowledge area and Table 4.17 presents the international standards that support this knowledge area.

Table 4.16: The structure of the "software engineering process" knowledge area.

Software process definition	
– Software process measurement	– Software process infrastructure
Software life cycles	
– Categories of software processes – Software life cycle models	– Software process adaptation – Practical considerations
Software process assessment and improvement	
– Software process assessment models – Software process assessment methods	– Software process improvement models – Continuous and staged software process ratings
Software measurement	
– Software process and product measurement – Quality measurement results	– Software information models – Software process measurement techniques

Table 4.17: The main international standards that support the software engineering process knowledge area.

Supporting standards – early versions	Supporting standards – updated versions
IEEE Std. 12207-2008 (a.k.a. ISO/IEC 12207:2008) Standard for Systems and Software Engineering – Software Life Cycle Processes Link: https://www.iso.org/Standard/43447.html	ISO/IEC/IEEE 12207:2017 Systems and Software Engineering – Software Life Cycle Processes Link: https://www.iso.org/Standard/63712.html
IEEE Std. 15288-2008 (a.k.a. ISO/IEC 15288:2008) Standard for Systems and Software Engineering – System Life Cycle Processes Link: https://www.iso.org/Standard/43564.html	ISO/IEC/IEEE 15288:2015 Systems and Software Engineering – System Life Cycle Processes Link: https://www.iso.org/Standard/63711.html
ISO/IEC/IEEE 15289:2011 Systems and Software Engineering – Content of Life-Cycle Information Products (Documentation) Link: https://www.iso.org/Standard/54388.html	Update One: ISO/IEC/IEEE 15289:2015 Systems and Software Engineering – Content of Life-Cycle Information Items (Documentation) Link: https://www.iso.org/Standard/66422.html

Table 4.17 (continued)

Supporting standards – early versions	**Supporting standards – updated versions**
	Update Two: ISO/IEC/IEEE 15289:2017 Systems and Software Engineering – Content of Life-Cycle Information Items (Documentation) Link: https://www.iso.org/Standard/71950.html Update Three: ISO/IEC/IEEE 15289:2019 Systems and Software Engineering – Content of Life-Cycle Information Items (Documentation) Link: https://www.iso.org/Standard/74909.html
IEEE Std. 24748.2-2012 Guide – Adoption of ISO/IEC TR 24748-2:2011 Systems and Software Engineering – Life Cycle Management – Part 2: Guide to the Application of ISO/IEC 15288 (System Life Cycle Processes) Link: https://www.iso.org/Standard/54994.html	ISO/IEC/IEEE 24748-2:2018 Systems and Software Engineering – Life Cycle Management – Part 2: Guidelines for the Application of ISO/IEC/IEEE 15288 (System Life Cycle Processes) Link: https://www.iso.org/Standard/70816.html Latest: https://www.iso.org/Standard/84661.html
IEEE Std. 24748.3-2012 Guide – Adoption of ISO/IEC TR 24748-3:2011 Systems and Software Engineering – Life Cycle Management – Part 3: Guide to the Application of ISO/IEC 12207 (Software Life Cycle Processes) Link: https://www.iso.org/Standard/54995.html	ISO/IEC/IEEE 24748-3:2020 Systems and Software Engineering – Life Cycle Management – Part 3: Guidelines for the Application of ISO/IEC/IEEE 12207 (Software Life Cycle Processes) Link: https://www.iso.org/Standard/77698.html
IEEE Std. 24748.1-2011 Guide – Adoption of ISO/IEC TR 24748-1:2010 Systems and Software Engineering – Life Cycle Management – Part 1: Guide for Life Cycle Management Link: https://www.iso.org/Standard/50502.html	ISO/IEC TS 24748-1:2016 Systems and Software Engineering – Life Cycle Management – Part 1: Guidelines for Life Cycle Management Link-Update I: https://www.iso.org/Standard/69655.html ISO/IEC/IEEE 24748-1:2018 Systems and Software Engineering – Life Cycle Management – Part 1: Guidelines for Life Cycle Management Link-Update II: https://www.iso.org/Standard/72896.html Latest: https://www.iso.org/Standard/84709.html
IEEE Std. 1517-2010 Standard for Information Technology – System and Software Life Cycle Processes – Reuse Processes Link: https://ieeexplore.ieee.org/document/5551093	No updates

Table 4.17 (continued)

Supporting standards – early versions	Supporting standards – updated versions
IEEE Std. 24774-2012 Guide – Adoption of ISO/IEC TR 24474:2010 Systems and Software Engineering – Life Cycle Management – Guidelines for Process Description Link: https://ieeexplore.ieee.org/document/6190704	ISO/IEC/IEEE 24774:2021 Systems and Software Engineering – Life Cycle Management – Specification for Process Description Link: https://www.iso.org/Standard/78981.html
ISO/IEC TR 29110-5-1-2:2011 Software Engineering – Lifecycle Profiles for Very Small Entities (VSEs) – Part 5-1-2: Management and Engineering Guide: Generic Profile Group: Basic Profile Link: https://www.iso.org/Standard/51153.html	Last reviewed and confirmed in 2022
IEEE Std. 1074-2006 Standard for Developing a Software Project Life Cycle Process Link: https://ieeexplore.ieee.org/document/1665059	No updates
IEEE Std. 828-2012 Standard for Configuration Management in Systems and Software Engineering Link: https://ieeexplore.ieee.org/document/6170935	No updates
IEEE Std. 14764-2006 (a.k.a. ISO/IEC 14764:2006) Standard for Software Engineering – Software Life Cycle Processes – Maintenance Link: https://www.iso.org/Standard/39064.html	ISO/IEC/IEEE 14764:2022 Software Engineering – Software Life Cycle Processes – Maintenance Link: https://www.iso.org/Standard/80710.html
ISO/IEC 15026-4:2012 Systems and Software Engineering – Systems and Software Assurance – Part 4: Assurance in the Life Cycle Link: https://www.iso.org/Standard/59927.html	ISO/IEC/IEEE 15026-4:2021 Systems and Software Engineering – Systems and Software Assurance – Part 4: Assurance in the Life Cycle Link: https://www.iso.org/Standard/74396.html
IEEE Std. 15939-2008 Standard Adoption of ISO/ IEC 15939:2007 Systems and Software Engineering – Measurement Process Link: https://www.iso.org/Standard/44344.html	ISO/IEC/IEEE 15939:2017 System and Software Engineering – Measurement Process Link: https://www.iso.org/Standard/71197.html
ISO/IEC 15940:2006 Information Technology – Software Engineering Environment Services Link: https://www.iso.org/Standard/29573.html	ISO/IEC 15940:2013 Systems and Software Engineering – Software Engineering Environment Services Link: https://www.iso.org/Standard/60100.html

Table 4.17 (continued)

Supporting standards – early versions	Supporting standards – updated versions
IEEE Std. 16085-2006 (a.k.a. ISO/IEC 16085:2006) Standard for Systems and Software Engineering – Software Life Cycle Processes – Risk Management Link: https://www.iso.org/Standard/40723.html	ISO/IEC/IEEE 16085:2021 System and Software Engineering – Life Cycle Processes – Risk Management Link: https://www.iso.org/Standard/74371.html
ISO/IEC/IEEE 16326:2009 Systems and Software Engineering – Life Cycle Processes – Project Management Link: https://www.iso.org/Standard/41977.html	ISO/IEC/IEEE 16326:2019 Systems and Software Engineering Life-Cycle Processes – Project Management Link: https://www.iso.org/Standard/75276.html
ISO/IEC/IEEE 29148:2011 Systems and Software Engineering – Life Cycle Processes – Requirements Engineering Link: https://www.iso.org/Standard/45171.html	ISO/IEC/IEEE 29148:2018 Systems and Software Engineering – System Life Cycle Processes – Requirements Engineering Link: https://www.iso.org/Standard/72089.html
ISO/IEC 20000-1:2011 Information Technology – Service Management – Part 1: Service Management System Requirements Link: https://www.iso.org/Standard/51986.html	ISO/IEC 20000-1:2018 Information Technology – Service Management – Part 1: Service Management System Requirements Link: https://www.iso.org/Standard/70636.html

4.2.9 The software engineering models and methods knowledge area

Software engineering models and methods are concerned with engineering and structuring software solutions. Structuring software solutions make them systematic, repeatable, and SUCCESSFUL.

Software engineering models provide:
- an organized and structured approach for problem solving;
- a notation for modeling software solutions; and
- procedures for constructing software models.

While software engineering methods provide an approach for specifying, designing, constructing, testing, verifying, and validating the final software solutions.

Table 4.18 presents the structure of the software engineering models and methods knowledge area and Table 4.19 presents the international standards that support this knowledge area.

Table 4.18: The structure of the "software engineering models and methods" knowledge area.

Modeling	
– Modeling principles – Properties and expression of models	– Syntax, semantics, and pragmatics – Preconditions, post-conditions, and invariants
Types of models	
– Information modeling – Behavioral modeling	– Structure modeling
Analysis of models	
– Analyzing for completeness – Analyzing for consistency – Analyzing for correctness	– Traceability – Interaction analysis
Software engineering methods	
– Heuristic methods – Formal methods	– Prototyping methods – Agile methods

Table 4.19: The main international standards that support the software engineering models and methods knowledge area.

Supporting standards – early versions	Supporting standards – updated versions
ISO/IEC/IEEE 26515:2012 Systems and Software Engineering-Developing User Documentation in an Agile Environment Link: https://ieeexplore.ieee.org/document/6170923/versions#versions	ISO/IEC/IEEE 26515:2018 Systems and Software Engineering – Developing Information for Users in an Agile Environment Link: https://ieeexplore.ieee.org/document/8584455
IEEE Std. 1320.1-1998 Standard for Functional Modeling Language-Syntax and Semantics for IDEF0 Link: https://ieeexplore.ieee.org/document/749110	No updates
IEEE Std. 1320.2-1998 Standard for Conceptual Modeling Language – Syntax and Semantics for IDEF1X97 (IDEF-object) Link: https://ieeexplore.ieee.org/document/8883275	31320-2-2012-ISO/IEC/IEEE International Standard-Information Technology – Modeling Languages – Part 2: Syntax and Semantics For IDEF1X97 (IDEF object) Link: https://ieeexplore.ieee.org/document/6357338
ISO/IEC 19501:2005 Information Technology – Open Distributed Processing-Unified Modeling Language (UML) Version 1.4.2 Link: https://www.iso.org/Standard/32620.html	Last reviewed and confirmed in 2019

Table 4.19 (continued)

Supporting standards – early versions	Supporting standards – updated versions
ISO/IEC 19505:2012 [two parts] Information Technology – Object Management Group Unified Modeling Language (OMG UML) Link-Part I: https://www.iso.org/Standard/32624.html Link-Part II: https://www.iso.org/Standard/52854.html	Last reviewed and confirmed in 2017
ISO/IEC 19506:2012 Information Technology – Object Management Group Architecture-Driven Modernization (ADM) – Knowledge Discovery Meta-Model (KDM) Link: https://www.iso.org/Standard/32625.html	Last reviewed and confirmed in 2017
ISO/IEC 19507:2012 Information Technology – Object Management Group Object Constraint Language (OCL) Link: https://www.iso.org/Standard/57306.html	Last reviewed and confirmed in 2022
ISO/IEC 15940:2006 Information Technology – Software Engineering Environment Services Link: https://www.iso.org/Standard/29573.html	ISO/IEC 15940:2013 Systems and Software Engineering – Software Engineering Environment Services Link: https://www.iso.org/Standard/60100.html
IEEE Std. 14102-2010 Standard Adoption of ISO/IEC 14102:2008 Information Technology – Guideline for the Evaluation and Selection of CASE Tools Link: https://ieeexplore.ieee.org/document/5567083	Last reviewed and confirmed in 2020
IEEE Std. 14471-2010 Guide-Adoption of ISO/IEC TR 14471:2007 Information Technology – Software Engineering – Guidelines for the Adoption of CASE Tools	No updates
IEEE Std. 1175.1-2002 Guide for CASE Tool Interconnections – Classification and Description Link: https://ieeexplore.ieee.org/document/1178882	No updates
IEEE Std. 1175.2-2006 Recommended Practice for CASE Tool Interconnection-Characterization of Interconnections Link: https://ieeexplore.ieee.org/document/4068343	No updates
IEEE Std. 1175.3-2004 Standard for CASE Tool Interconnections – Reference Model for Specifying Software Behavior Link: https://ieeexplore.ieee.org/document/1350464	No updates

4.2.10 The software quality knowledge area

Software quality is concerned with the software engineering perspective of culture and ethics, value and cost of quality, quality models and quality characteristics, quality improvement, software safety, software quality assurance, software quality control, software and systems verification and validation, software reviews and audits, software quality requirements, software defects characterization, software quality management techniques, software quality measurements, and software quality tools.

Table 4.20 presents the structure of the software quality knowledge area and Table 4.21 presents its supporting international standards.

Table 4.20: The structure of the "software quality" knowledge area.

Software quality fundamentals	
– Software engineering culture and ethics – Value and costs of quality	– Models and quality characteristics – Software quality improvement – Software safety
Software quality management processes	
– Software quality assurance – Verification and validation	– Reviews and audits
Practical considerations	
– Software quality requirements – Defect characterization	– Software quality management techniques – Software quality measurement

Table 4.21: The main international standards that support the software quality knowledge area.

Supporting Standards – early versions	Supporting standards – updated versions
IEEE Std. 90003-2008 Guide – Adoption of ISO/IEC 90003:2004 Software Engineering – Guidelines for the Application of ISO 9001:2000 to Computer Software Link: https://ieeexplore.ieee.org/document/4690902	90003-2018 Guide – ISO/ IEC/IEEE International Standard-Software Engineering – Guidelines for the Application of ISO 9001:2015 to Computer Software Link: https://ieeexplore.ieee.org/document/8559961
IEEE Std. 730-2002 Standard for Software Quality Assurance Plans Link: https://ieeexplore.ieee.org/document/1040117	IEEE Std. 730-2014 Standard for Software Quality Assurance Processes Link: https://ieeexplore.ieee.org/document/6835311

Table 4.21 (continued)

Supporting Standards – early versions	Supporting standards – updated versions
ISO/IEC 25000:2005 Software Engineering – Software Product Quality Requirements and Evaluation (SQuaRE) – Guide to SQuaRE Link: https://www.iso.org/Standard/35683.html	ISO/IEC 25000:2014 Systems and Software Engineering-Systems and Software Quality Requirements and Evaluation (SQuaRE) – Guide to SQuaRE Link: https://www.iso.org/Standard/64764.html
ISO/IEC 25010:2011 Systems and Software Engineering-Systems and Software Quality Requirements and Evaluation (SQuaRE) – System and Software Quality Models Link: https://www.iso.org/Standard/35733.html	Last reviewed and confirmed in 2017
ISO/IEC 25060 Through 25064 Software Engineering – Software Product Quality Requirements Evaluation (SQuaRE) – Common Industry Format (CIF) for Usability Link: https://www.iso.org/Standard/35786.html	ISO/TR 25060:2023 Systems and Software Engineering – Systems and Software Quality Requirements and Evaluation (SQuaRE) – General Framework for Common Industry Format (CIF) for Usability-Related Information Link: https://www.iso.org/Standard/83763.html
IEEE Std. 982.1-2005 Standard for Dictionary of Measures of the Software Aspects of Dependability Link: https://ieeexplore.ieee.org/document/1634994	No updates
IEEE Std. 1633-2008 Recommended Practice for Software Reliability Link: https://ieeexplore.ieee.org/document/4554206	IEEE Std. 1633-2016 Recommended Practice for Software Reliability Link: https://ieeexplore.ieee.org/document/7827907
IEEE Std. 1061-1998 Standard for Software Quality Metrics Method Link: https://ieeexplore.ieee.org/document/749159	No updates
IEEE Std. 1012-2012 Standard for System and Software Verification and Validation Link: https://ieeexplore.ieee.org/document/6204026	IEEE Std. 1012-2016 Standard for System, Software and Hardware Verification and Validation Link: https://ieeexplore.ieee.org/document/8055462
IEEE Std. 1028-2008 Standard for Software Reviews and Audits Link: https://ieeexplore.ieee.org/document/4601584	No updates
IEEE Std. 1044-2009 Standard for Classification for Software Anomalies Link: https://ieeexplore.ieee.org/document/5399061	No updates

Table 4.21 (continued)

Supporting Standards – early versions	Supporting standards – updated versions
IEEE Std. 15026.1-2011 Trial-Use Standard Adoption of ISO/IEC TR 15026-1:2010 Systems and Software Engineering-Systems and Software Assurance – Part 1: Concepts and Vocabulary Link: https://www.iso.org/Standard/50520.html	Update I: ISO/IEC 15026-1:2013 Systems and Software Engineering – Systems and Software Assurance – Part 1: Concepts and Vocabulary Link: https://www.iso.org/Standard/62526.html Update II: ISO/IEC/IEEE 15026-1:2019 Systems and Software Engineering – Systems and Software Assurance – Part 1: Concepts and Vocabulary Link: https://www.iso.org/Standard/73567.html
IEEE Std. 15026.2-2011 Standard Adoption of ISO/IEC 15026-2:2011 Systems and Software Engineering-Systems and Software Assurance – Part 2: Assurance Case Link: https://www.iso.org/Standard/52926.html	ISO/IEC/IEEE 15026-2:2022 Systems and Software Engineering – Systems and Software Assurance – Part 2: Assurance Case Link: https://www.iso.org/Standard/80625.html
ISO/IEC 15026-3:2011 Systems and Software Engineering-Systems and Software Assurance – Part 3: System Integrity Levels Link: https://www.iso.org/Standard/57107.html	ISO/IEC 15026-3:2015 Systems and Software Engineering-Systems and Software Assurance – Part 3: System Integrity Levels Link: https://www.iso.org/Standard/64842.html ISO/IEC/IEEE DIS 15026-3 Systems and Software Engineering-Systems and Software Assurance – Part 3: System Integrity Levels Link: https://www.iso.org/Standard/84444.html
ISO/IEC 15026-4:2012 Systems and Software Engineering-Systems and Software Assurance – Part 4: Assurance in the Life Cycle Link: https://www.iso.org/Standard/59927.html	ISO/IEC/IEEE 15026-4:2021 Systems and Software Engineering-Systems and Software Assurance – Part 4: Assurance in the Life Cycle Link: https://www.iso.org/Standard/74396.html
IEEE Std. 1228-1994 Standard for Software Safety Plans Link: https://ieeexplore.ieee.org/document/9097571	No updates
IEEE Std. 829-2008 Standard for Software and System Test Documentation Link: https://ieeexplore.ieee.org/document/4578383	No updates
IEEE Std. 1008-1987 Standard for Software Unit Testing Link: https://ieeexplore.ieee.org/document/27763	No updates

Table 4.21 (continued)

Supporting Standards – early versions	Supporting standards – updated versions
IEEE Std. 26513-2010 Standard Adoption of ISO/ IEC 26513:2009 Systems and Software Engineering – Requirements for Testers and Reviewers of Documentation Link: https://ieeexplore.ieee.org/document/5712772	26513-2017 ISO/IEC/IEEE International Standard – Systems and Software Engineering – Requirements for Testers and Reviewers of Information for Users Link: https://ieeexplore.ieee.org/document/8085435
ISO/IEC/IEEE 29119 [four parts] (Draft) Software and Systems Engineering – Software Testing Link: https://www.iso.org/Standard/60245.html	ISO/IEC/IEEE 29119-4:2021 Software and Systems Engineering – Software Testing – Part 4: Test Techniques Link: https://www.iso.org/Standard/79430.html

4.2.11 The software engineering professional practice knowledge area

Software engineering professional practice is concerned with the knowledge, skills, and attitudes that software engineers must have to practice software engineering in a professional, responsible, and ethical manner.

As indicated in Table 4.22, the software engineering professional practice composes professionalism, group of dynamics and psychology, and communication skills, whereas Table 4.23 presents its supporting international standards.

Table 4.22: The software engineering professional practice overview.

Professionalism	Group of dynamics and psychology	Communication skills
– Accreditation – Certification and licensing – Codes of ethics and professional conduct – Nature and role of professional societies – Economic impact of software – Employment contracts – Legal issues – Documentation – Tradeoff analysis	– Dynamics of working in teams – Individual cognition – Dealing with problem complexity – Interacting with stakeholders – Dealing with uncertainty and ambiguity – Dealing with multicultural environments	– Reading, understanding, writing, summarizing, and presentation skills. – Team and group communication skills

Table 4.23: The main international standards that support the software engineering professional practice knowledge area.

Supporting standards – early versions	Supporting standards – updated versions
ISO/IEC TR 19759:2005 Software Engineering – Guide to the Software Engineering Body Knowledge (SWEBOK) Link: https://www.iso.org/Standard/33897.html	ISO/IEC TR 19759:2015 Software Engineering-Guide to the Software Engineering Body Knowledge (SWEBOK) Link: https://www.iso.org/Standard/67604.html
ISO/IEC 24773:2008 Software Engineering – Certification of Software Engineering Professionals Link: https://www.iso.org/Standard/41543.html	ISO/IEC 24773-1:2019 Software and Systems Engineering – Certification of Software and Systems Engineering Professionals – Part 1: General Requirements Link: https://www.iso.org/Standard/69724.html

4.2.12 The software engineering economics knowledge area

Software engineering economics SEE is concerned with making business decisions related to software engineering. Economics is the study of values, costs, resources, and their relationship under certain circumstances.

The SEE provides a way to study the attributes of software and software processes in relevance to economic measures. SEE is concerned with aligning software technical decisions with the organizational business goals.

As indicated in Table 4.24, the SEE knowledge area composes the SEE fundamentals; SEE life cycle; SEE analysis methods; and SEE practical considerations.

Table 4.24: The structure of the "software engineering economics" knowledge Area.

Software engineering economics fundamentals		
– Finance – Accounting – Controlling – Decision-making process	– Valuation – Inflation – Depreciation – Taxation – Cash flow	– Time-value of money – Efficiency – Effectiveness – Productivity
Life cycle economics		
– Product – Project – Program – Portfolio – Product life cycle	– Project life cycle – Proposal – Investment decision – Planning horizon – Price and pricing	– Cost and costing – Performance measurement – Termination decisions – Replacement and retirement decisions

Table 4.24 (continued)

Economic analysis methods	
– For-profit decision analysis – Minimum acceptable rate of return – Return on investment – Return on capital employed – Cost-benefit analysis	– Cost-effectiveness analysis – Break-even analysis – Multiple attribute evaluation – Optimization analysis
Practical considerations	
– The "good enough" principle – Friction-free economy	– Ecosystems – Offshoring and outsourcing

4.2.13 The computing foundation knowledge area

Computing foundation includes knowledge about computers and their underlying principles of hardware and software that serve as a framework for software engineering.

The computing foundations knowledge area includes many topics such as problem-solving techniques, abstraction, programming language basics, debugging tools and techniques, data structure and representation, algorithms and complexity, system basic concepts, computer organization, operating systems basics, compiler basics, database basics, data management, network communication basics, parallel and distributed computing, basic developer human factors, basic developer human factors, and secure software development and maintenance.

4.2.14 The mathematical foundations knowledge area

The mathematical foundations knowledge area includes various topics such as the sets, relations, functions, basic logic, proof techniques, basic counting, graphs and trees, discrete probability, finite state machines, grammars, numerical precision, accuracy, errors, number theory, and algebraic structures.

4.2.15 The engineering foundations knowledge area

Engineering foundations knowledge area includes various topics such as the empirical methods and experimental techniques, statistical analysis, measurement, engineering design, modeling, simulation, prototyping, standards, and root-cause analysis.

4.3 Software engineering relevant certifications

Table 4.25 presents many examples of the software engineering international certifications.

Table 4.25: Examples of software engineering international certifications.

- **Amazon Web Services (AWS) Certification:**
  - AWS certified DevOps Engineer – Professional
- **IEEE-Computer Society Certification:**
  - Certified Software Development Professional (CSDP)
- **National Initiative for Cybersecurity Careers and Studies (NICCS) Certifications:**
  - Certified Secure Software Lifecycle Professional (CSSLP)
- **International Software Testing Qualifications Board (ISTQB):**
  - Core foundation (Level 1): Certified Tester Foundation Level (CTFL)
  - Core Advanced (Level 2): Certified Tester Advanced Level Test Analyst (CTAL-TA) and Certified Tester Advanced Level Test Manager (CTAL-TM)
  - Specialist (Level 3): Certified Tester AI Testing (CT-AI) and Certified Tester Game Testing (CT-GaMe)
- **The International Council of Electronic Commerce Consultants (EC-Council):**
  - Certified Ethical Hacker (CEH)
- **The International Information System Security Certification Consortium (ISC):**
  - Certified Information Systems Security Professional (CISSP)

- **Project Management Institute (PMI):**
  - Project Management Professional (PMP)
  - Certified Associate in Project Management (CAPM)
  - PMI Agile Certified Practitioner (PMI-ACP)
  - Professional in Business Analysis (PMI-PBA)
  - Program Management Professional (PgMP)
  - Portfolio Management Professional (PfMP)
  - PMI Risk Management Professional (PMI-RMP)
  - PMI Scheduling Professional (PMI-SP)
  - PMI Project Management Ready
  - Disciplined Agile Certifications (Disciplined Agile SCRUM Master (DASM))
  - Disciplined Agile Senior SCRUM Master (DASSM)
  - Disciplined Agile Coach (DAC)
  - Disciplined Agile Value Stream Consultant (DAVSC)

Table 4.25 (continued)

- **SCRUM.ORG Certifications:**
  - Professional SCRUM Master I (PSM I)
  - Professional SCRUM Master II (PSM II)
  - Professional SCRUM Master III (PSM III)
  - Professional SCRUM Product Owner I (PSPO I)
  - Professional SCRUM Product Owner II (PSPO II)
  - Professional SCRUM Product Owner III (PSPO III)
  - Professional SCRUM Development (PSD)
  - Professional Agile Leadership (PAL)
  - Professional Agile Leadership – Evidence-Based Management (PAL EBM)
  - Scaled Professional SCRUM (SPS)
  - Professional SCRUM with KANBAN (PSK)
  - Professional SCRUM with User Experience (PSU)

4.4 Software project management

The IEEE-CS software engineering management knowledge area represents a software-specific project management approach. This can be referred to as the IEEE-CS SPM approach. It is strongly associated with all other 14 software engineering knowledge areas, which are the software requirements, software design, software construction, software testing, software maintenance, software configuration management, software engineering process, software engineering models and methods, software quality, software engineering professional practice, SEE, computing foundation, mathematical foundation, and engineering foundation.

The IEEE-CS SPM approach composes the following phases:

- **Initiation and scope definition phase**. This phase is meant to quickly review RFPs and decide either to go or not to go with the software project opportunity associated with that RFPs.
- **Software project planning phase**. This phase is meant to carefully review RPFs and create a detailed plan that defines the work needed to SUCCESSFULLY carry on and manage the software project associated with that RFPs.
- **Software project enactment phase**. This phase is meant to execute, monitor, and control the plan created during the earlier software project planning phase.
- **Review and evaluation phase**. This phase is meant to review and evaluate the work throughout the project life cycle and ensure that things are going SATISFACTORILY. The same thing applies to the project work related to its schedule, cost, and quality assurance and control.
- **Closure phase**. This phase is meant to determine the work that must be SUCCESSFULLY completed in order to declare the project completion.

- **Software engineering measurement phase**. This phase is meant to measure and assess the software solution after it is operational.

Table 4.14 presents the “breakdown structure of the IEEE-CS software engineering management knowledge area.”

4.4.1 The initiation and scope definition phase

The initiation and scope definition phase is used to determine the software requirements by using elicitation methods and by assessing the software project feasibility. This phase includes the following processes:

- **Determination and negotiation of requirements**. It is essential to negotiate the software project requirements with the stakeholders and finalize them.
- **Feasibility study and analysis**. It is essential to conduct a feasibility study and analyze its results to define the project objectives and determine whether the project is feasible or not. This depends on many constraints such as the:
  - availability of the needed technologies, and human and physical resources; and
  - surrounding environmental, social, and political considerations.
- **Determination and endorsement of the process for requirements’ reviews and revisions**. It is essential that the stakeholders agree on the proposed process for reviewing the CHANGE-REQUESTS.

4.4.2 The software project planning phase

The software planning phase is meant to define the work needed to SUCCESSFULLY carry on and manage the software project that is being planned for. First of all, it is essential that a software development method is selected for the project based on the:

- project scope and requirements; and
- project risks, threats, and opportunities.

This phase includes the following processes:

- **Process planning**. This process is used to reach out a meaningful and applicable plan for carrying on and managing the software project in hand. This implies the necessity of having in place a detailed and comprehensive SPM plan.
- **Determination of deliverables**. This process is used to determine the set of deliverables that should be handed over to the customer throughout the project life cycle.
- **Estimations of effort, schedule, and cost**. As part of the targeted SPM plan, this process is used to conduct reasonable estimations with respect to the:
  - software project team efforts;
  - project work activities and tasks that should be carried on by the project team;

  - project schedule;
  - project itemized costs;
  - project overall cost; and
  - project price that the customer will be asked to pay in return to SUCCESSFULLY carrying on the project.
- **Resource allocations**. This process is used to fairly allocate the project's needed resources. Resources include physical and human resources. Physical resources include equipment, materials, and offices, whereas human resources include the project manager and team Members.
- **Risk management**. This process is used to:
  - identify the project risks, threats, and opportunities;
  - assess these risks, threats, and opportunities with respect to their occurrence probabilities and their anticipated impacts; and
  - define corrective actions and responses to be carried on reactively or proactively during the project life cycle.
- **Quality management**. This process is used to:
  - identify the software project quality requirements, including the project quality qualitative/quantitative requirements; and
  - define how to ensure that these quality requirements are satisfied at the end of the project.
- **Plan management**. This process is used to plan for managing the SPM plan including all of its subsidiary plans. Subsidiary plans include the scope management plan, risk management plan, quality management plan, resource management plan, and stakeholders' engagement plan.

4.4.3 The software project enactment phase

The software project enactment phase is used to execute the SPM plan and all of its subsidiary plans including the scope management plan, risk management plan, quality management plan, resource management plan, and stakeholders' engagement plan.

This phase includes the following:

- Implementation of plans. The SPM plan and all of its subsidiary plans are executed, monitored, and controlled during this software project enactment phase.
- Software acquisition and supplier contact management. All acquisitions of third-party software licenses and contracts with subcontractors are managed during this software project enactment phase.
- Implementation of measurement process. The process for conducting various measurements is defined and determined during this software project enactment phase. Conducting these measurements takes place during the following review and evaluation phase.

- Monitoring, controlling, and reporting processes. These processes are meant to monitor, control, and report the execution of the software project plan and its subsidiary plans throughout this software project enactment phase.

4.4.4 The review and evaluation phase

The review and evaluation phase is used to evaluate the project progress and achievements throughout the project life cycle, but at predetermined times, and/or as needed.

This phase includes the following:

- Determining satisfaction of requirements. During this review and evaluation phase, it is essential to ensure that the software project predetermined requirements are well taken care of and satisfied throughout the project life cycle.
- Reviewing and evaluating performance. During this review and evaluation phase, it is essential to regularly review and evaluate the:
  - performance of the project team members in terms of their abilities to do their assigned duties throughout the project life cycle;
  - effectiveness and appropriateness of the used methods, processes, and procedures throughout the project life cycle; and
  - performance and quality of the project context in terms of satisfying the project objectives and achieving the project requirements, outcomes, deliverables, and milestones throughout the project life cycle.

4.4.5 The closure phase

The closure phase is used to ensure that the SPM plan and all of its subsidiary plans were SUCCESSFULLY and SATISFACTORILY executed and completed.

This phase includes the following:

- Determining closure. This process is used to determine that the project is ready for closure after ensuring that all of its activities were SUCCESSFULLY and SATISFACTORILY completed, and that the project objectives and requirements were achieved.
- Closure activities. This process is used to execute all necessary closure activities after it is determined that the project is ready for closure. The main closure activities that should be carried on in coordination with the project stakeholders are:
  - archiving the project documentation, reports, and material;
  - destroying sensitive information, software, and tools;
  - updating all involved databases and repositories such as the organizational lessons learned ones;
  - allowing the prespecified warranty period; and
  - collecting the final payment.

4.4.6 The software engineering measurement phase

The software engineering measurement phase is used to define the activities needed to implement a software measurement process that is carried on after the targeted software solution is deployed at the customer's side and becomes operational.

This phase includes the following:

- **Establishing and sustaining measurement commitment**. This process includes:
  - defining the requirements for the targeted measurements;
  - defining the scope of measurements;
  - identifying the team commitment to measurement; and
  - identifying the needed resources for measurements.
- **Planning the measurement process**. This process includes:
  - characterizing the organizational unit that provides the context for measurements;
  - identifying the needed information for measurements;
  - selecting the right measures for measurements;
  - defining the data collection, analysis, and reporting procedures;
  - selecting the CRITERIA for evaluating the information results from measurements;
  - assigning the resources necessary to carry on measurements; and
  - acquiring and deploying all necessary supporting tools.
- **Performing the measurement process**. This process includes:
  - integrating measurement procedures with software processes;
  - collecting data for measurements; and
  - communicating results with the project team and stakeholders.
- **Evaluating measurements**. This process includes:
  - evaluating the measurement process and its produced information against a predefined evaluation CRITERIA;
  - determining the strengths and weaknesses of the produced information and the used measurement process;
  - identifying any potential improvement in accordance with the measurement resulted information; and
  - communicating any proposed improvements to the stakeholders.

4.4.7 A comparative view: the IEEE-CS software project management approach versus the PMI project management approach

The main differences between the IEEE-CS SPM approach and the PMI phased project management approach are as follows:

- The initiation and scope definition phase of the IEEE-CS SPM approach pays more attention to the scope definition, feasibility study, process definitions, and process reviews.

- The software project planning phase of the IEEE-CS SPM approach is specific to software.
- The software project enactment phase of the IEEE-CS SPM approach is a merger of the execution phase, and monitoring and control phase of the PMI phased project management approach.
- The reviews and performance phase of the IEEE-CS SPM approach does not exist as a phase in the PMI phased project management approach, but rather it may exist there as a process.
- The closure phase of the IEEE-CS SPM approach is similar to the closure phase of the PMI phased project management approach. However, it is not the end of the story in the case of the IEEE-CS SPM approach as it is followed by another phase as indicated in the following point.
- The software engineering measurement phase of the IEEE-CS SPM approach does not exist in the PMI phased project management approach. It is simply added in the IEEE-CS SPM approach to conduct additional measurements of the performance and the behavior of the software solution after it is deployed and brought into an operational mode at the customer side. The main purpose of such measurements is to proactively predict and prevent potential problems and issues in the software solution before they happen.

Chapter 5
Agile and SCRUM software project management

This chapter presents an executive summary of the SCRUM.ORG guide to SCRUM project management [4]. It elaborates on the Agile software project management methods in general, then on the SCRUM software project management method:

- **SCRUM principles.** These overview the SCRUM principles of empirical process control (e.g., *transparency*, *inspection*, and *adaptation*); self-organization; collaboration; value-based prioritization; time-boxing; and iterative development.
- **SCRUM organization.** This overviews the product owner; SCRUM master; SCRUM team; SCRUM of SCRUMs (SoS); SCRUM portfolios, programs, and projects; Tukman's theory of team development stages; Maslow's theory of needs hierarchy; and SCRUM leadership styles.
- **SCRUM business justification.** This overviews how SCRUM projects can be justified.
- **SCRUM quality.** This overviews how quality is perceived in SCRUM.
- **SCRUM change.** This overviews how change is perceived in SCRUM.
- **SCRUM risk.** This overviews the risk identification; risk assessment; risk prioritization; risk mitigation; risk communication; risk minimization; and risks in SCRUM portfolios and programs.
- **SCRUM initiate phase:** This overviews the SCRUM initiate phase processes, which are the Create Project Vision process; Identify SCRUM Master and Stakeholders' process; Form SCRUM Team process; Develop Epics process; Create Prioritized Product Backlog process; and Conduct Release Planning process.
- **SCRUM plan and estimate phase.** This overviews the SCRUM Plan and Estimate processes, which are the Create USER-STORIES process; Approve/Estimate/Commit USER-STORIES process; Create Task process; Estimate Task process; and Create Sprint Backlog process.
- **SCRUM implement phase.** This overviews the SCRUM Implement Phase processes, which are the Create Deliverables process; Conduct Daily Standup process; and Groom Prioritized Product Backlog process.
- **SCRUM Review and Retrospect Phase.** This overviews the SCRUM Review and Retrospect Phase processes, which are the Convene SCRUM of SCRUMs process; Demonstrate and Validate Sprint process; and Sprint Retrospect process.
- **SCRUM release phase.** This overviews the SCRUM Release Phase processes, which are the Ship Deliverable process and Retrospect Project process.

The context of this chapter serves the course learning outcomes of CLO1 and CLO4 that were specified in the Preface.

https://doi.org/10.1515/9783111206868-005

5.1 Agile overview

Agile refers to moving forward and responding easily and quickly. Agile development models handle the weaknesses of the traditional project management methods with respect to their ability of handling the increasing environmental demands expected by organizations.

5.1.1 The Agile manifesto

The Manifesto for Agile software development was introduced in 2001 for the purpose of finding better ways for software development. This involves concepts and actions such as the following:
- Having efficient and effective interactions among the team members
- Ongoing delivery of working software components
- Collaboration with the customer through involving the customer in the project from day 1
- Welcoming and endorsing CHANGE-REQUESTS received from the customer throughout the project life cycle
- Paying less attention to processes, tools, and comprehensive documentation

The Agile Manifesto focuses on:
- The individuals and their interactions rather than focusing on tools and processes
- Continuously delivering working software rather than focusing on developing comprehensive documentation
- Collaborating with the customer and ensuring their collaboration rather than focusing on negotiating the project contract
- Welcoming changes and responding to them rather than focusing on complying with the project plans

5.1.2 The Agile main principles

The Agile principles are as follows:
- The highest priority should be satisfying the customer through continuous delivery of valuable software.
- CHANGE-REQUESTS should be welcomed throughout the project life cycle.
- Incremental and ongoing delivery of working software components.
- Ongoing collaboration should be there between the business people and developers.
- Projects must be carried on around motivated individuals to get the projects completed.

- Face-to-face conversation is the most efficient and effective approach to share information with and within the project development team.
- The primary measure of progress should be the outcome working software components.
- Agility can be enhanced through following up on technical and design excellence and developments.
- Increasing the amount of completed work is essential and that can be achieved through using simple development and management methods.
- Team self-organization helps the emergence of the best architectures and designs.
- Regularly, the team becomes more and more effective and efficient.

5.1.3 The Agile main methods

Table 5.1 presents the main Agile methods other than the SCRUM one.

Table 5.1: The main Agile software project management methods other than the SCRUM one.

(a) KANBAN method
– The KANBAN method indicates the use of visual tools to track and improve production. – The KANBAN method was introduced by Taiichi Ohno, the founder of the Toyota Production Systems (TPS). – Examples of the KANBAN visual tools are the task cards, SCRUMBOARDS, and BURNDOWN-CHARTs.
(b) Lean KANBAN method
– The lean KANBAN method merges the use of the KANBAN visualization methods and the lean principles to create an incremental visual management system. – The lean KANBAN optimizes an organizational system to produce valuable results using its resources and needs, while reducing waste. – The lean KANBAN method involves increasing the productivity by reducing delays, early detection of errors, and reducing the total amount of effort required to finish a task.
(c) Crystal methods
– The crystal methods of software development were introduced during the early 1990s by Alistair Cockburn. They are people-centric, lightweight, and easy to adapt. Their development processes and tools can be tuned to cope with the project requirements. – The crystal methods include the Crystal Clear, Crystal Orange Web, Crystal Diamond, Crystal Yellow, Crystal Red, Crystal Sapphire, Crystal Orange, and Crystal Maroon. – The crystal methods have four roles: executive sponsor, lead designer, developers, and experienced users.

Table 5.1 (continued)

(d) Dynamic Systems Development Method (DSDM)

- The Dynamic Systems Development Method)DSDM((framework was first published in 1995.
- The DSDM represents the project work quality and effort in terms of cost and time.
- The DSDM adjusts the project deliverables to prioritize the project deliverables into four categories: MUST-HAVE, SHOULD-HAVE, COULD-HAVE, and WILL-NOT-HAVE.
- The DSDM has six phases: pre-project phase, feasibility phase, foundations phase, exploration-and-engineering phase, deployment phase, and benefit-assessment phase.

(e) Extreme Programming (XP) method

- The Extreme Programming (XP) method was introduced in Chrysler Corporation and in the 1990s it became familiar.
- The XP makes it possible to control the cost of changing software.
- The XP method facilitates the incremental development, flexible scheduling, automated code testing, verbal communications, evolving design, and close collaboration.
- The XP method involves communications, simplicity, feedback, and courage.
- The XP method has four roles, which are customer, developer, tracker, and coach.

(f) Feature-Driven Development (FDD) method

- The Feature-Driven Development (FDD) method was introduced in 1997 by Jeff DeLuca. The FDD operates on the principle of completing a project by breaking it down into small functions that can be delivered in less than 2 weeks.
- The FDD method has two principles, which are the software development is a human activity, and software development is a client-valued functionality.
- The FDD method has six roles: project manager, development manager, class owners, chief architect, chief programmers, and domain experts.
- The FDD method is an iterative approach that includes developing an overall model, building a feature list, and planning, designing, and building feature by feature.

(g) Test-Driven Development (TDD) method

- The Test-Driven Development (TDD) Method is known also as Test-First Development. It was introduced by Kent Beck as a software development method that involves writing automated test code first and developing the least amount of code necessary to pass that test later.
- The entire project is broken down into small and client-valued features to be developed in a very short development cycle.
- The TDD method has two levels: acceptance TDD and developer TDD.

(h) Adaptive Software Development (ASD) method

- The Adaptive Software Development (ASD) method was based on the rapid application development work of Jim Highsmith and Sam Bayer.
- The ASD method involves constant adaptation of processes to the work in hands, provision of solutions to problems surfacing in large projects, and iterative and incremental development with continuous prototyping.
- The ASD method has three phases: speculate phase, collaborate phase, and learn phase.

Table 5.1 (continued)

(i) Agile Unified Process (AUP) method
– The Agile Unified Process (AUP) method evolved from IBM's Rational Unified Process. – The AUP combines industry-endorsed Agile methods such as TDD, Agile modeling, Agile change management, and database refactoring to deliver a quality working product. – The AUP method has four phases: inception, elaboration, construction, and transition.
(j) Domain-Driven Design method
– The Domain-Driven Design method is an Agile development approach for handling complex designs with implementation linked to an evolving model. It was conceptualized in 2004 by Eric Evans. Domain is defined as an area of activity to which the user applies a program or functionality.

5.2 SCRUM overview

A SCRUM project is a collaborative effort to create a new product, solution, or service in accordance with the project vision statement.

There are many constraints that impact projects such as time-related constraints, cost-related constraints, scope-related constraints, quality-related constraints, resource-related constraints, and organizational-specific constraints. Such constraints make projects difficult to plan, execute, manage, and succeed.

SCRUM is a well-known Agile method that can be considered as an adaptive, iterative, fast, flexible, and effective method. It is designed to continuously deliver value throughout a project.

The SCRUM framework is structured such that it supports the development of products, solutions, and/or services regardless the projects' complexities.

Figure 5.1 shows that a SCRUM starts with a stakeholders' meeting to have the project vision created by the product owner who then develops a prioritized product backlog that includes a list of prioritized business and project requirements presented in the form of USER-STORIES.

A sprint lasts for 1–6 weeks and begins with a sprint planning meeting to select from the Highest-Priority USER-STORIES in the prioritized product backlog, and then include the selected ones in that sprint's backlog.

During each sprint, the SCRUM team creates the predefined deliverables or product increments of that sprint. Also, daily standup meetings are conducted to enable the team members to discuss their daily progress.

By the end of each sprint, a sprint review meeting is held to demonstrate the deliverables to the product owner and stakeholders.

The product owner accepts the deliverables only if they meet the predefined ACCEPTANCE-CRITERIA. If accepted, the deliverables will then be shipped to the customer and implemented for their use.

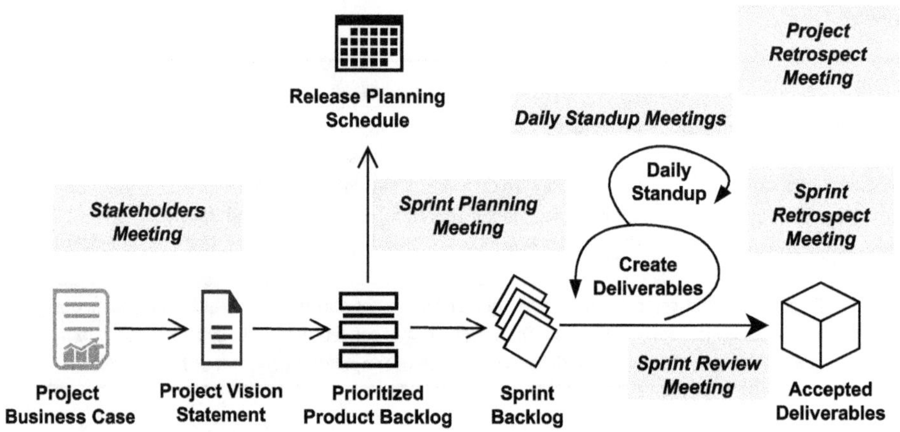

Figure 5.1: SCRUM flow diagram: a one sprint flow diagram and meetings.

The sprint cycle ends with a sprint retrospect meeting to enable the SCRUM team to discuss ways to improve processes and performance before they move on with the following sprint.

Using the SCRUM method in software development and in software project management has many benefits such as ensuring and promoting the adaptability, transparency, continuous feedback, continuous improvement, continuous delivery of effective deliverables and value, efficient development process, motivation among employees, and fast problem solving throughout the project life cycle.

In reference to Figure 5.2, SCRUM is divided into three areas: SCRUM principles, SCRUM aspects, and SCRUM processes. These areas are addressed in detail in the following sections.

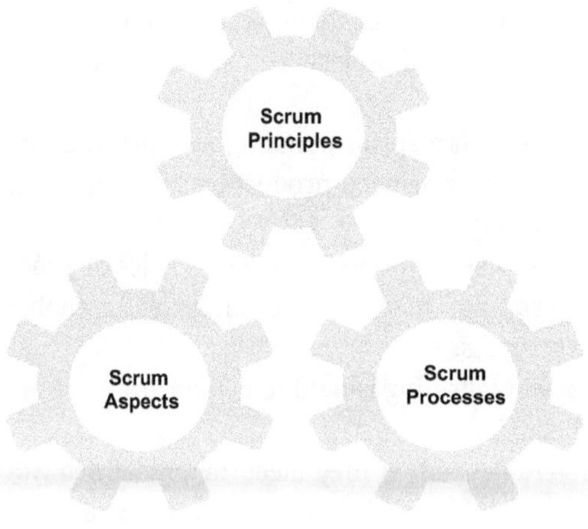

Figure 5.2: SCRUM areas: principles, aspects, and processes.

5.3 SCRUM principles

In reference to Figure 5.3, the SCRUM principles are the empirical process control, self-organization, collaboration, value-based prioritization, time-boxing, and iterative development.

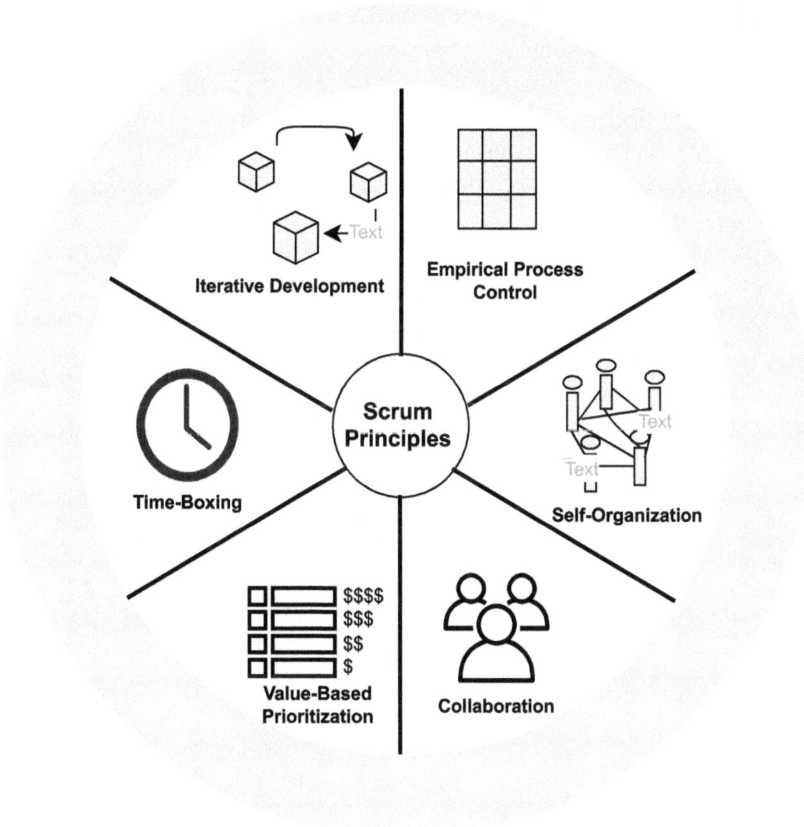

Figure 5.3: SCRUM principles.

5.3.1 Empirical process control

This principle describes the SCRUM philosophy based on the concepts of transparency, inspection, and adaptation.

5.3.1.1 Transparency

Transparency indicates that all aspects of any of the SCRUM processes can be observed by anyone. In reference to Figure 5.4, the transparency can be achieved through the following:

- **Project vision statement.** This statement is viewable by all stakeholders and the team.
- **Prioritized product backlog's USER-STORIES**. These USER-STORIES are viewable by everyone within and outside the team.
- **Release planning schedule.** This can be coordinated across multiple SCRUM teams.
- **Team progress visibility.** This can be achieved through the use of INFORMATION-RADIATORS such as SCRUMBOARDS, and BURNDOWN-CHARTS.
- **Daily standup meetings.** During these meetings, all team members can present what they have done yesterday, what they will do today, and any issues they are facing with.
- **Sprint review meetings.** During these meetings, the team can demonstrate their deliverables to the product owner and stakeholders seeking their approval before the team can proceed with deploying these deliverables at the customer site.

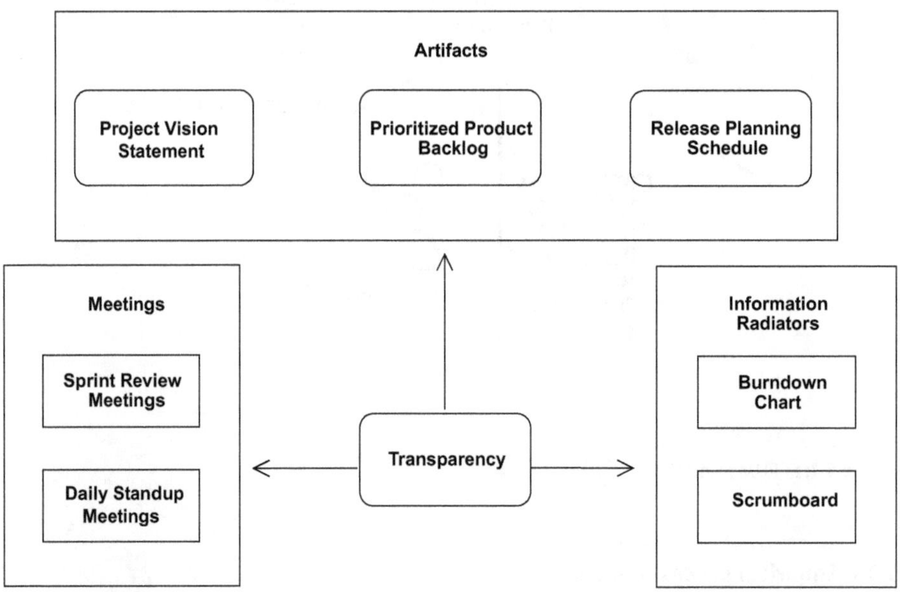

Figure 5.4: Transparency in SCRUM.

5.3.1.2 Inspection

In reference to Figure 5.5, inspection can be achieved through the:

- Use of a common SCRUMBOARD and other INFORMATION-RADIATORS that show the progress of the SCRUM team with the current sprint.

- Collection of feedback from the customer and other stakeholders during the Develop Epics process, Create Prioritized Product Backlog process, and conduct release planning process.
- Inspection and approval of the deliverables by the product owner and the customer during the Demonstrate and Validate Sprint process.

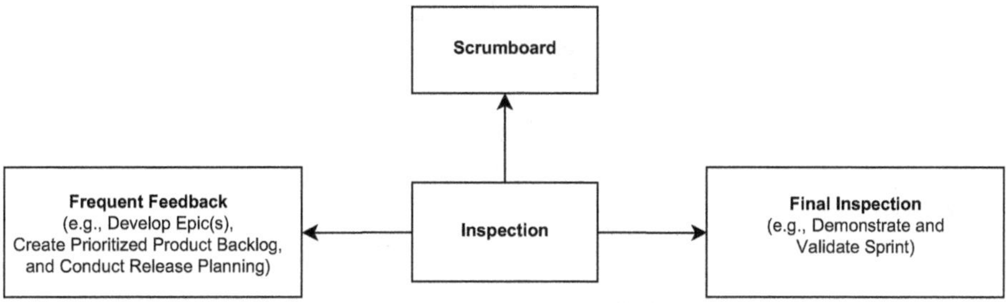

Figure 5.5: Inspection in SCRUM.

5.3.1.3 Adaptation

In reference to Figure 5.6, the SCRUM team and stakeholders adapt by improving their work in the following ways:

- During the daily standup meetings, the team members discuss issues that challenge the completion of their tasks and activities seeking support and guidance from others.
- Identified risks are used as inputs to many of the SCRUM processes.
- In the Sprint Retrospect process, improvements are decided based on the outputs of the Demonstrate and Validate Sprint process.
- In the project retrospect meeting, participants identify lessons learned and perform reviews while looking for opportunities to improve their processes.

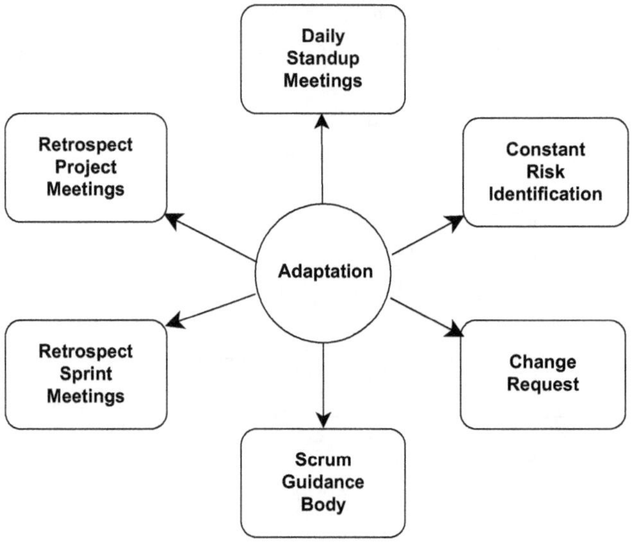

Figure 5.6: Adaptation in SCRUM.

5.3.2 Self-organization

This principle is used to indicate that if the project team members are focused and self-organized, then they can creatively and innovatively produce high-value deliverables.

The SCRUM appropriate leadership style is the Servant-Leadership style that indicates achieving results by focusing on the SCRUM team's needs. Figure 5.7 shows the main objectives of any SCRUM self-organizing teams.

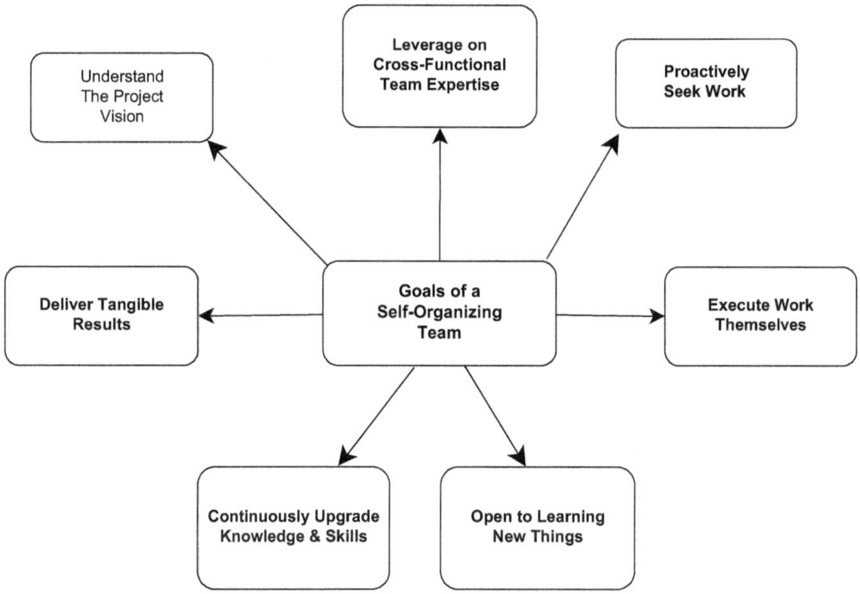

Figure 5.7: Goals of self-organizing SCRUM team.

5.3.3 Collaboration

This principle involves the dimensions of collaboration, which are the awareness, articulation, and appropriation.

- **Awareness:** awareness indicates that each member of a SCRUM team should be aware of the duties, tasks, and activities that are carried on by each member in his/her team.
- **Articulation:** articulation indicates that during a sprint, the work must be divided carefully among the team members in a way that their individual outcomes can be easily integrated as presented.
- **Appropriation:** appropriation indicates that resources must be used appropriately such that no or minimal efforts and time are wasted.

5.3.4 Value-based prioritization

This principle indicates that the SCRUM team should focus on delivering the highest possible business values throughout the project life cycle.

In reference to Figure 5.8, the product owner has to interpret the project stakeholders" needs while prioritizing the USER-STORIES in the prioritized product backlog. Such prioritization is based on the following factors:

- The risks associated with each USER-STORY.
- The value and impact of each USER-STORY.
- The dependencies among the USER-STORIES.

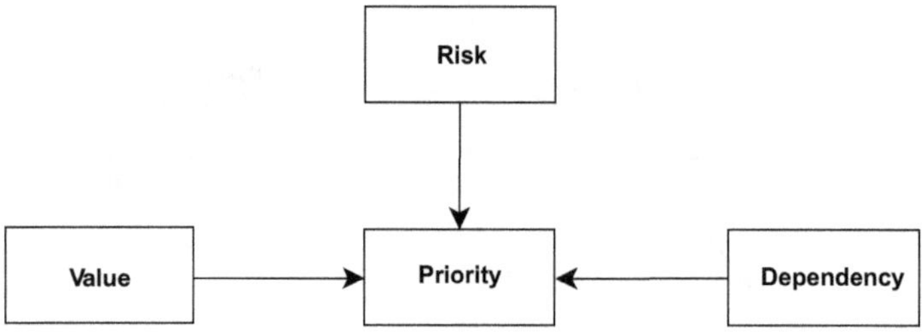

Figure 5.8: Value-based prioritization of USER-STORIES in a SCRUM product prioritized backlog.

5.3.5 Time-boxing

This principle indicates that the time constraints in SCRUM and how they can be used to manage the project planning and execution.

In reference to Figure 5.9, the SCRUM time-boxings are as follows:

- **Sprints time-boxing.** A sprint is a time-boxed iteration that lasts for 1–6 weeks. During a sprint, the SCRUM master guides and facilitates the SCRUM team while creating the deliverables through converting the USER-STORIES in that sprint into shippable functionalities. A typical sprint time-boxing is 4 weeks.
- **Daily standup meetings time-boxing.** This refers to a 15 min daily meeting, during which the team members meet and report the progress of their portions of the project. Each member typically reports to the rest of the team on what he/she completed yesterday, on what he/she is planning to complete today, and on what challenges or issues he/she is currently experiencing.
- **Sprint planning meetings' time-boxing.** This refers to an 8-h single meeting conducted prior to each sprint. It is divided into two submeetings as follows:
  - An objective definition submeeting, during which the product owner explains the highest priority USER-STORIES in the prioritized product backlog to the SCRUM team, before the SCRUM team defines the sprint objective.
  - A task estimation submeeting, during which the SCRUM team decides on "how" to complete the selected prioritized product backlog items to fulfill the sprint objective(s).
- **Sprint review meetings' time-boxing.** This refers to a 4-h meeting conducted prior to the end of each sprint, during which the SCRUM team demonstrates the

deliverables of the current sprint to the product owner, before he/she accepts or rejects them in accordance with the ACCEPTANCE-CRITERIA.

- **Sprint retrospect meetings' time-boxing.** This refers to a 4-h meeting prior to the end of each sprint after shipping the deliverables of that sprint to the customer. During a sprint retrospect meeting, the SCRUM team reviews and reflects on the current and about-to-complete sprint in terms of the processes followed, tools used, collaborative efforts, and communication mechanisms used. Accordingly, the team decides on what went well during that sprint and what did not go well. That way they learn and make improvements in the following sprints.

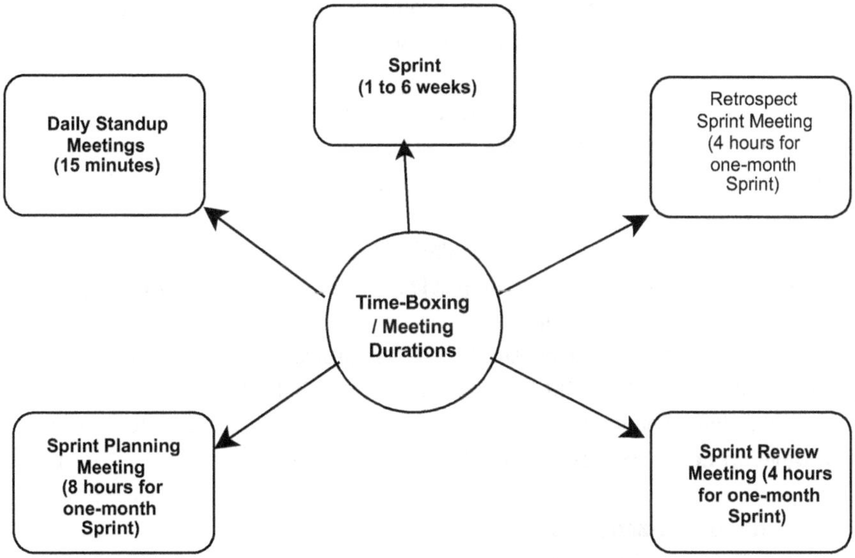

Figure 5.9: SCRUM time-boxings.

5.3.6 Iterative development

This principle describes the iterative development, and shows how to manage changes and build products that satisfy the customer's needs.

5.4 SCRUM organization

SCRUM organization refers to the SCRUM organizational structure in terms of the SCRUM's portfolios, programs, and projects. Also, it refers to the various SCRUM core and optional roles. Core rules are the product owner, SCRUM master, and SCRUM team. Optional roles are the stakeholders, customer, users, sponsor, vendors, SCRUM Guidance Body (SGB), chief product owner, and chief SCRUM master.

5.4.1 Product owner

The product owner is expected to:
- clearly and effectively communicate the product functional requirements (e.g., *USER-STORIES*) to the SCRUM team;
- clearly and precisely define the product's ACCEPTANCE-CRITERIA;
- clearly and certainly ensure that the product's deliverables (e.g., at the sprint level and at the project level) satisfy the product's ACCEPTANCE-CRITERIA.

The product owner's responsibilities across the SCRUM processes are as follows:
- **The Create Project Vision process.** During this process, the product owner is expected to define the project vision and to facilitate the creation of the project charter and its budget.
- **The identify SCRUM master and stakeholders' process.** During this process, the product owner is expected to help select a SCRUM master for the project and to identify the project stakeholders.
- **The Form SCRUM Team process.** During this process, the product owner is expected to help select the SCRUM team members, to help develop a collaboration plan, and to help develop the team building plan jointly with the SCRUM master.
- **The Develop Epics process.** During this process, the product owner is expected to create a set of epics that can be used to develop and promote the targeted product.
- **The Create Prioritized Product Backlog process.** During this process, the product owner is expected to prioritize the USER-STORIES and place them into the prioritized product backlog.
- **The Conduct Release Planning process.** During this process, the product owner is expected to create the release planning schedule and to help determine the sprint's length.
- **The Create USER-STORIES process.** During this process, the product owner is expected to help create the USER-STORIES and to define the ACCEPTACE-CRITERIA for USER-STORIES.
- **The Approve, Estimate, and Commit USER-STORIES process.** During this process, the product owner is expected to approve the USER-STORIES and to encourage the SCRUM team to commit to them.
- **The create tasks process.** During this process, the product owner is expected to explain the USER-STORIES to the SCRUM team while the team is deciding on the tasks and activities needed to complete each of them.

- **The estimate tasks process.** During this process, the product owner is expected to guide the SCRUM team and help them estimate and decide on the efforts required to carry on the tasks and activities needed to complete each USER-STORY.
- **The Create Sprint Backlog process.** During this process, the product owner is expected to clarify the USER-STORIES to the SCRUM team while the team is creating the sprint backlog.
- **The Create Deliverables process.** During this process, the product owner is expected to clarify the business requirements to the SCRUM team and provide them with the details of the required deliverables out of each sprint.
- **The Groom Prioritized Product Backlog process.** During this process, the product owner is expected to groom the prioritized product backlog in accordance with the feedback and recommendations made during the retrospect meetings at the end of each sprint.
- **The Demonstrate and Validate Sprint process.** During this process, the product owner is expected to accept or reject the proposed deliverables of each sprint, to provide feedback to the SCRUM master and SCRUM team, and accordingly to update the release plan and the prioritized product backlog.
- **The ship deliverable process.** During this process, the product owner is expected to facilitate the deployment of the product releases in coordination with the customer.
- **The Retrospect Project process.** During this process, the product owner is expected to participate in all sprint retrospect meetings as well as in the overall project retrospect meeting.

It is important here to highlight that the product owner must be a SCRUM expert, knowledgeable in the project business domain, possessing excellent communication skills, capable of handling uncertainties, and possessing strong negotiation skills. He/she must also be approachable, proactive, decisive, pragmatic, and goal-oriented.

5.4.2 SCRUM master

The SCRUM master is expected to
- be a servant-leader coaching and motivating his/her team;
- facilitate a protective and productive work environment for his/her team;
- resolve and eliminate issues and external influencing factors; and
- guard the SCRUM principles, aspects, and processes.

The SCRUM master's responsibilities across the SCRUM processes are as follows:
- **The identify SCRUM master and stakeholders' process.** During this process, the SCRUM master is expected to help the product owner identify the project stakeholders.

- **The Form SCRUM Team process.** During this process, the SCRUM master is expected to facilitate the selection of the SCRUM team members and the creation of the SCRUM team collaboration and building plan; and to ensure that the project has backup resources throughout its life cycle.
- **The Develop Epics process.** During this process, the SCRUM master is expected to help the product owner create the project epics.
- **The Create Prioritized Product Backlog process.** During this process, the SCRUM master is expected to help the product owner create the prioritized product backlog and define the deliverables ACCEPTANCE-CRITERIA and the project DONE-CRITERIA.
- **The Conduct Release Planning process.** During this process, the SCRUM master is expected to coordinate the creation of the release planning schedule and to determine the sprint's length.
- **The Create USER-STORIES process.** During this process, the SCRUM master is expected to help the product owner create and implement the USER-STORIES.
- **The Approve, Estimate, and Commit USER-STORIES process.** During this process, the SCRUM master is expected to facilitate the SCRUM team meetings to estimate and plan for the implementation of the USER-STORIES.
- **The create tasks process.** During this process, the SCRUM master is expected to help the SCRUM team identify and create the task list for the current sprint.
- **The estimate tasks process.** During this process, the SCRUM master is expected to help the SCRUM team estimate the needed effort to complete the current sprint.
- **The Create Sprint Backlog process.** During this process, the SCRUM master is expected to help the SCRUM team select the right USER-STORIES from the prioritized product backlog for inclusion in the current sprint's backlog.
- **The Create Deliverables process.** During this process, the SCRUM master is expected to support the SCRUM team and help them create the current sprint's deliverables, and update the SCRUMBOARD and the impediment log document.
- **The Conduct Daily Standup Meetings process.** During this process, the SCRUM master is expected to ensure that both of the SCRUMBOARD and the impediment log document are updated.
- **The Groom Prioritized Product Backlog process.** During this process, the SCRUM master is expected to facilitate the review meetings to review the prioritized product backlog.
- **The Convene SCRUM of SCRUMs process.** During this process, the SCRUM master is expected to participate in the SoS meetings, and discuss and resolve any issues that affect his/her SCRUM team.
- **The Demonstrate and Validate Sprint process.** During this process, the SCRUM master is expected to enable the SCRUM team to demonstrate the deliverables of the current sprint to the product owner and to solicit the endorsement of these deliverables, before shipping them to the customer.

- **The Sprint Retrospect process.** During this process, the SCRUM master is expected to improve the work environment of the incoming sprints based on the lessons learned from the current sprints in the project.
- **The Project Retrospect process.** During this process, the SCRUM master is expected to improve the work environment of the future projects based on the lessons learned from the current project.

It is important here to highlight that the SCRUM master must be a SCRUM expert, servant-leader, moderator, problem solver, approachable, motivator, perceptive, mentor, and introspective.

5.4.3 SCRUM team

The SCRUM team is expected to develop the product, solution, and/or service in accordance with the USER-STORIES in the current sprint's backlog. The SCRUM team's responsibilities across the SCRUM processes are as follows:
- **The Form SCRUM Team process.** During this process, the SCRUM team is expected to help create the collaboration plan and create the team building plan.
- **The Develop Epics process.** During this process, the SCRUM team is expected to clearly understand the epics.
- **The prioritized product backlog process.** During this process, the SCRUM team is expected to clearly understand the USER-STORIES in the prioritized product backlog.
- **The Conduct Release Planning process.** During this process, the SCRUM team is expected to consider and agree on the sprint's length as proposed by the SCRUM master.
- **The Create USER-STORIES process.** During this process, the SCRUM team is expected to help the product owner create the project USER-STORIES.
- **The Approve, Estimate, and Commit USER-STORIES process.** During this process, the SCRUM team is expected to estimate the effort and time required for each task.
- **The create tasks process.** During this process, the SCRUM team is expected to decide on and develop the task list for each of the USER-STORIES, while taking into consideration any dependencies among them.
- **The estimate tasks process.** During this process, the SCRUM team is expected to estimate the effort and time for each task.
- **The Create Sprint Backlog process.** During this process, the SCRUM team is expected to select the right USER-STORIES from the prioritized product backlog and include them in the incoming sprint's backlog.
- **The Create Deliverables process.** During this process, the SCRUM team is expected to create the deliverables, identify risks, and implement necessary risk mitigation plans, and keep the impediment log document updated.
- **The Conduct Daily Standup process.** During this process, the SCRUM team is expected to keep the BURNDOWN-CHART, SCRUMBOARD, and impediment log docu-

ment updated, discuss any issues raised by the participating team members, identify risks, and submit CHANGE-REQUESTS whenever that is necessary.
- **The Groom Prioritized Backlog process.** During this process, the SCRUM team is expected to participate in the review meetings of the prioritized product backlog.
- **The Convene SCRUM of SCRUMs process.** During this process, the SCRUM team is expected to provide the SCRUM master with the necessary inputs on issues that need to be discussed and resolved at the SoS meetings.
- **The Demonstrate and Validate Sprints process.** During this process, the SCRUM team is expected to demonstrate the completed deliverables to the product owner and seek his/her approval and endorsement.
- **The Sprint Retrospect process.** During this process, the SCRUM team is expected to provide the product owner and the SCRUM master with their suggestions for improvements that would benefit the incoming sprints.
- **The Project Retrospect process.** During this process, the SCRUM team is expected to participate in the project retrospect meeting.

It is important here to highlight that the SCRUM team members must be knowledgeable in the SCRUM areas (e.g., principles, aspects, and processes), collaborative, self-organized, motivated, proactive, technical expert, team player, responsive, goal-oriented, and introspective.

5.4.4 SCRUM of SCRUMs (SoS) structure

If a project needs a team of more than nine members, then it will be required to divide that project into multiple SCRUMs rather than having it all in one SCRUM.

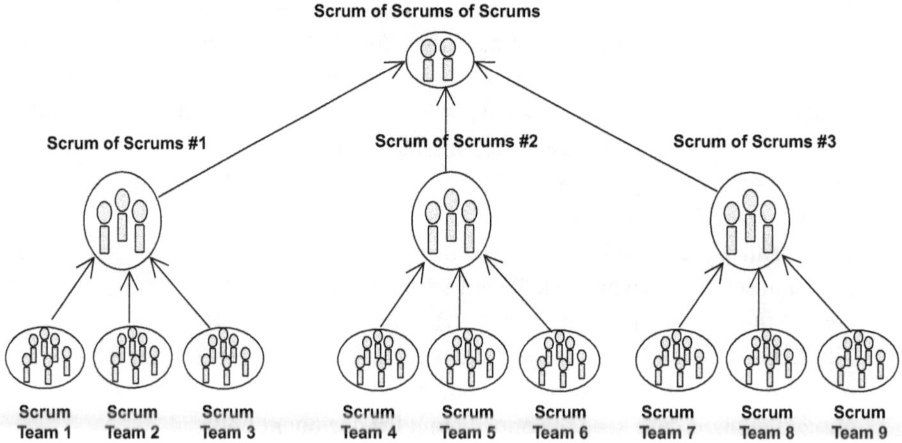

Figure 5.10: A multi-SCRUM project.

Figure 5.10 presents a two-layered SoS project structure, in which nine SCRUM teams are distributed across three SoS groups, which are put into a SoS of SCRUMs group. In this SCRUM structure, the SCRUM project population would be as follows:

- **SCRUM teams' members:** There should be nine SCRUM teams, each of up to nine members. Thus, the count of the overall SCRUM teams' members cannot exceed 81.
- **Product owners/SCRUM masters:** There are nine SCRUM teams. Thus, there should be nine product owners and nine SCRUM masters.
- **Chief product owners/chief SCRUM masters.** There are three SoS and one SoS of SCRUMs. Thus, there should be four chief product owners and four chief SCRUM master. The chief product owner is expected to coordinate and supervise multiple product owners when the project is large and has multiple SCRUMs. Also, he/she is expected to interface with the program product owner to ensure that the project is in full alignment with the program goals and objectives. The chief SCRUM master is expected to gather and communicate information across the SCRUM teams within the same project.

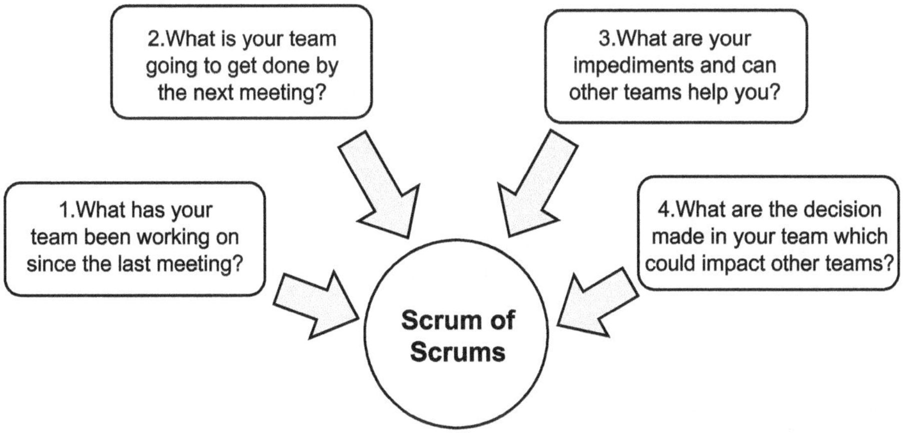

Figure 5.11: Typical questions exchanged among the product owners and the SCRUM masters during a SCRUM of SCRUMs meeting.

Figure 5.11 presents the type of questions that may be asked during a SoS meeting. This may include information about what each team has completed since the last SoS meeting, what each team is planning to complete by the next SoS meeting, what obstacles that each team has been facing, and what decisions that each team has taken and may affect the work being carried out by the other teams.

5.4.5 SCRUM portfolios and programs

Figure 5.12 shows the duties and roles that are associated with a SCRUM portfolio, a SCRUM program, and a SCRUM project. Table 5.2 provides an executive summary of various SCRUM roles and responsibilities.

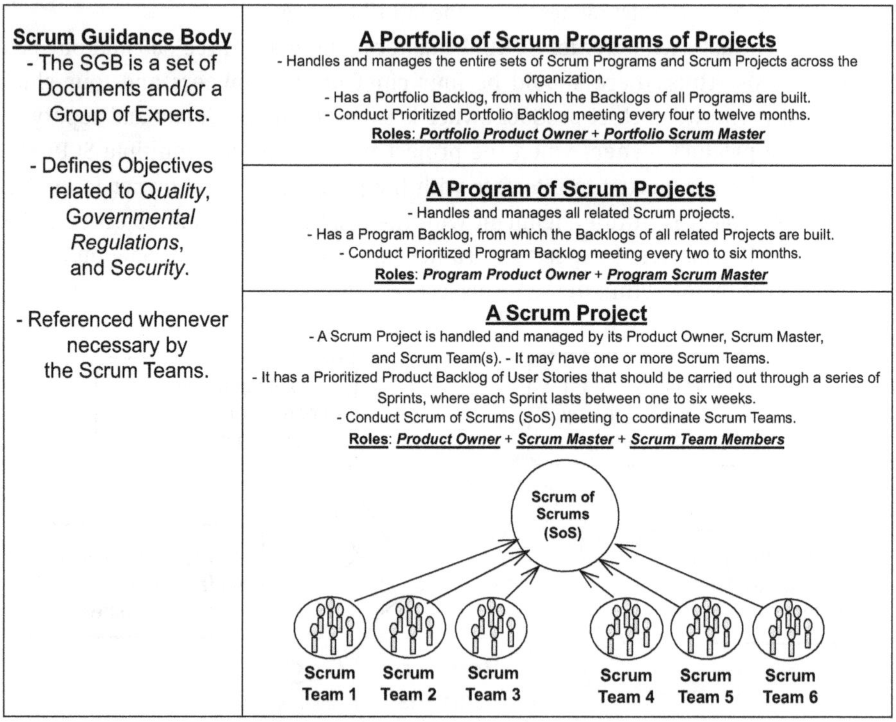

Figure 5.12: SCRUM portfolios, SCRUM programs, and SCRUM projects.

Table 5.2: Summary of various SCRUM roles and responsibilities.

Roles directly associated with portfolios

- **Portfolio product owner.** A portfolio product owner is responsible for defining the strategic objectives and priorities for the SCRUM portfolio he/she is handling.
- **Portfolio SCRUM master.** A portfolio SCRUM master is responsible for solving problems and issues, and conducting the portfolio meetings.

Roles directly associated with programs

- **Program product owner.** A program product owner is responsible for defining the portfolio strategic objectives and priorities for the SCRUM program he/she is handling.
- **Program SCRUM master.** A program SCRUM master is responsible for solving the program problems and issues, and conducting the program meetings.

Table 5.2 (continued)

Roles directly associated with projects
– **Chief product owner.** A chief product owner is mainly responsible for coordinating the collaboration among the product owners in a multi-SCRUM project.
– **Chief SCRUM master.** A chief SCRUM master is mainly responsible for coordinating the collaboration among the SCRUM masters in a multi-SCRUM project.
– **Product owner.** A product owner is mainly responsible for defining the project's overall requirements, starting and monitoring the project, identifying the SCRUM master and SCRUM team members, providing the project's financial resources, deriving the product vision, ensuring the product delivery, ensuring the transparency and clarity of the prioritized product backlog USER-STORIES, deciding on the release content, deciding on the USER-STORIES ACCEPTANCE-CRITERIA, inspecting the deliverables, and deciding on the sprint's length.
– **SCRUM master.** A SCRUM master is mainly responsible for ensuring that the SCRUM processes are endorsed and followed by all team members, ensuring that the product development is going smoothly, coordinating the release planning meeting, and scheduling for all necessary meetings.
– **SCRUM team.** The SCRUM team is mainly responsible for ensuring that the project deliverables are created in accordance with the product requirements, and assuring that the product owner and SCRUM master that the work is being performed in accordance with the plan.
– **Stakeholders.** Stakeholders include customers, users, and sponsors. They interface with the product owner, SCRUM master, and SCRUM team and provide them with their inputs.

5.4.6 SCRUM-relevant human resources theories

5.4.6.1 Tukman's theory of team development stages

Figure 5.13 depicts the Tukman's theory of team development stages, which indicates that the process of developing project teams would go through the following four stages:

- **Forming stage.** During this stage, the roles and responsibilities of the individual team members are not clear nor yet defined.
- **Storming stage.** During this stage, there exist much confusion and less agreements among team members.
- **Norming stage.** During this stage, the team would be in a good shape and can work together on the project.
- **Performing stage.** During this stage, the team would be working together consistently and coherently with high productivity.

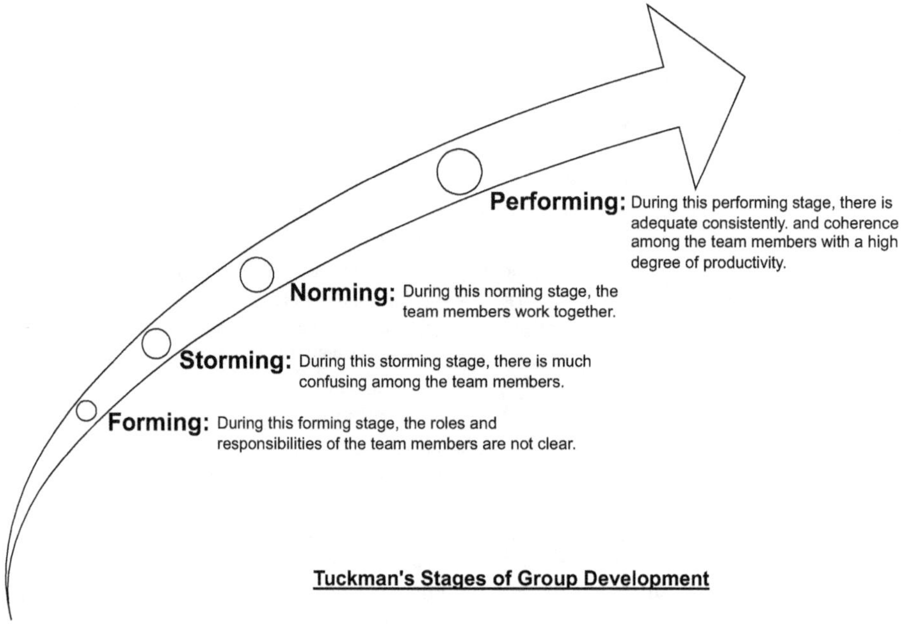

Figure 5.13: Tukman's theory of group (team) development stages.

5.4.6.2 Maslow's theory of needs hierarchy

Figure 5.14 is self-explanatory depicting the Maslow's theory of needs hierarchy.

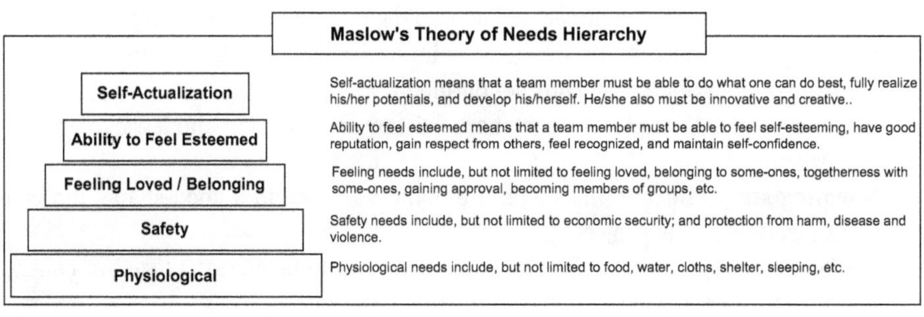

Figure 5.14: Maslow's theory of needs hierarchy.

5.4.6.3 Theory X and theory Y

The following two theories were proposed by Douglas McGregor:

– **Theory X leadership style.** Theory X leaders would assume that their team members are unmotivated and would avoid work whenever that is possible. This style of leadership is not suitable for SCRUM projects.

- **Theory Y leadership style.** Theory Y leaders would assume that their team members are self-motivated and seek work, and would stand for challenge and accept more responsibilities. This style of leadership is indeed suitable for SCRUM projects.

5.4.7 SCRUM conflict management methods

The main conflict management methods are as follows:

- **Win-win conflict management method.** With this method, team members can cooperatively deal directly with problems to resolve any disagreements. SCRUM projects need work environments that encourage the SCRUM team members to openly discuss problems and work through them to reach win-win outcomes.
- **Lose-win conflict management method.** In case a team member feels that his/her contributions are not recognized or that he/she is not well-treated, then he/she may lose focus and stop his/her effective contribution to the project and agree (e.g., even if he/she is in disagreement) to whatever he/she is told to do. This method is not appropriate for SCRUM projects.
- **Lose-lose conflict management method.** In some situations, team members attempt to search for solutions that bring only some degree of satisfaction to them and to the other disputing parties. This approach involves "give and take" to satisfy every team member. SCRUM meetings are there to ensure solving problems through mutual discussions.
- **Win-lose conflict management method.** A SCRUM master may think that he/she is a de facto manager and impose his/her viewpoint on others. This method results in a win-lose outcome. This method is not recommended for SCRUM projects.

5.4.8 SCRUM leadership styles

The main leadership styles that may be followed in SCRUM projects are as follows:

- **Servant-leadership style**. Servant-leaders listen and heal their team members. Also, they are expected to be persuasive, aware of what is going to happen and of what is going on. This is the most suitable leadership style for the SCRUM master.
- **Delegating leadership style**. When time is limited, leaders of this style would delegate some of their planning and decision-making responsibilities to some of their competent team members.
- **Autocratic leadership style**. Leaders of this style would make decisions on their own with little without involving their team members.
- **Directing leadership style**. Leaders of this style would instruct their team members what tasks are required, and when and how they should be performed.
- **Laissez-faire leadership style**. Leaders with this style would leave their team members unsupervised with little or no interference in their work.

- **Coaching and supportive leadership style**. Leaders with this style support and monitor their team members by listening to them, assisting and encouraging them, and maintaining positive outlooks when encountering uncertainties.
- **Task-oriented leadership style**. Leaders with this style ensure that all tasks are completed before deadlines.
- **Assertive leadership styles**. Leaders with this style establish authority with respect.

5.5 SCRUM business justification

SCRUM business justification refers to the second SCRUM area, which provides reasonable justification for a SCRUM project before it is endorsed and carried out, by simply answering a question like "Why is this project needed?"

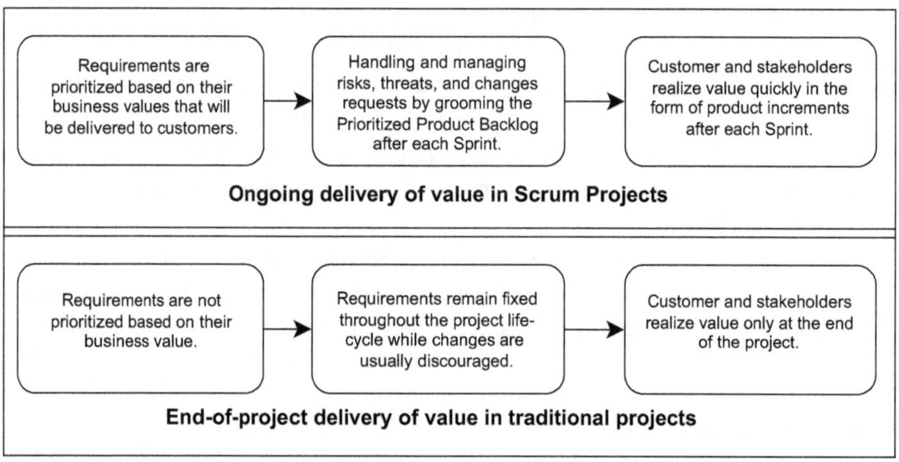

Figure 5.15: SCRUM delivery of value versus traditional project delivery.

In reference to Figure 5.15, the ongoing delivery of value deliverables is the most important justification for using the SCRUM method to carry out projects instead of the traditional methods.

The following is a summary of the responsibilities of various SCRUM roles in justifying the business need for SCRUM projects:

- **Portfolio product owner.** A portfolio product owner is responsible for delivering value and creating business justification for his/her portfolio. He/she is also responsible for providing value guidance and approving business justification for programs.

- **Program product owner.** A program product owner is responsible for delivering value and creating business justification for his/her program. Also, he/she is responsible for providing value guidance and approving business justification for projects under his/her program.
- **SCRUM master.** A SCRUM master is mainly responsible for facilitating the creation of the project deliverables, and for managing risks and CHANGE-REQUESTS.
- **SCRUM team.** The SCRUM team is mainly responsible for creating the project deliverables, demonstrating them to the product owner, and implementing them for the customer.
- **Stakeholders.** Stakeholders include customers, users, and sponsors. Sponsors are responsible for funding their SCRUM projects and monitoring them to confirm that they realized their benefits. Customers and users are involved in defining the prioritized list of USER-STORIES in the prioritized product backlog.

Typically, the process for justifying the business needs for a SCRUM project goes through the following steps:
- **Step#1:** Assessing and presenting a business case for that project.
- **Step#2:** Continuously justifying the value of that project.
- **Step#3:** Continuously confirming the benefits realization of that project.

Also, the process for justifying the business need for a SCRUM project is typically based on the following factors:
- **Project reasoning.** This includes the factors that make a project justifiable and meaningful.
- **Business needs.** This includes the needed business outcomes of a project in accordance with that project vision statement.
- **Project benefits.** This includes the expected benefits from a project such as a measurable improvement in the product.
- **Major risks.** This includes any uncertainties and/or events that were not planned for in a project, but may impact the SUCCESS of that project.
- **Project timescales.** This refers to the duration of a project.
- **Project costs.** This refers to the costs of investments and development of a project.

5.6 SCRUM quality

SCRUM quality refers to ensuring that the deliverables and product of a SCRUM project satisfy the predefined ACCEPTANCE-CRITERIA of that project, and that the business value of that project is achieved in accordance with the customer expectations. SCRUM quality management involves quality planning, quality control, and quality assurance.

The factors used to determine the scope quality requirements for a SCRUM project are as follows:

- The business needs that should be fulfilled by that project.
- The ability of the customer to satisfy these business needs.
- The stakeholders' current and potential future business needs.

The following summarizes the responsibilities of the SCRUM roles with respect to securing the SCRUM quality:
- **Portfolio product owner.** A portfolio product owner is responsible for setting the ACCEPTANCE-CRITERIA for the portfolio and for reviewing the portfolio deliverables.
- **Portfolio SCRUM master**. A portfolio SCRUM master is responsible for ensuring that a sustainable focus is maintained on the quality of features of the overall portfolio.
- **Program product owner**. A program product owner is responsible for setting the ACCEPTANCE-CRITERIA for the program and for reviewing the program deliverables.
- **Program SCRUM master**. A program SCRUM master is responsible for ensuring that a sustainable focus is maintained on the quality of features of the overall program.
- **Product owner.** A product owner is responsible for stating and clearly defining the business needs and requirements in the prioritized product backlog, for ensuring that the deliverables satisfy the quality requirements, for setting the ACCEPTANCE-CRITERIA for the entire project, for facilitating the creation of the USER-STORIES' ACCEPTANCE-CRITERIA, and for reviewing and validating the deliverables.
- **SCRUM master.** A SCRUM master is responsible for estimating the obstacles that may impact the quality of deliverables and processes, for ensuring that a sustainable focus is maintained on the quality of the project features, and for ensuring that the SCRUM processes are followed by all of the SCRUM team members.
- **SCRUM team.** A SCRUM team is responsible for developing and maintaining all sprints' deliverables, for encouraging effective communications such that the business needs and requirements are made clear and understandable, for sharing knowledge and familiarizing the team members with these features, for benefiting from the experience of others, and for making appropriate and smooth changes to deliverables.
- **SGB.** An SGB provides a framework for developing the ACCEPTANCE-CRITERIA.
- **Stakeholders.** The stakeholders review and accept the deliverables and the outcome product.

5.7 SCRUM change

SCRUM change refers to ensuring that the CHANGE-REQUESTS received from the stakeholders are welcomed and endorsed. It is important to notice that the stakeholders can request changes at any time throughout the project life cycle, and it is almost impossible for them to precisely define their business requirements' upfront.

Figure 5.16 presents an example of how CHANGE-REQUESTS are received, approved, implemented, and integrated in the final outcome product.

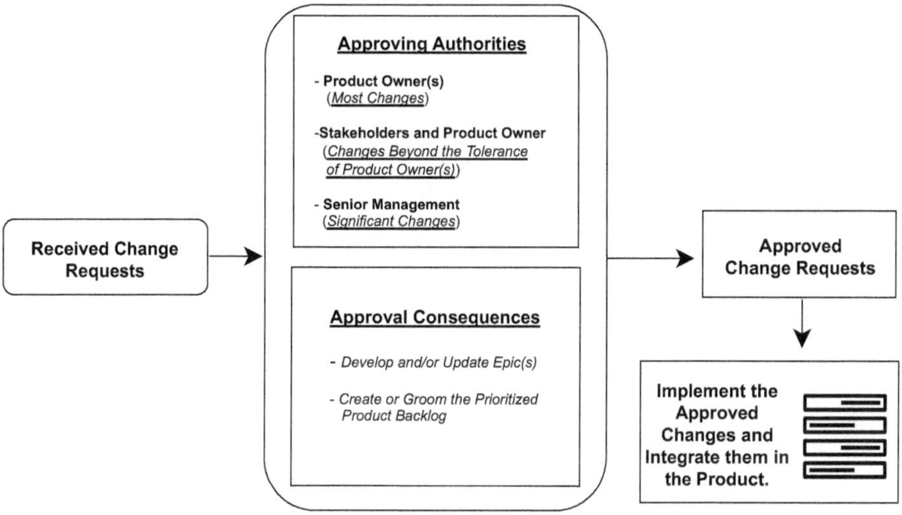

Figure 5.16: A sample SCRUM change approval/endorsement process.

Figure 5.17 shows how a prioritized product backlog is updated in accordance with the approved CHANGE-REQUESTS.

Figure 5.18 shows how changes are incorporated in the SCRUM portfolios and programs.

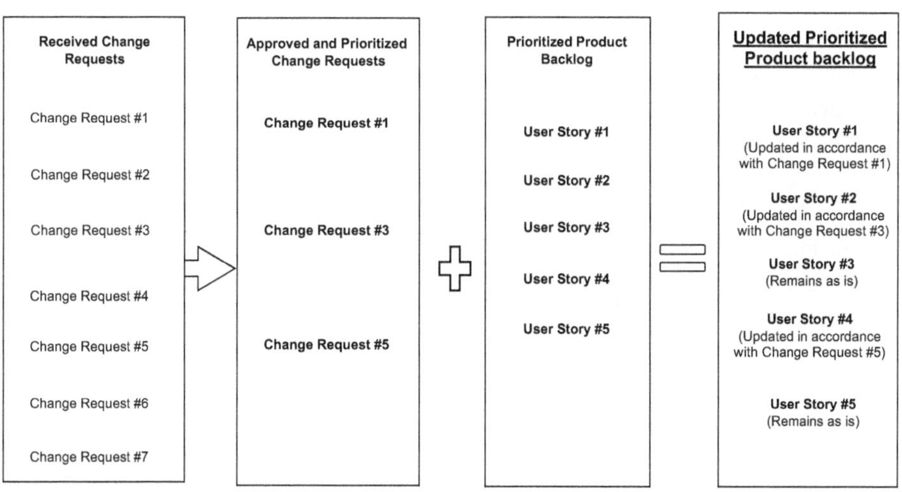

Figure 5.17: Updating the prioritized product backlog in accordance with the approved changes.

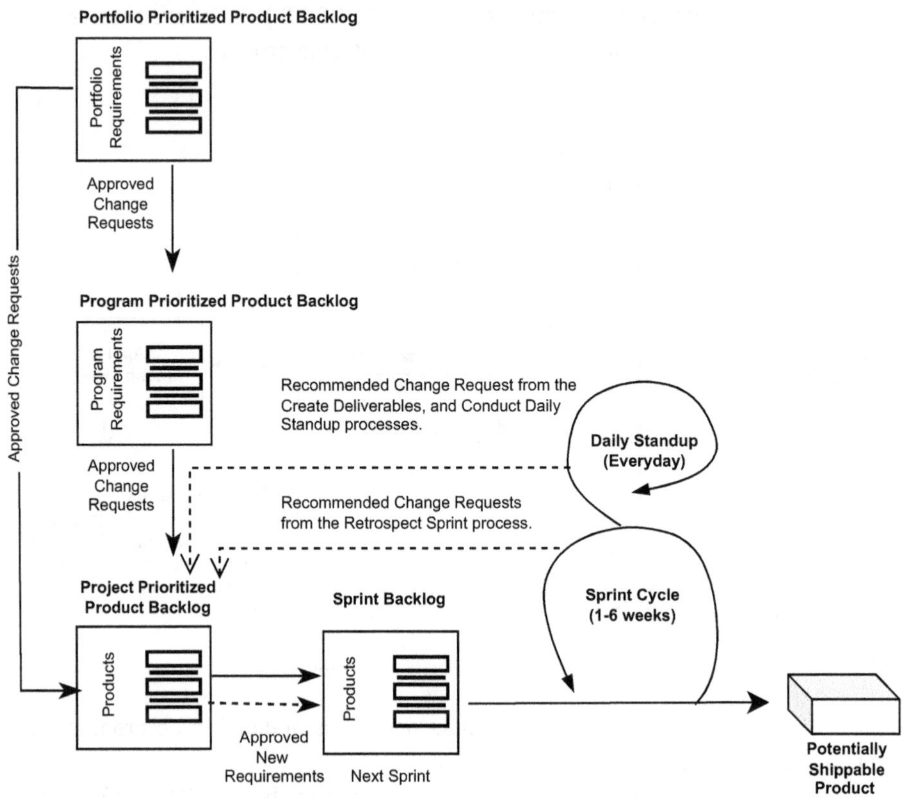

Figure 5.18: Integrating changes in SCRUM portfolios and SCRUM programs.

The following is a summary of the responsibilities of the SCRUM roles in carrying out the approved CHANGE-REQUESTS in SCRUM projects:

- **Portfolio product owner.** A portfolio product owner is responsible for providing the portfolio CHANGE-REQUESTS, and for approving the amended, removed, and/or added products in accordance with the portfolio requirements.
- **Portfolio SCRUM master.** A portfolio SCRUM master is responsible for facilitating the identification, assessment, and management of the portfolio CHANGE-REQUESTS.
- **Program product owner.** A program product owner is responsible for providing the program CHANGE-REQUESTS, and for approving the amended, removed, and added products in accordance with the program requirements.
- **Program SCRUM master.** A program SCRUM master is responsible for facilitating the identification, assessment, and management of the program CHANGE-REQUESTS.
- **Product owner.** A product owner is responsible for providing the project CHANGE-REQUESTS; for assessing the impact of the portfolio, program, and/or project CHANGE-REQUESTS; and for prioritizing the USER-STORIES in the prioritized product backlog.

- **SCRUM master.** A SCRUM master is responsible for facilitating the identification, assessment, and escalation of the SCRUM team CHANGE-REQUESTS.
- **SCRUM team.** The SCRUM team is responsible for suggesting changes during the two processes of Create Deliverables and Conduct Daily Standup.
- **Stakeholders.** A stakeholder is responsible for providing CHANGE-REQUESTS and for helping prioritize these CHANGE-REQUESTS.
- **SGB.** The SGB provides an overall guidance for the change management procedures throughout the project life cycle.

5.8 SCRUM risk

A risk is an unexpected and/or uncertain event that can impact the project objectives. If the risk's impact is negative, then that risk is referred to as a threat. If the risk's impact is positive, then that risk is referred to as an opportunity.

The difference between a risk and an issue is simple. A risk is a future unexpected and/or uncertain event with no current impact, but it needs to be planned for encountering it, just in case it happens in the future. While an issue is a present event that has current impact and needs immediate action to encounter it.

Risk management procedure is composed of five steps: risk identification, risk assessment, risk prioritization, risk mitigation, and risk communication.

5.8.1 Risk identification

The SCRUM team is expected to identify all risks that may impact the project and its objectives.

Risk identification can be achieved through:
- reviewing the lessons learned from the outcomes of the sprint/project retrospect meeting outcomes, from the risk checklists, and from the risk prompt lists;
- brainstorming;
- deriving a risk breakdown structure.

5.8.2 Risk assessment

The SCRUM team is expected to assess all risks that may impact the project value and objectives. A harsh risk that cannot be contained would be considered a strong cause for not proceeding with the project. A business justification is possible as long as all risks can be handled and contained.

Risk assessment can be achieved through:

- **Risk meetings.** The product owner can easily prioritizes risks by meeting with the SCRUM team and the concerned stakeholders.
- **Probability trees**. In reference to Figure 5.19, a probability tree composes all risks such that their outcomes are represented each with a branch in that tree. Then, the values of these outcomes are summed together to calculate the overall anticipated impact of any risk to the project.

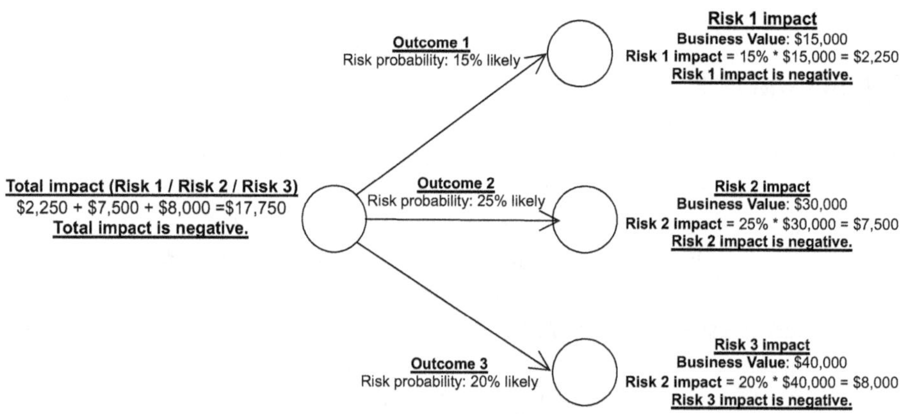

Figure 5.19: A sample risk assessment probability tree.

- **Pareto analysis**. This is used to rank risks by their impact magnitudes. Figure 5.20 presents an example of the Pareto analysis.

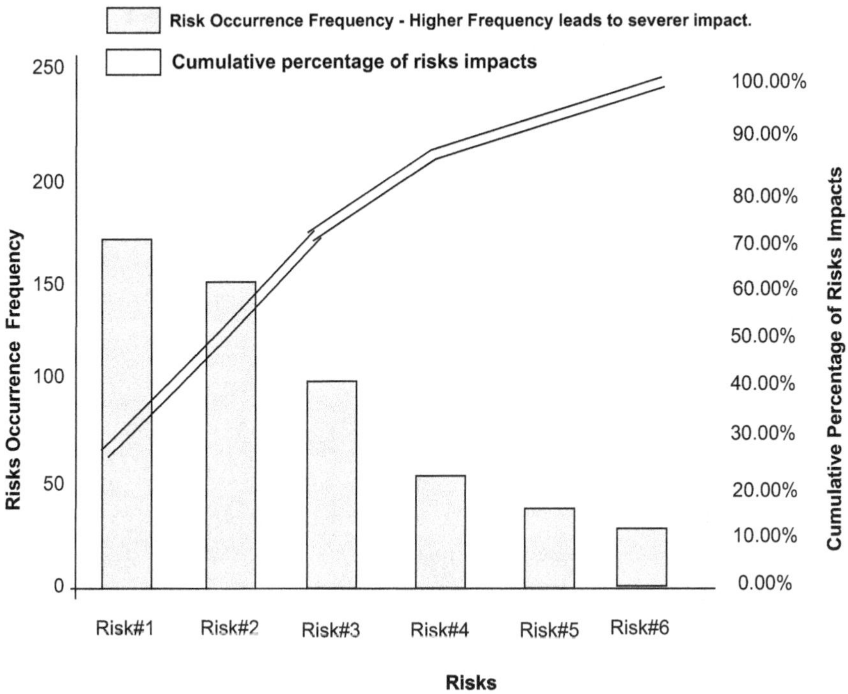

Figure 5.20: A sample Pareto analysis.

- **Probability impact grid.** This is used to rank risks by their severity. A risk severity is calculated by multiplying its probability by its impact rating value. For example, if the risk probability is 60% and its impact rating value is 0.7, then the risk severity equals (0.6 × 0.7) and that evaluates to 0.42. Figure 5.21 presents an example on this matter.
- **Expected monitory value (EMV).** This is used to rank risks by their EMVs. A risk EMV is calculated by multiplying the risk impact in dollars with its probability in percentage. For example, if the estimated negative impact value of a risk is $2,500 and its probability is 60%, then its EMV equals ($2,500 × 0.6) and that evaluates to $1,500.

Risks Severity: Risk Severity = Risk Probability * Risk Impact

Probability	Threats			Opportunities		
0.90	0.09	0.27	0.72	0.72	0.27	0.09
0.70	0.075	0.21	0.56	0.56	0.21	0.075
0.50	0.05	0.15	0.40	0.40	0.15	0.05
0.30	0.03	0.09	0.24	0.24	0.09	0.03
0.10	0.01	0.03	0.08	0.08	0.03	0.01
Impact	Low 0.1	Medium 0.3	High 0.8	High 0.8	Medium 0.3	Low 0.1

Low PI value   Moderate PI value   High PI value

Figure 5.21: Risk severity: a sample probability impact grid.

5.8.3 Risk prioritization

Risks are taken into consideration while creating or updating a prioritized product backlog during the Create Prioritized Product Backlog or the Groom Prioritized Product Backlog processes.

Prioritizing risks is essential as their consideration is associated with the USER-STORIES which may be impacted by these risks. Thus, sometimes, the prioritized product backlog is sometimes called risk-adjusted prioritized product backlog. Figure 5.22 presents a sample process for prioritizing risks.

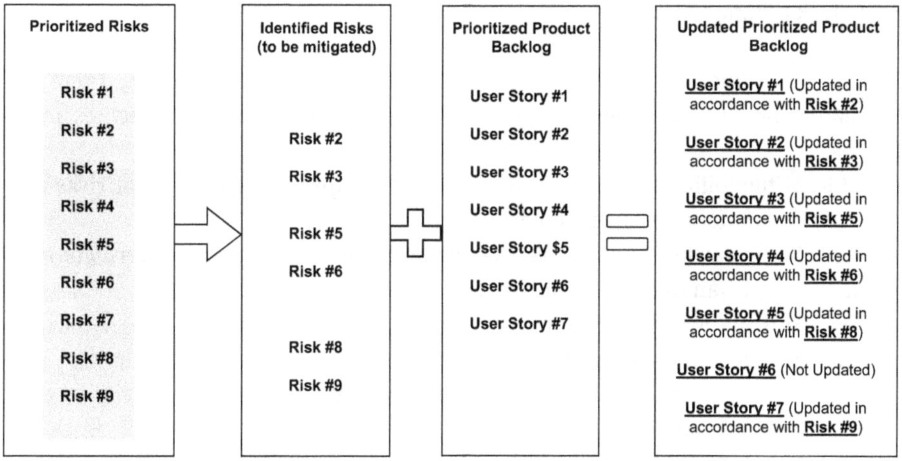

Figure 5.22: A sample process for risk prioritization.

5.8.4 Risk mitigation

SCRUM allows for early detection of failures; therefore, it has a natural mitigation feature built in. Risks can be mitigated by implementing a number of proactive and/or reactive responses. When a risk event occurs, a reactive response (e.g., as a plan B) may be formulated and used. Risks are accepted when their probabilities and/or impacts are too low.

5.8.5 Risk communication

It is extremely important to communicate information related to the risks with the project stakeholders. This includes the potential impact of each risk and the plans for responding to them. Risk communication should take place in parallel with the four sequential steps of risk identification, risk assessment, risk prioritization, and risk mitigation.

The SCRUM team may need to discuss during the daily standup meetings the identified risks with the SCRUM master. The product owner needs to prioritize these risks and communicate them to the SCRUM team.

5.8.6 Risk minimization in SCRUM

The SCRUM framework minimizes risks using the Agile iterative process. There are SCRUM practices that help managing risks effectively. Examples of these practices are as follows:

- SCRUM flexibility can reduce the probability, impact, and severity of risks that are relevant to the business environment.
- SCRUM regular feedback can reduce the probability, impact, and severity of risks that are relevant to the stakeholders' expectations.
- SCRUM team ownership can reduce the probability, impact, and severity of risks that are relevant to the sprints' tasks, resources, and schedule estimations.
- SCRUM transparency can reduce the probability, impact, and severity of risks that are hard to detectable.
- SCRUM iterative delivery can reduce the probability, impact, and severity of risks relevant to investments.

5.8.7 Risks in SCRUM portfolios and programs

When the portfolio risks are identified, the portfolio product owner must assess the probability and impact of each risk, prioritize these risks, and determine the right responses. The portfolio product owner must communicate these risks to the relevant stakeholders, program's teams, and projects' teams.

In reference to Figure 5.23, when the program risks are identified, the program product owner must enter them into the program risk-adjusted prioritized product backlog, assess their probabilities and impacts in order to prioritize them, and determine the programs' appropriate responses. The program product owner must communicate the risks to relevant stakeholders and the project teams.

The following is a summary of the responsibilities of various SCRUM roles in handling risks in SCRUM projects:

- **Portfolio product owner**. A portfolio product owner is responsible for capturing and assessing the portfolio's risks, and for prioritizing and communicating them to the stakeholders, and the program and project teams.
- **Portfolio SCRUM master.** A portfolio SCRUM master is responsible for helping in identifying, assessing, and communicating the portfolio's potential risks.
- **Program product owner.** A program product owner is responsible for capturing and assessing the program's potential risks, and for prioritizing and communicating them to the stakeholders, and to the project teams.
- **Program SCRUM master.** A program SCRUM master is responsible for helping in identifying, assessing, and escalating the program's potential risks.

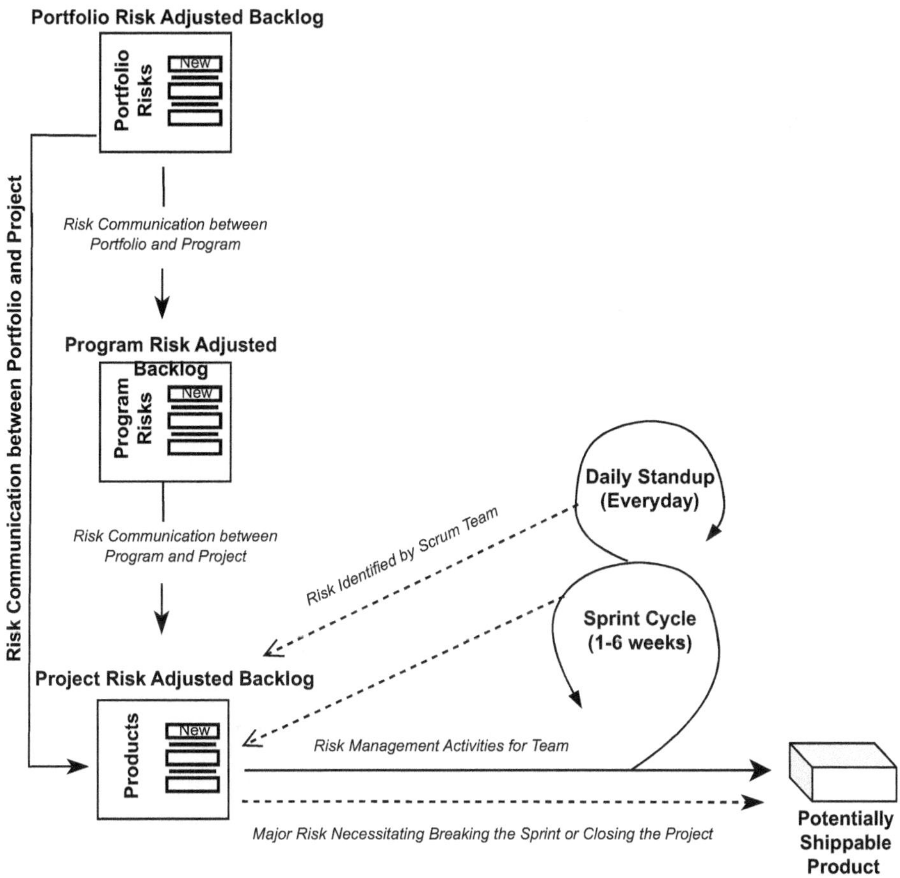

Figure 5.23: Risks in SCRUM portfolios and programs.

- **Product owner.** A product owner is responsible for capturing and assessing the project's potential risks; for prioritizing and communicating them to the stakeholders, program and portfolio teams; and for ensuring that the project's risk levels are acceptable.
- **SCRUM master.** A SCRUM master is responsible for helping in identifying and escalating the project's potential risks to the SCRUM team.
- **SCRUM team.** A SCRUM team is responsible for identifying the project's potential risks during the product development and for implementing their management activities as advised by the product owner.
- **Stakeholders.** Stakeholders are responsible for interacting with the SCRUM team to provide them with their inputs with respect to managing the project's potential risks.
- **SGB.** The SGB provides guidance with respect to the procedures of managing the potential risks.

5.9 SCRUM initiate phase

With reference to Figure 5.24(a), the SCRUM initiate phase has six processes: Create Project Vision, Identify SCRUM Master and Stakeholders, Form SCRUM Team, Develop Epics, Create Prioritized Product Backlog, and Conduct Release Planning.

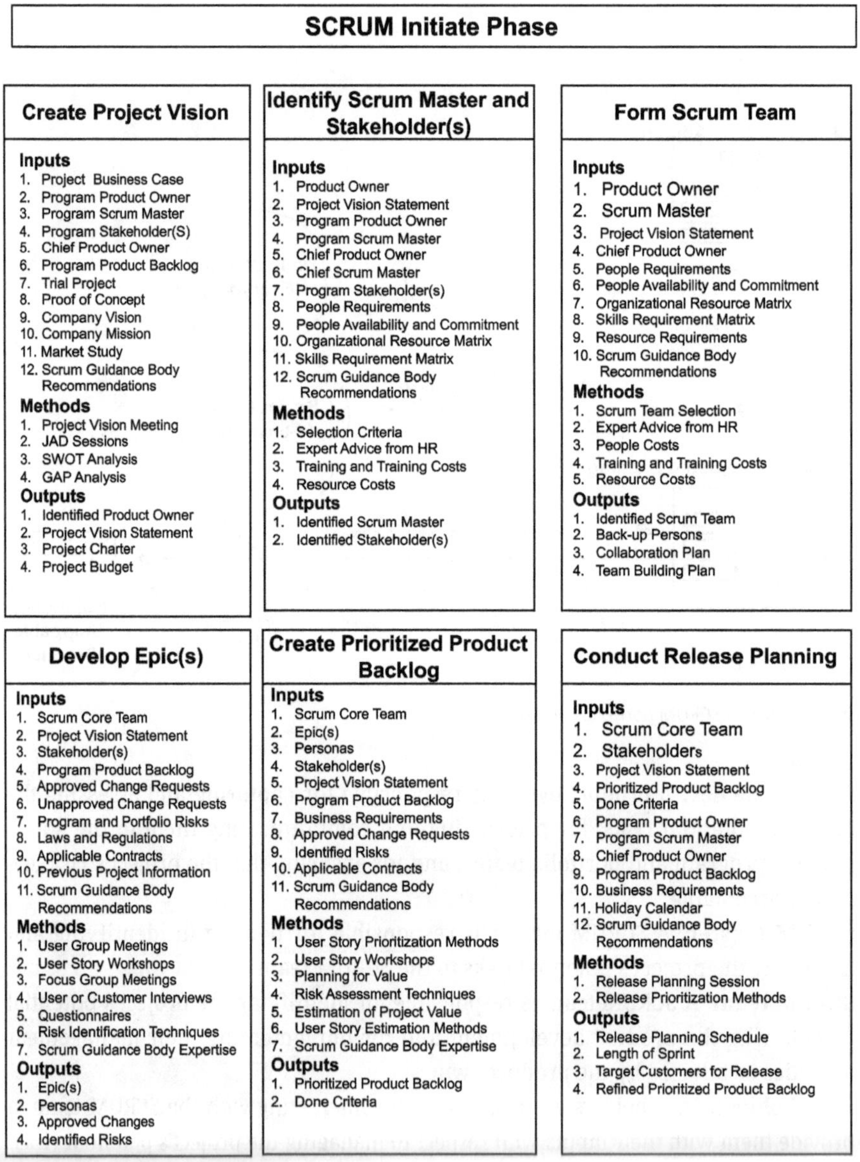

Figure 5.24: (a) SCRUM initiate phase processes and (b) SCRUM initiate phase data flow diagram and interactions across its processes.

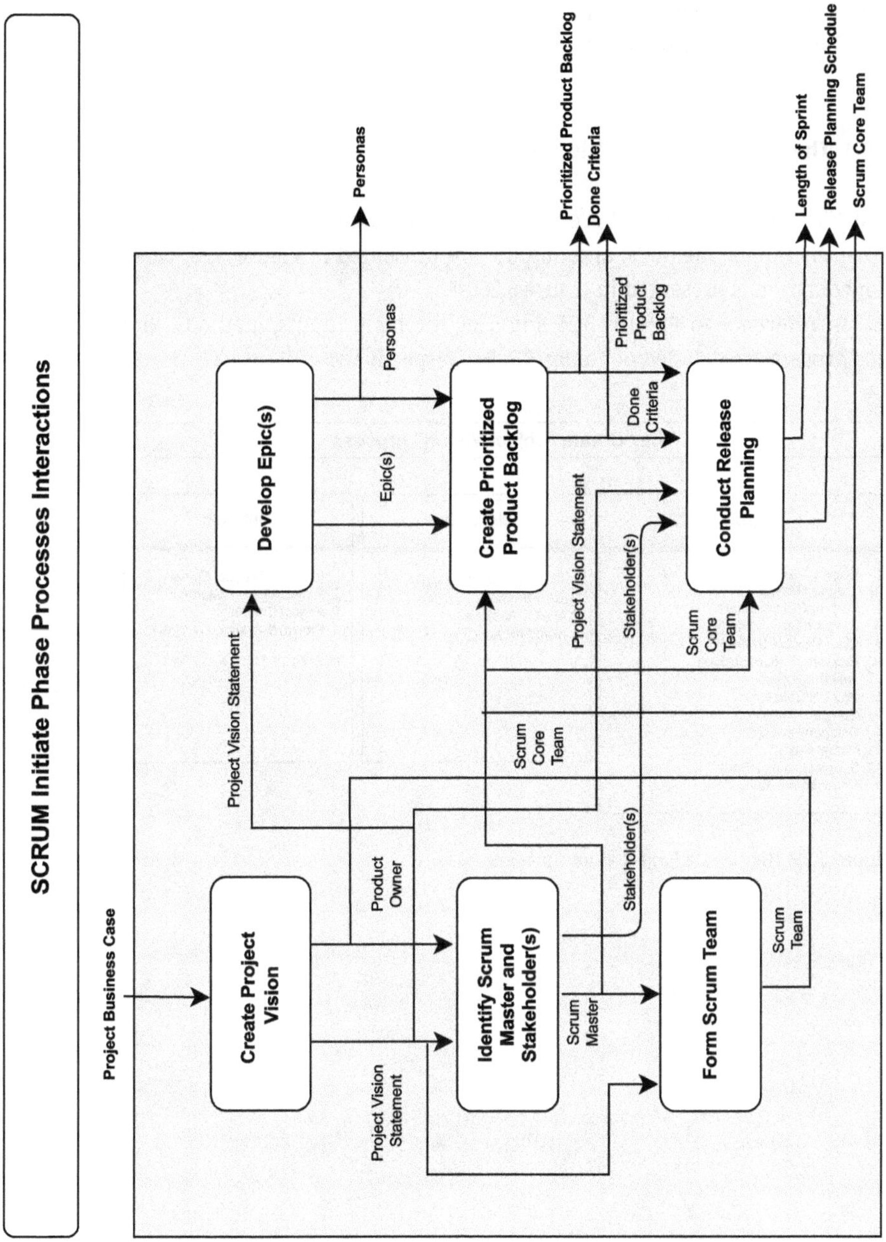

Figure 5.24: (b) SCRUM Initiate Phase Processes interactions.

Figure 5.24(b) presents the high-level data flow diagram of the SCRUM initiate phase and the interactions among its processes.

5.9.1 The "Create Project Vision" process

The "Create Project Vision" process is used to create the project business vision statement. Its inputs, methods, and outputs are presented in Figure 5.25, whereas its data flow diagram is presented in Figure 5.26(a).

In reference to Figures 5.25 and 5.26, the main inputs, methods, and outputs of this process are detailed in Tables 5.3–5.5, respectively.

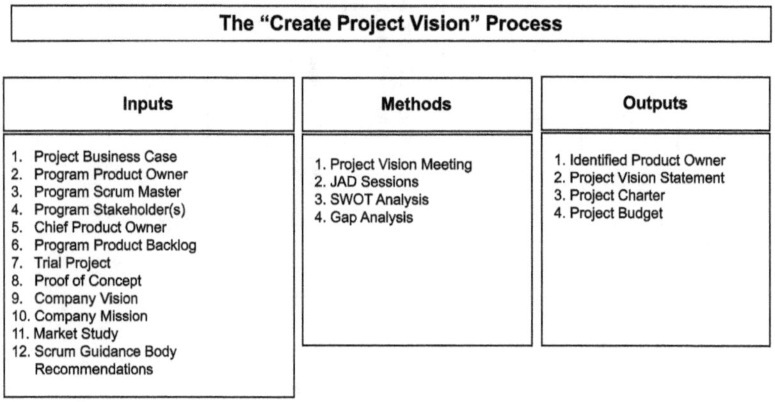

Figure 5.25: The "Create Project Vision" process.

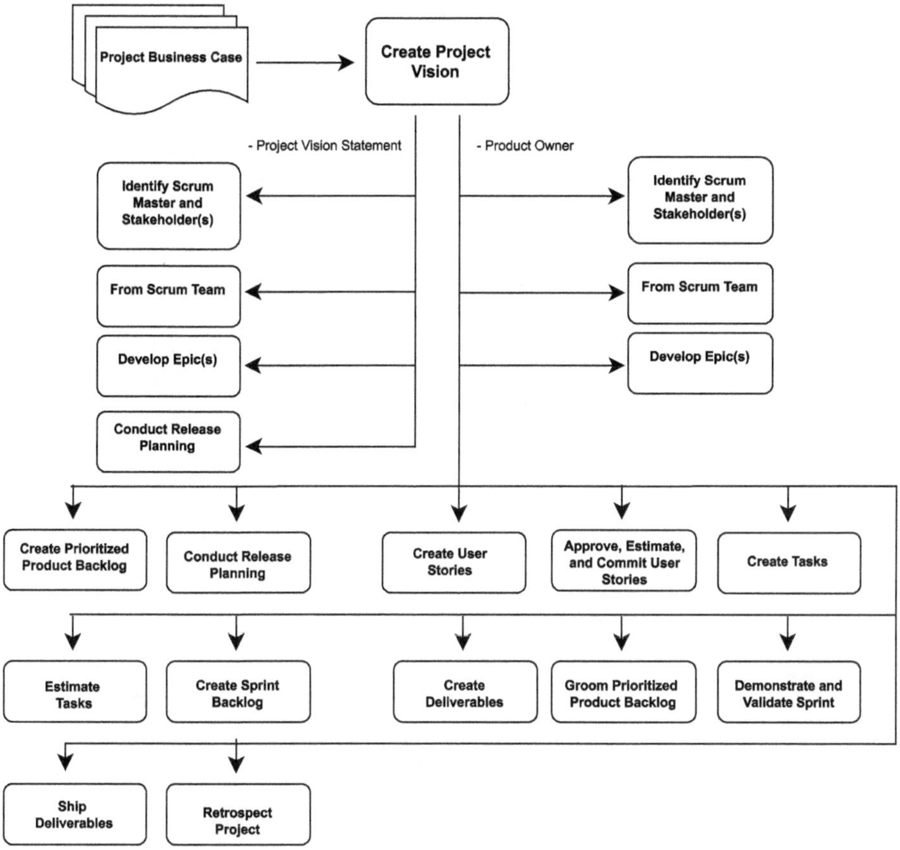

Figure 5.26: (a) The "Create Project Vision" process data flow diagram. (b) The gap analysis process.

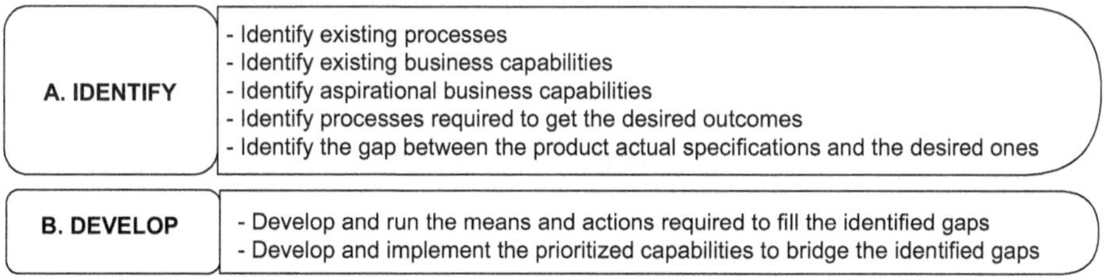

Figure 5.26: (b) The gap analysis process.

Table 5.3: Inputs to the "Create Project Vision" process.

Input	Details
Project business case	A SCRUM project starts with presenting a project business case to the project stakeholders and sponsors. The stakeholders must understand the project expected business benefits, while the sponsors must confirm their financial support and resources for the project.
Inputs from the program product owner	A program product owner works with his/her superior portfolio product owner toward aligning his/her program with the objectives of its superior portfolio. He/she is also responsible for ensuring that the vision, objectives, outcomes, and releases of his/her program's projects are in alignment with those of that program. Also, he/she must help appoint product owners for his/her program's projects.
Inputs from the program stakeholders	A program's stakeholders include the customers, users, and sponsors of that program. They can help define the project business vision; interface with their superior portfolio's stakeholders to ensure the alignment of the program with the portfolio's objectives; help appoint stakeholders for the program's projects; and ensure that the vision, objectives, outcomes, and releases of his/her program's projects are in alignment with those of the program.
Inputs from the program SCRUM master	A program SCRUM master ensures that a conductive working environment is facilitated to his/her SCRUM teams to enable them to SUCCESSFULLY complete their projects. He/she is involved in teaching SCRUM BEST-PRACTICES to all members in his/her program; guiding SCRUM masters of his/her program's projects; working with the SCRUM Guidance Body (SGB) on defining quality, governmental regulations, and security-related objectives; ensuring an effective use of the SCRUM processes across his/her program; working with his/her superior portfolio SCRUM master to align his/her program with the objectives of that portfolio and on ensuring the alliance of the projects' visions, objectives, outcomes, and releases with those of the program; and helping to appoint SCRUM master for the projects.
Inputs from the chief product owner	A chief product owner is responsible for coordinating the work of the product owners when a project is large enough to be divided into multiple SCRUMs. He/she prepares and maintains the overall prioritized product backlog for all of the project SCRUMs.
Program product backlog	A program product owner is responsible for developing his/her program's product backlog containing a prioritized list of business requirements at the program's level. These will be refined by the projects' product owners into prioritized product backlogs for these projects.
Trial project	To help predict and evaluate the actual project working environment and design before starting it, a trial demo could be carried out to evaluate its viability, cost, time, and risks.

Table 5.3 (continued)

Input	Details
Proof of concept	A proof of concept, or say a prototype, is used to demonstrate that the current project's backbone concept is financially and technically feasible. This can also help understand the project requirements and assess the design decisions.
Company vision	A company vision helps and directs the product owner while he/she is preparing the project vision statement.
Company mission	A company mission helps the formulation of that company's strategies and the company's overall decision-making and taking process. A project vision statement must be in full alignment with the company's vision and mission statements.
Market study	A market study refers to the research activities carried out during a project initiation phase. This includes gathering, and analyzing the data of the customers' preferences for products, such as data on market trends, market segmentation, and marketing processes. A market study may also include an analytical study of the competitors in terms of their strengths and weaknesses to better position the project upcoming product.
Inputs from the SCRUM Guidance Body (SGB)	A SCRUM Guidance Body (SGB) consists of a group of documents and/or a group of experts who are typically involved in defining quality, governmental regulations, and security-related objectives to guide the work carried out by the product owner, SCRUM master, and SCRUM team. An SGB can also help capture the BEST-PRACTICES that should be used across all SCRUM projects in the organization.

Table 5.4: Methods and tools used during the "Create Project Vision" process.

Method	Details
Project vision meeting	A project vision meeting takes place between that project product owner and the program stakeholders, program product owner, program SCRUM master, and chief product owner (*if applicable*). It is used to develop an effective project vision statement through identifying the business context, business requirements, and stakeholder expectations.
JAD sessions	A joint application design (JAD) session is a method used to gather requirements, speed up the Create Project Vision process through enabling the stakeholders and probably others to agree on the project specifications, particularly its scope and requirements. Also, it helps increase the users' participation, speed up the products' development, and improve the product specifications and quality.

Table 5.4 (continued)

Method	Details
SWOT analysis	SWOT analysis is a structured project planning method used to assess the strengths, weaknesses, opportunities, and threats of a project. It helps the project stakeholders and decision-makers to finalize the project lists of processes, methods and tools, objectives, priorities, changes, and risks. The strengths and weaknesses are internal factors, while the opportunities and threats are external ones.
Gap analysis	In reference to Figure 5.26(b), the gap analysis process is used to compare the product's actual state with the desired one. It involves finding out and articulating the difference between current business capabilities and the desired ones.

Table 5.5: Outputs of the "Create Project Vision" process.

Output	Details
Identified product owner	An identified project product owner is one of the main outcomes of this process.
Project vision statement	The main output of this process is a project vision statement that explains the business and project needs and requirements.
Project charter	A project charter is a formal document that authorizes the project. It provides the project team with written permission and authority to start their work on the project.
Project budget	A project budget includes the expenses and costs of human and physical resources including the project team and materials. It is signed by the sponsors to ensure that sufficient funds are available. Once it is signed, the product owner and the SCRUM master will help manage the project budget on regular basis. Also, they should help ensure that the required human and physical resources are available.

5.9.2 The "Identify SCRUM Master and Stakeholders" process

The "Identify SCRUM Master and Stakeholders" process is used to identify the project SCRUM master and stakeholders, mainly by the project product owner. Its inputs, methods, and outputs are presented in Figure 5.27 and its data flow diagram is presented in Figure 5.28.

In reference to Figures 5.27 and 5.28, the main inputs, methods, and outputs of this process are detailed in Tables 5.6–5.8, respectively.

The "Identify SCRUM Master and Stakeholders" Process

Inputs

1. Product Owner
2. Project Vision Statement
3. Program Product Owner
4. Program Scrum Master
5. Chief Product Owner
6. Chief Scrum Master
7. Program Stakeholder(s)
8. People Requirements
9. People Availability and Commitment
10. Organizational Resource Matrix
11. Skills Requirement Matrix
12. Scrum Guidance Body Recommendations

Methods

1. Selection Criteria
2. Expert Advice from HR
3. Training and Training Costs
4. Resource Costs

Outputs

1. Identified Scrum Master
2. Identified Stakeholder(s)

Figure 5.27: The "Identify SCRUM Master and Stakeholders" process.

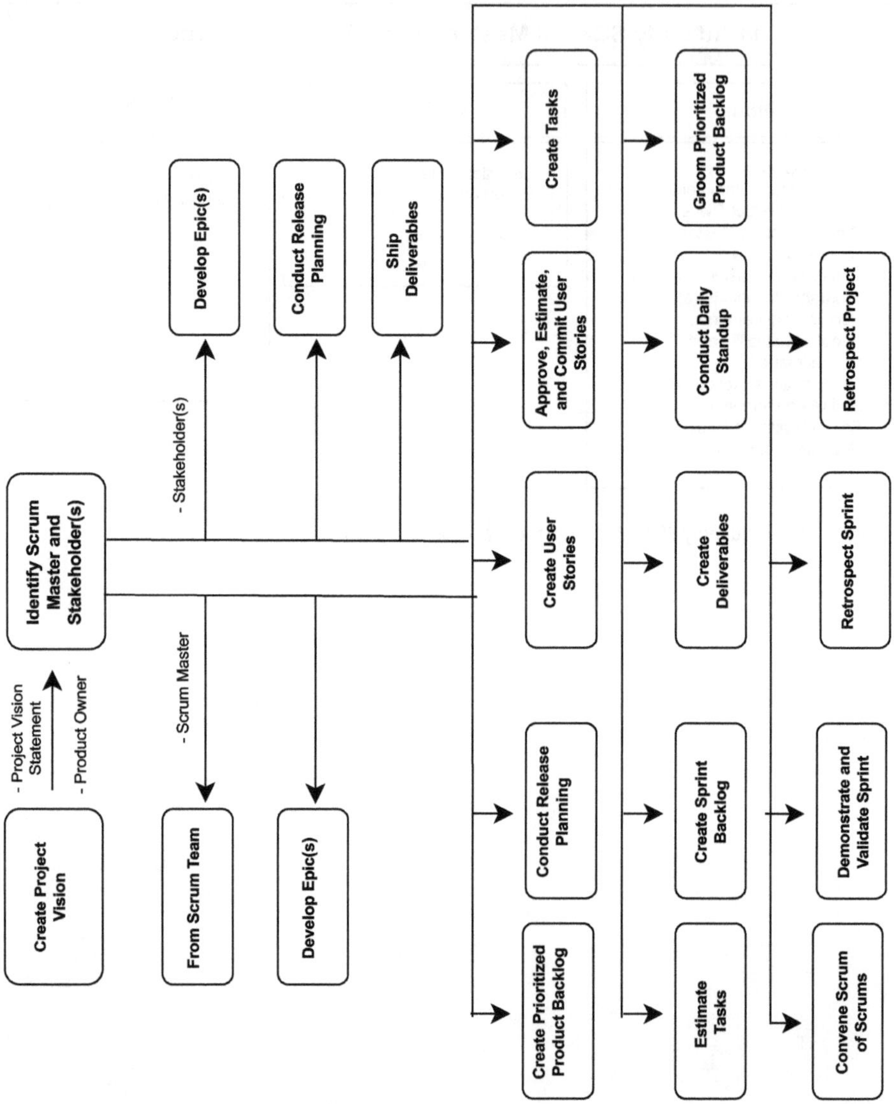

Figure 5.28: The "Identify SCRUM Master and Stakeholders" process data flow diagram.

Table 5.6: Inputs to the "Identify SCRUM Master and Stakeholders" process.

Input	Details
Product owner	As described earlier in Table 5.5 (e.g., identified product owner).
Project vision statement	As described earlier in Table 5.5.
Inputs from the program product owner	As described earlier in Table 5.3.
Inputs from the program SCRUM master	As described earlier in Table 5.3.
Inputs from the chief product owner	As described earlier in Table 5.3.
Inputs from the chief SCRUM master	As described earlier in Table 5.2.
Inputs from the program stakeholders	As described earlier in Table 5.3.
Needed human resources and their requirements	The project SCRUM master and stakeholders, and product owner should first identify the human resources needed to work on their project. Also, they should identify the requirements of these resources.
Human resources availability and commitments	Before selecting the project SCRUM master and stakeholders, the project product owner should confirm their availability. At the same time, only the confirmed and committed team members can be selected.
Organizational resource matrix	The organizational resource matrix combines a functional structure and a project-based structure. Matrix organizations create cross-functional teams by building a project team from different functional departments. Product owners direct these team members with respect to their assigned activities and tasks, while their functional managers perform all managerial activities such as their performance appraisals.
Skills requirement matrix	The skills requirement matrix, or say the competency matrix, is used to evaluate the gaps in the team's skills and accordingly determine their training needs.
Inputs from the SCRUM Guidance Body (SGB)	As described earlier in Table 5.3.

Table 5.7: Methods used during the "Identify SCRUM Master and Stakeholders" process.

Method	Details
SELECTION-CRITERIA	Identifying the right SCRUM master and stakeholders is a very critical task by the project product owner. The SELECTION-CRITERIA used to select a SCRUM master is based on the following three critical factors: – Problem-solving skills. A SCRUM master must be a problem solver. – Availability. A SCRUM master must be available throughout the project life cycle. He/she should be available at all meetings including the stakeholders' meeting, sprint planning meetings, daily standup meetings, sprint review meetings, sprint retrospect meetings, and project retrospect meeting. – Commitment. A SCRUM master must be committed to facilitating a conductive work environment for the SCRUM team. – Servant-leadership style. A SCRUM master must be of this leadership style. The details of this leadership style are available in Section 5.4.8.
Expert advice from the human resources department	An expert advice from the human resources department can help identify the SCRUM master and the stakeholder(s).
Training and training costs	The project product owner is responsible for identifying the gaps in the team's skills and their training needs. In addition to finding out these training needs, he/she is responsible for figuring out the costs for running the needed training sessions.
Resource costs	The project product owner is responsible for figuring out the costs for the required human and the physical resources.

Table 5.8: Outputs of the "Identify SCRUM Master and Stakeholders" process.

Outputs	Details
Identified SCRUM master	A SCRUM master is a servant-leader that facilitates a conductive work environment for the project SCRUM team; guides and teaches the SCRUM practices to the entire project team; and ensures that SCRUM processes are being followed.
Identified stakeholders	Stakeholders interface with the SCRUM team and influence the project throughout its life cycle.

5.9.3 The "Form SCRUM Team" process

The "Form SCRUM Team" process is used to identify the project SCRUM team members, mainly by the project product owner and the SCRUM master. Its inputs, methods, and outputs are presented in Figure 5.29 and its data flow diagram is presented in Figure 5.30.

In reference to Figures 5.29 and 5.30, the main inputs, methods, and outputs of this process are detailed in Tables 5.9–5.11, respectively.

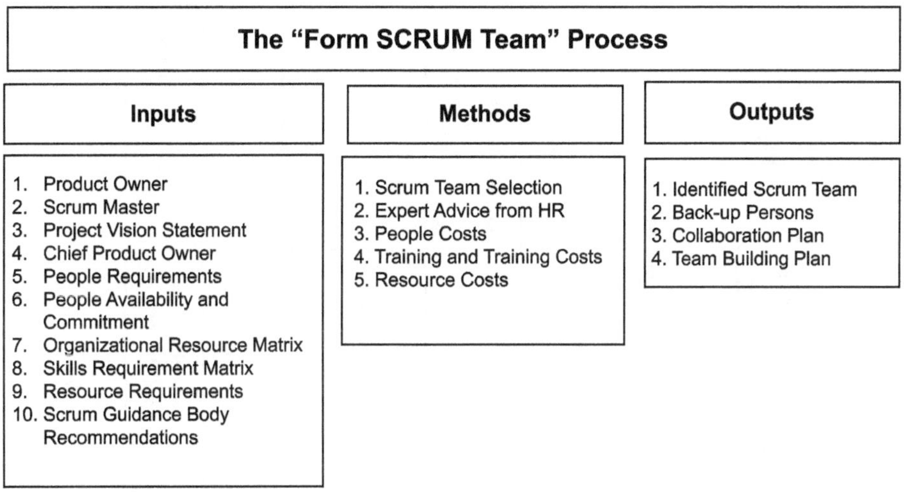

Figure 5.29: The "Form SCRUM Team" process.

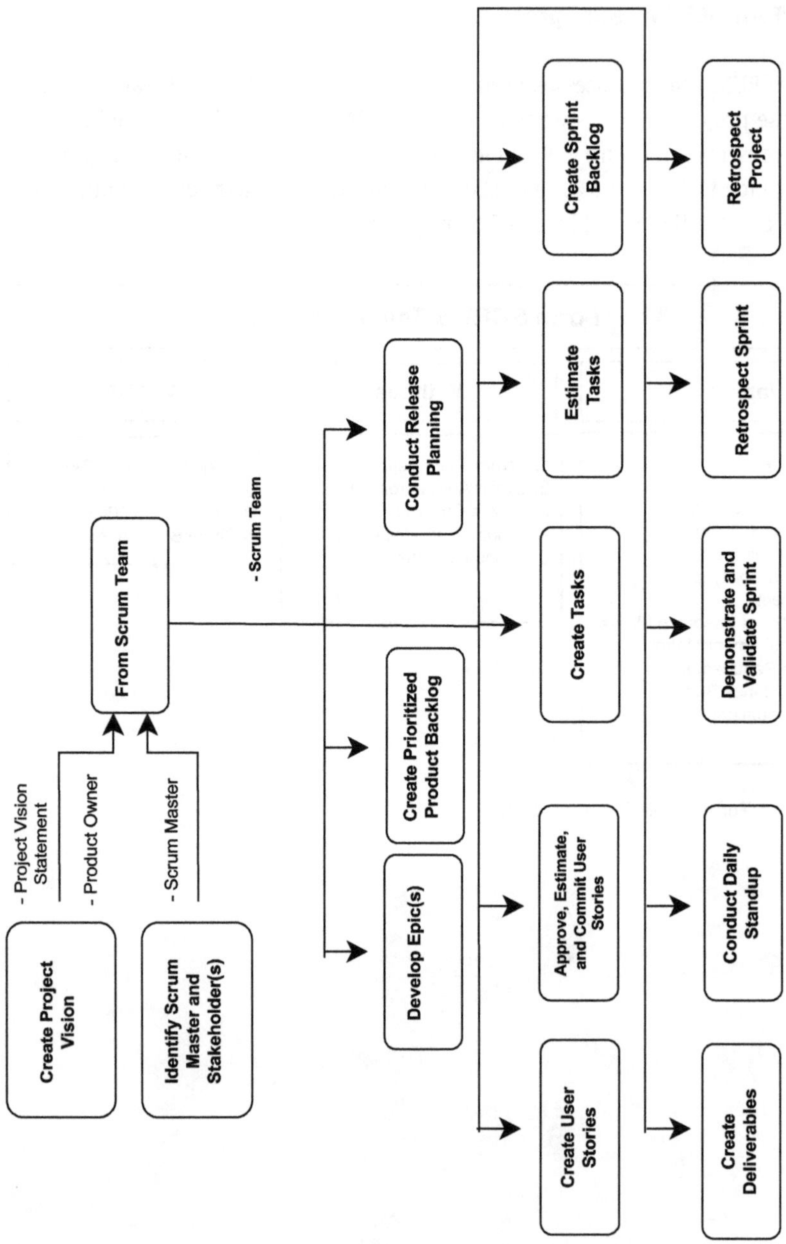

Figure 5.30: The "Form SCRUM Team" process data flow diagram.

Table 5.9: Inputs to the "Form SCRUM Team" process.

Input	Details
Inputs from the product owner	As described earlier in Table 5.5 (e.g., identified product owner).
Inputs from the SCRUM master	As described earlier in Table 5.8 (e.g., identified SCRUM master).
Project vision statement	As described earlier in Table 5.5.
Inputs from the chief product owner	As described earlier in Table 5.3.
Needed human resources and their requirements	As described earlier in Table 5.6.
Human resources availability and commitments	As described earlier in Table 5.6.
Organizational resource matrix	As described earlier in Table 5.6.
Skills requirement matrix	As described earlier in Table 5.6.
Resource requirements	These include all required resources excluding the human resources requirements, such as office infrastructure, meeting spaces, work equipment, collaboration tools, video conferencing, document repositories, and translation services.
Inputs from the SCRUM Guidance Body (SGB)	As described earlier in Table 5.3.

Table 5.10: Methods used during the "Form SCRUM Team" process.

Method	Details
SCRUM team SELECTION-CRITERIA	Selecting the right SCRUM team members is essential for the SUCCESS of SCRUM projects. The SELECTION-CRITERIA factors include the knowledge and expertise in various fields; having the ability to determine the SUCCESS of self-organizing teams; and being independent, self-motivated, customer-focused, responsible, collaborative independent thinking, and team players.
Expert advice from the human resources department	Expert advice from human resources department can help select the right SCRUM team members.
Human resources' cost assessment	The costs of the entire SCRUM project team members should be evaluated, analyzed, and approved.
Training and training costs	As described earlier in Table 5.7
Resource costs	The costs of the entire SCRUM project required physical resources that must be evaluated, analyzed, and approved.

Table 5.11: Outputs of the "Form SCRUM Team" process.

Output	Details
Identified SCRUM team	Identifying the SCRUM team is the responsibility of the project product owner, but in collaboration with the SCRUM master. A SCRUM team is expected to be cross-functional and self-organizing. The SCRUM team members are expected to: – understand the business and project requirements as specified by the project product owner; – estimate the USER-STORIES in terms of their cost and schedule; – create the project shippable deliverables including their development, testing, and quality assurance; and – decide the amount of work to commit to in a sprint.
Identified back-up persons	It is essential to create a backup for each SCRUM team member as availability and commitment of members may change due to illness, family emergency, and leaving the organization.
Collaboration plan	As collaboration is an essential SCRUM element, it is a must that a plan has to be created to enable the decision-makers, stakeholders, and team members to collaborate with each other.
Team building plan	The SCRUM master is responsible for creating a plan for building his/her SCRUM team. He/she must maintain the relationships among his/her team members as much positive, focused, and collaborative as possible.

5.9.4 The "Develop Epics" process

The "Develop Epics" process is used to develop and articulate promotional SUCCESS stories out of the upcoming product as well as large-size USER-STORIES. The project product owner is the sole responsible role for this process. However, the SCRUM master and the SCRUM team provide their inputs in this respect.

Figure 5.31 presents the inputs, methods, and outputs of this process and Figure 5.32 presents its data flow diagram. The main inputs, methods, and outputs of this process are detailed in Tables 5.12–5.14, respectively.

The "Develop Epics" Process		
Inputs	**Methods**	**Outputs**
1. Scrum Core Team 2. Project Vision Statement 3. Stakeholder(s) 4. Program Product Backlog 5. Approved Change Requests 6. Unapproved Change Requests 7. Program and Portfolio Risks 8. Laws and Regulations 9. Applicable Contracts 10. Previous Project Information 11. Scrum Guidance Body Recommendations	1. User Group Meetings 2. User Story Workshops 3. Focus Group Meetings 4. User or Customer Interviews 5. Questionnaires 6. Risk Identification Techniques 7. Scrum Guidance Body Expertise	1. Epic(s) 2. Personas 3. Approved Changes 4. Identified Risks

Figure 5.31: The "Develop Epics" process.

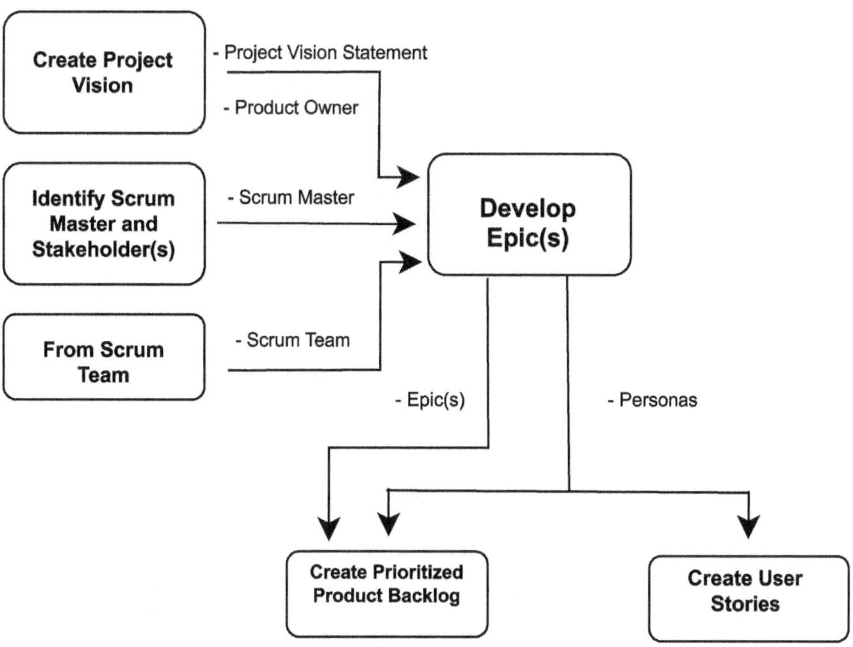

Figure 5.32: The "Develop Epics" process data flow diagram.

Table 5.12: Inputs to the "Develop Epics" process.

Input	Details
Inputs from the SCRUM team	The SCRUM team includes the product owner, SCRUM master, and the SCRUM team members. Their input can help carry out this process.
Project vision statement	As described earlier in Table 5.5.
Inputs from the stakeholders	The stakeholders' inputs can help carry out this process. Stakeholders are as described earlier in Table 5.8.
Program product backlog	As described earlier in Table 5.3.
Approved and unapproved CHANGE-REQUESTS	Approved CHANGE-REQUESTS could become inputs to the Develop Epics process. They can come from the program or portfolio, and/or from other SCRUM processes. Unapproved CHANGE-REQUESTS can come from the Create Deliverables, Conduct Daily Standup, and other SCRUM processes.
Program and portfolio risks	Portfolios and/or program risks can impact projects that belong to them. If a risk may affect a project, then that must be communicated to the product owner and SCRUM team of that project. Portfolios and program risks can be used as inputs to the Develop Epics process.
Laws and regulations	Laws are imposed on the organization by governmental entities, while regulations can be internally applicable with the company or external. Internal regulations relate to quality management systems, financial regulations, and staff regulations, whereas external regulations relate to governmental standards and requirements. Both of laws and regulations must be considered during the epics development.
Applicable contracts	A project contract defines the scope of work and the terms and conditions of that project. Epics should be developed while remembering the terms and conditions of the contract type being used. The most common types of SCRUM contracts are: – Incremental delivery contract. This includes regular inspection and check-points to help the customer and/or stakeholders decide on whether to accept the product development, stop it, or request modifications. – Joint venture contract. This is used when two or more parties get into a partnership to carry out the work of a project. – Development in phase contract. This makes funding available after each release is SUCCESSFULLY completed. – Incentive and penalty contract. This indicates that the supplier will be paid an agreed-upon financial incentive if the supplier delivers the product on time; otherwise, there will be a financial penalty imposed on that supplier.

Table 5.12 (continued)

Input	Details
Previous projects' information	Similar projects can provide valuable inputs for developing epics and assessing risk. Such inputs include the project manager's notes, project logs, and stakeholder comments.
Inputs from the SCRUM Guidance Body (SGB)	As described earlier in Table 5.3.

Table 5.13: Methods used during the "Develop Epics" process.

Method	Details
User group meetings	User group meetings involve stakeholders who provide the SCRUM team with early information about user expectations, and that can help formulate the ACCEPTANCE-CRITERIA for the product and can provide valuable inputs to the Develop Epics process.
USER-STORY workshops	USER-STORY workshops are facilitated by the SCRUM master. They can help the product owner prioritize the requirements and provide the SCRUM team with a shared perspective of the ACCEPTANCE-CRITERIA. Such workshops involve the SCRUM team and stakeholders.
Focus group meetings	Focus group meetings are guided sessions for the SCRUM teams to provide their opinions, perceptions, and/or ratings of a product. Focus group meetings are sometimes conducted to resolve issues that arise during the Develop Epics process.
User or customer interviews	Engaging stakeholders is important to get their inputs that help develop epics. Interviewing users and customers help ensure that epics' requirements align with the project vision. Such interviews help identify and understand the stakeholders' expectations. Interviews also help gather opinions, facts, and feedback about the partially developed product.
Questionnaires	Questionnaires and surveys represent a cost-effective method for getting quantitative and/or qualitative statistical inputs from the users and customers. During the Develop Epics process, the product owner and/or the SCRUM master may run a questionnaire or a survey to get relevant information from the stakeholders and/or the SCRUM team.
Risk identification methods	As described earlier in Section 5.8.1.
SCRUM Guidance Body (SGB) Expertise	An SGB expertise can be in the form of documented rules, regulations, and/or standards and BEST-PRACTICES for creating epics.

Table 5.14: Outputs used during the "Develop Epics" process.

Output	Details
Epics	The Develop Epics process involves creating fictional characters that represent the users and stakeholders who cannot use the end product directly. Fictional characters are best referred to as a persona that are created to figure out the needs of the targeted users. A product owner can use a persona to prioritize features during the process of creating the prioritized product backlog. Creating a persona involves assigning a fictional title and a picture to the fictional character. A persona includes specific attributes such as age, gender, education, environment, interests, and goals.
Approved changes	The product owner may approve unapproved CHANGE-REQUESTS during the Develop Epics process. Such changes can then be prioritized and implemented.
Identified risks	New risks may be identified during the Develop Epics process, and can then contribute to the development of the prioritized product backlog.

5.9.5 The "Create Prioritized Product Backlog" process

The "Create Prioritized Product Backlog" process is used to create the project Prioritized Product Backlog. The project product owner is the sole responsible role for this process. However, the SCRUM master and the SCRUM team provide their inputs in this respect.

Figure 5.33 presents the inputs, methods, and outputs of this process and Figure 5.34 presents its data flow diagram. The main inputs, methods, and outputs of this process are detailed in Tables 5.15–5.17, respectively.

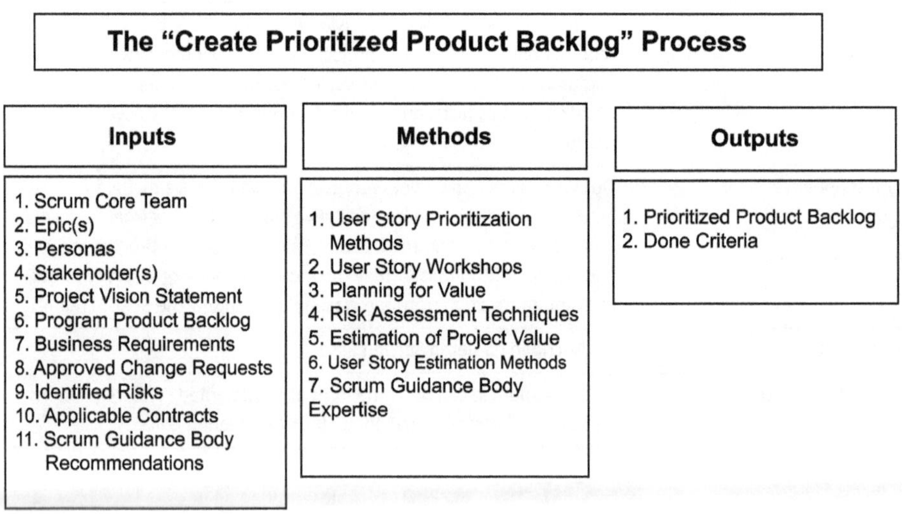

Figure 5.33: The "Create Prioritized Product Backlog" process.

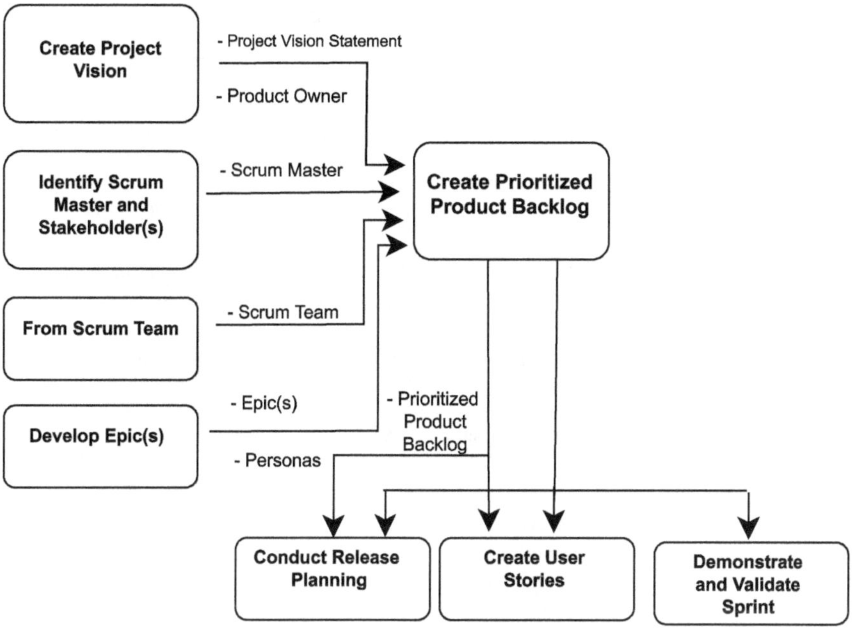

Figure 5.34: The "Create Prioritized Product Backlog" process data flow diagram.

Table 5.15: Inputs to the "Create Prioritized Product Backlog" process.

Input	Details
Inputs from the SCRUM team	As described earlier in Table 5.9.
Epics	As described earlier in Table 5.11.
Personas	As described earlier in Table 5.11.
Inputs from the stakeholders	As described earlier in Table 5.9.
Project vision statement	As described earlier in Table 5.5.
Program product backlog	As described earlier in Table 5.3.
Business requirements	The insights gained through the user and customer Interviews, questionnaires, JAD sessions, gap analysis, and SWOT analysis can help improve the business requirements and can help create the prioritized product backlog.
Approved CHANGE-REQUESTS	As described earlier in Table 5.9.
Identified risks	As described earlier in Table 5.11.
Inputs from the SCRUM Guidance Body (SGB)	As described earlier in Table 5.3.

Table 5.16: Methods used during the "Create Prioritized Product Backlog" process.

Method	Details
USER-STORY prioritization methods	USER-STORIES are prioritized into the Prioritized Product Backlog using the following methods: – MoSCoW prioritization scheme. This method derives its name from the first letters of the phrases "MUST HAVE," "SHOULD HAVE," "COULD HAVE," and "WILL NOT HAVE." – Paired comparison. This method prepares a list of all of the USER-STORIES in the prioritized product backlog, before comparing them with the other USER-STORIES in the list to decide on which of the two is more important. At the end, a prioritized list of USER-STORIES is generated.
USER-STORY workshops	As described earlier in Table 5.10.
Risk assessment techniques	As described earlier in Section 5.8.2.
USER-STORY estimation methods	The approve, estimate, and commit USER-STORIES process can be used to create estimates for epics while creating the prioritized product backlog. Sample tools include the user group meetings, planning poker, and points for cost estimation.
SCRUM Guidance Body Expertise	As described earlier in Table 5.10.

Table 5.17: Outputs of the "Create Prioritized Product Backlog" process.

Output	Details
Prioritized Product Backlog	The Prioritized Product Backlog is developed by the product owner. It contains a prioritized list of business and project requirements written in the form of USER-STORIES. The Prioritized Product Backlog is based on the primary factors of value, risk or uncertainty, and dependencies.
DONE-CRITERIA	DONE-CRITERIA is devised by the product owner as a set of rules and guidelines that can be used to determine if the product is acceptable in terms of satisfying the business and project requirements. A USER-STORY is considered done when it is presented to and approved by the product owner based on the DONE-CRITERIA and the USER-STORY ACCEPTANCE-CRITERIA.

5.9.6 The "Conduct Release Planning" process

The "Conduct Release Planning" process is used to devise the project release plan that indicates when the product increments/deliverables will be delivered to the customer.

Figure 5.35 presents the inputs, methods, and outputs of this process and Figure 5.36 presents its data flow diagram. The main inputs, methods, and outputs of this process are detailed in Tables 5.18–5.20, respectively.

The "Conduct Release Planning" Process

Inputs	Methods	Outputs
1. Scrum Core Team 2. Stakeholder(s) 3. Project Vision Statement 4. Prioritized Product Backlog 5. Done Criteria 6. Program Product Owner 7. Program Scrum Master 8. Chief Product Owner 9. Program Product Backlog 10. Business Requirements 11. Holiday Calendar 12. Scrum Guidance Body Recommendations	1. Release Planning Sessions 2. Release Prioritization Methods	1. Release Planning Schedule 2. Length of Sprint 3. Target Customers for Release 4. Refined Prioritized Product Backlog

Figure 5.35: The "Conduct Release Planning" process.

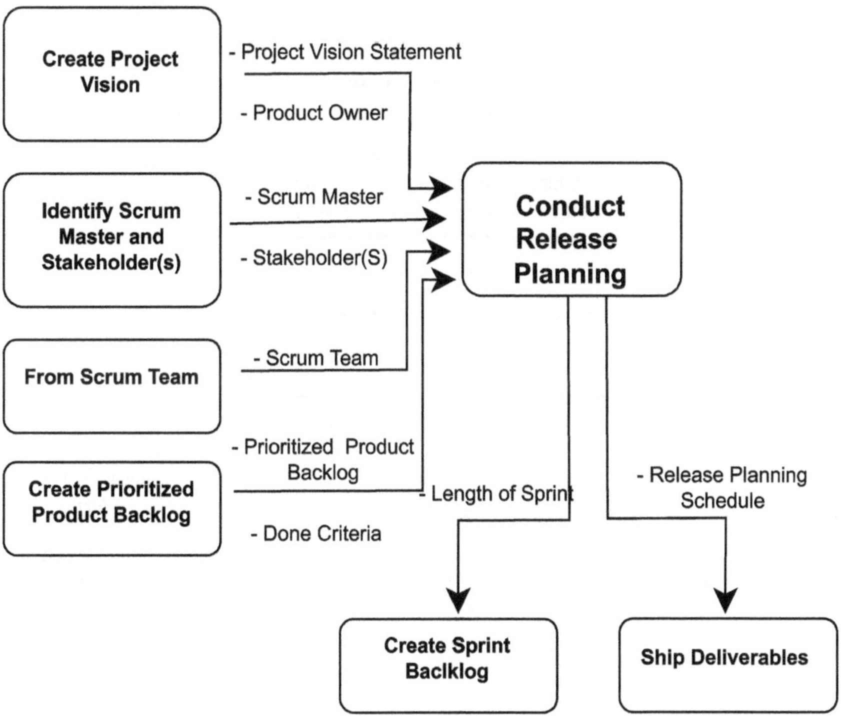

Figure 5.36: The "Conduct Release Planning" process data flow diagram.

Table 5.18: Inputs to the "Conduct Release Planning" process.

Input	Details
Inputs from the SCRUM team	As described earlier in Table 5.9.
Inputs from the stakeholders	As described earlier in Table 5.9.
Project vision statement	As described earlier in Table 5.5.
Prioritized Product Backlog	As described earlier in Table 5.17.
DONE-CRITERIA	As described earlier in Table 5.17.
Inputs from the program product owner	As described earlier in Table 5.3.
Inputs from the program SCRUM master	As described earlier in Table 5.3.
Inputs from the chief product owner	As described earlier in Table 5.3.
Program product backlog	As described earlier in Table 5.3.
Business requirements	As described earlier in Table 5.15.
Holiday calendar	The SCRUM team members keep track of their key dates and availability through the use of a shared calendar that can help plan for and execute the sprints.
Inputs from the SCRUM Guidance Body (SGB)	As for the conduct release planning process, the SGB can help provide information about the rules, regulations, standards, and BEST-PRACTICES for developing the release plan.

Table 5.19: Methods used during the "Conduct Release Planning" process.

Method	Details
Release planning sessions	A product owner conducts a series of release planning sessions in order to develop a release plan that defines when the product increments will be delivered to the customer. The main objective of a SCRUM release planning meeting is to share with the SCRUM team the product's delivery schedule.
Release prioritization methods	Developing a release plan needs the use of some predefined release prioritization methods that are industry and organization specific and determined by the organization's senior management.

Table 5.20: Outputs of the "Conduct Release Planning" process.

Output	Details
Release planning schedule	A release planning schedule is a key output of the Conduct Release Planning process. It shows the deliverables that will be released to the customers, and when.
Length of sprint	The product owner and the SCRUM team decide on the sprint's length, and that length remains unchanged throughout the project life cycle. A sprint's length can be anywhere between 1 and 6 weeks. However, practically, it is kept at 4 weeks.
Target customers for release	A release plan specifies the targeted customers and users such as the stakeholders who may choose to limit certain releases to certain users.
Refined Prioritized Product Backlog	The Prioritized Product Backlog may be refined during the Conduct Release Planning process.

5.10 SCRUM plan and estimate phase

In reference to Figure 5.37(a), the SCRUM plan and estimate phase involves five processes: Create USER-STORIES; Approve, Estimate, and Commit USER-STORIES; Create Tasks; Estimate Tasks; and Create Sprint Backlog.

Figure 5.37(b) presents the high-level data flow diagram of the SCRUM plan and estimate phase and the interaction among the five processes.

SCRUM Plan and Estimate Phase Processes

Create User Stories

Inputs
1. Scrum Core Team
2. Prioritized Product Backlog
3. Done Criteria
4. Personas
5. Stakeholder(s)
6. Epic(s)
7. Business Requirements
8. Laws and Regulations
9. Applicable Contracts
10. Scrum Guidance Body Recommendations

Methods
1. User Story Writing Expertise
2. User Story Workshops
3. User Group Meetings
4. Focus Group Meetings
5. Customer or User Interviews
6. Questionnaires
7. User Story Estimation Methods
8. Scrum Guidance Body Expertise

Outputs
1. User Stories
2. User Story Acceptance Criteria
3. Updated Prioritized Product Backlog
4. Updated or Refined Personas

Approve, Estimate, and Commit User Stories

Inputs
1. Scrum Core Team
2. User Stories
3. User Story Acceptance Criteria
4. Scrum Guidance Body Recommendations

Methods
1. User Group Meetings
2. Planning Poker
3. Fist of Five
4. Points for Cost Estimation
5. Other Estimation Techniques
6. Scrum Guidance Body Expertise

Outputs
1. Approved, Estimated, and Committed User Stories

Create Tasks

Inputs
1. Scrum Core Team
2. Approved, Estimated and Committed User Stories

Methods
1. Task Planning Meetings
2. Index Cards
3. Decomposition
4. Dependency Determination

Outputs
1. Task List
2. Updated Approved, Estimated, and Committed User Stories
3. Dependencies

Estimate Tasks

Inputs
1. Scrum Core Team
2. Task List
3. User Story Acceptance Criteria
4. Dependencies
5. Identified Risks
6. Scrum Guidance Body Recommendations

Methods
1. Task Estimation Meetings
2. Estimation Criteria
3. Planning Poker
4. Fist of Five
5. Other Task Estimation Techniques

Outputs
1. Effort Estimated Task List
2. Updated Task List

Create Sprint Backlog

Inputs
1. Scrum Core Team
2. Effort Estimated Task List
3. Length of Sprint
4. Previous Sprint Velocity
5. Dependencies
6. Team Calendar

Methods
1. Sprint Planning Meetings
2. Sprint Tracking Tools
3. Sprint Tracking Metrics

Outputs
1. Sprint Backlog
2. Sprint Burndown Chart

Figure 5.37: (a) SCRUM Plan and Estimate Phase processes. (b) SCRUM Plan and Estimate Phase data flow diagram and interactions across its processes.

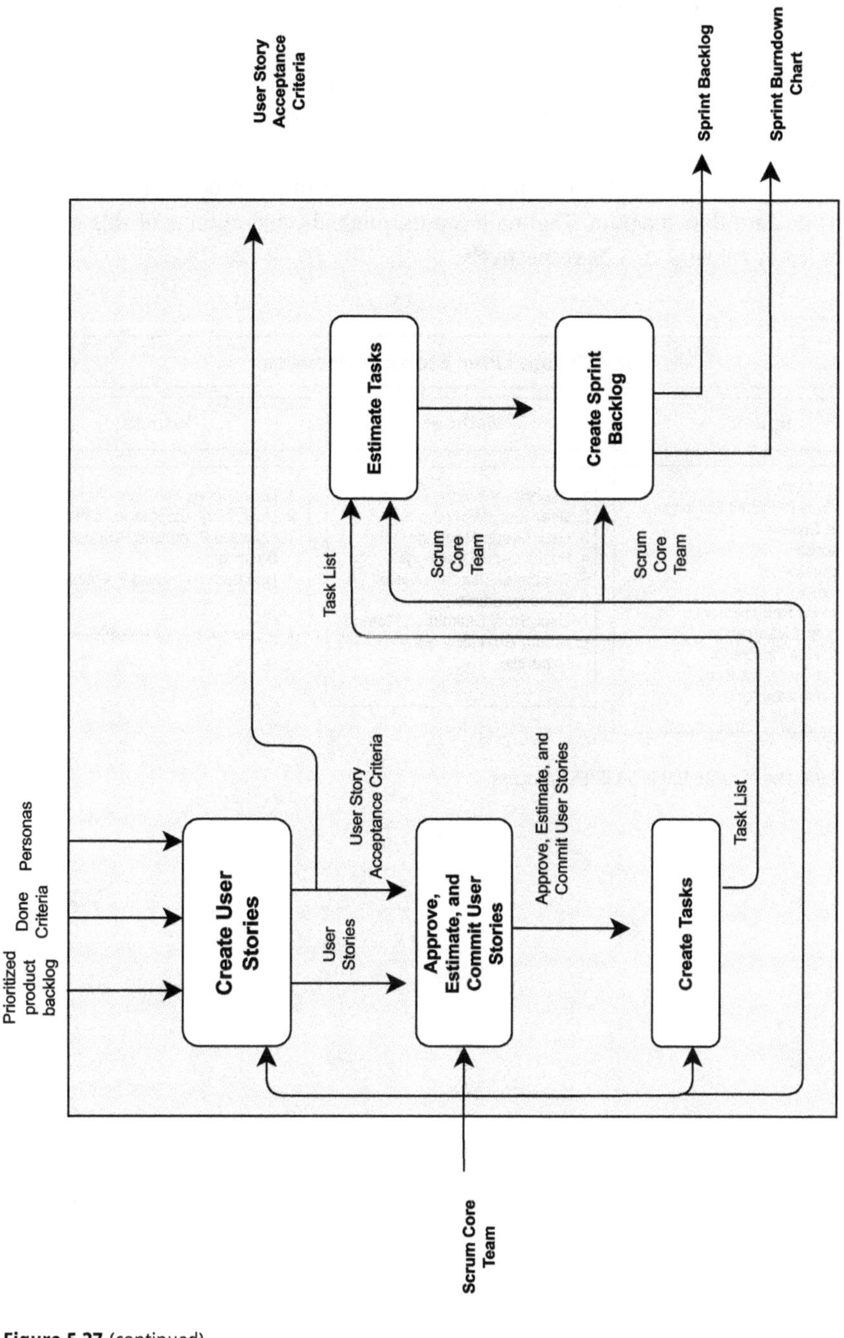

Figure 5.37 (continued)

5.10.1 The "Create USER-STORIES" process

The "Create USER-STORIES" process is used to define how the project's product owner can create the project USER-STORIES before he/she prioritize them into the project Prioritized Product Backlog.

Figure 5.38 presents the inputs, methods, and outputs of this process and Figure 5.39 presents its data flow diagram. The main inputs, methods, and outputs of this process are detailed in Tables 5.21–5.23, respectively.

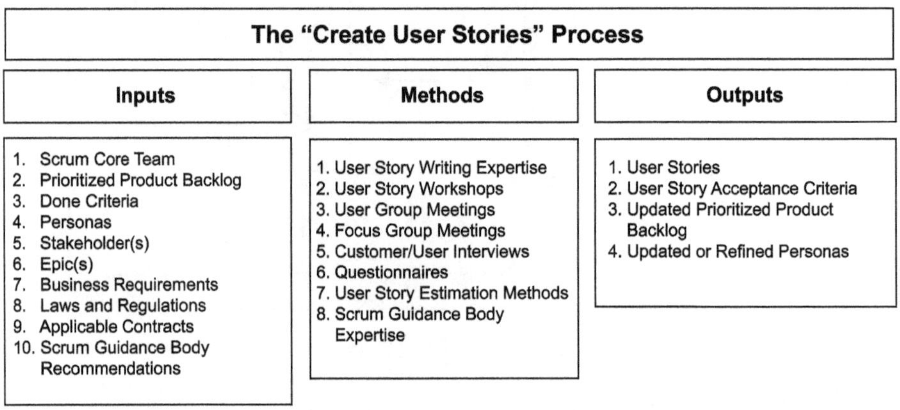

Figure 5.38: The "Create USER-STORIES" process.

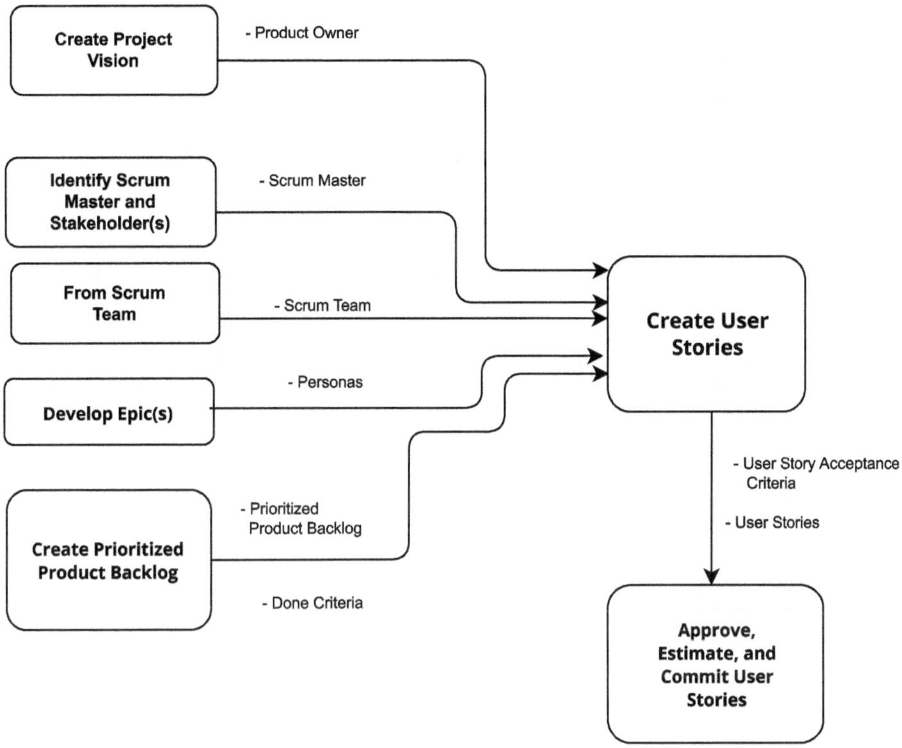

Figure 5.39: The "Create USER-STORIES" process data flow diagram.

Table 5.21: Inputs to the "Create USER-STORIES" process.

Input	Details
Inputs from the SCRUM team	As described earlier in Table 5.9.
Prioritized Product Backlog	As described earlier in Table 5.17.
DONE-CRITERIA	As described earlier in Table 5.17.
Personas	As described earlier in Table 5.11.
Inputs from the stakeholders	As described earlier in Table 5.9.
Epics	As described earlier in Table 5.11.
Business requirements	As described earlier in Table 5.15.
Laws and regulations	As described earlier in Table 5.9.
Applicable contracts	As described earlier in Table 5.9.
Inputs from the SCRUM Guidance Body (SGB)	The SGB inputs to the Create USER-STORIES process include information on the rules, regulations, standards, and BEST-PRACTICES that are relevant to USER-STORIES.

Table 5.22: Methods used during the "Create USER-STORIES" process.

Method	Details
USER-STORY writing expertise	A product owner is responsible for developing the USER-STORIES before he/she prioritize them into the Prioritized Product Backlog. He/she must be able to create and/or refine the USER-STORIES in away such that they can be easily approved, estimated, and committed to by the SCRUM team. To ensure that a product owner has the right writing skills, they must involve him/her in a USER-STORY writing workshop.
USER-STORY workshops	As described earlier in Table 5.10.
User group meetings	As described earlier in Table 5.10.
Focus group meetings	As described earlier in Table 5.10.
User or customer interviews	As described earlier in Table 5.10.
Questionnaires	As described earlier in Table 5.10.
USER-STORY estimation methods	As described earlier in Table 5.13.
SCRUM Guidance Body Expertise	As described earlier in Table 5.10.

Table 5.23: Outputs of the "Create USER-STORIES" process.

Output	Details
USER-STORIES	USER-STORIES express the business and project requirements in short and simple statements. Some of them may be too large to handle within a single sprint and these are referred to as epics that should be decomposed into smaller USER-STORIES. The Prioritized Product Backlog is continuously updated (e.g., due to changes in the business and project requirements) as a result of having new, updated, and/or deleted USER-STORIES.
USER-STORIES ACCEPTANCE-CRITERIA	Every USER-STORY has an associated ACCEPTANCE-CRITERIA that provides clarity to the team on what is expected of a USER-STORY. The product owner devises and communicates the ACCEPTANCE-CRITERIA to the SCRUM team.
Updated Prioritized Product Backlog	The Prioritized Product Backlog can be updated with the estimates for the USER-STORIES, epics, and USER-STORY ACCEPTANCE-CRITERIA.
Updated or refined personas	Although all personas are initially created in the Develop Epics process, the SCRUM team may decide that some of them need refinement.

5.10.2 The "Approve, Estimate, and Commit USER-STORIES" process

The "Approve, Estimate, and Commit USER-STORIES" process is used to select USER-STORIES from the prioritized product backlog and approve them into the current sprint backlog; estimate their tasks, needed resources, and time; and having the SCRUM team to commit to work toward working out these USER-STORIES to create their anticipated deliverables.

Figure 5.40 presents the inputs, methods, and outputs of this process and Figure 5.41 presents its data flow diagram. The main inputs, methods, and outputs of this process are detailed in Tables 5.24–5.26, respectively.

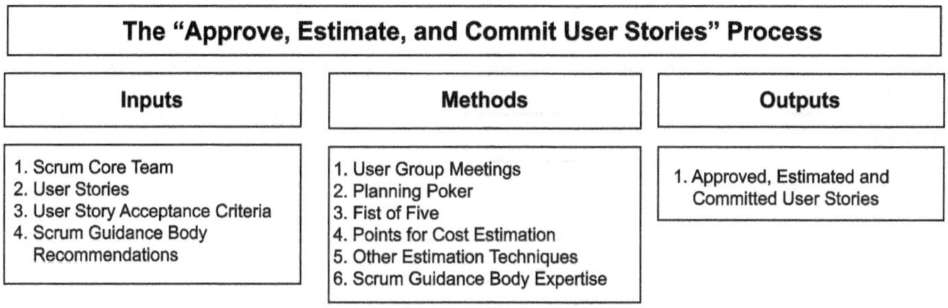

Figure 5.40: The "Approve, Estimate, and Commit USER-STORIES" process.

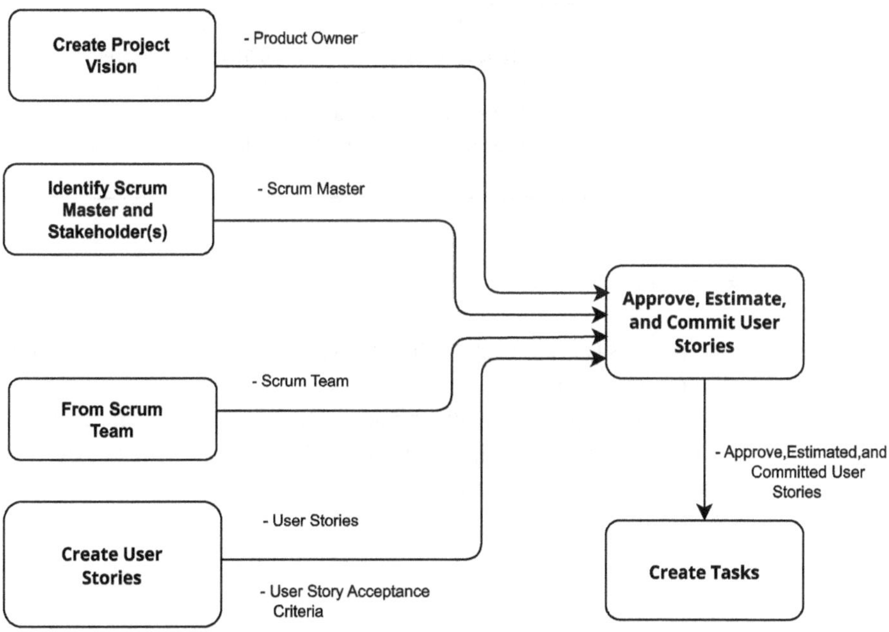

Figure 5.41: The "Approve, Estimate, and Commit USER-STORIES" process data flow diagram.

Table 5.24: Inputs to the "Approve, Estimate, and Commit USER-STORIES" process.

Input	Details
Inputs from the SCRUM team	As described earlier in Table 5.9.
USER-STORIES	As described earlier in Table 5.23.
USER-STORIES ACCEPTANCE-CRITERIA	As described earlier in Table 5.23.
Inputs from the SCRUM Guidance Body (SGB)	The SGB inputs to the Approve, Estimate, and Commit USER-STORIES process include information on the rules, regulations, standards, and BEST-PRACTICES that are relevant to approving, estimating, and committing to USER-STORIES.

Table 5.25: Methods used during the "Approve, Estimate, and Commit USER-STORIES" process.

Method	Details
User group meetings	As described earlier in Table 5.10.
SCRUM Guidance Body Expertise	As described earlier in Table 5.10.

Table 5.26: Outputs of the "Approve, Estimate, and Commit USER-STORIES" process.

Output	Details
Approved, Estimated, and Committed USER-STORIES	The Create Prioritized Product Backlog and Create USER-STORIES processes give the USER-STORIES high-level estimates that are used by the product owner to approve the USER-STORIES for a sprint in accordance with the stakeholders' expectations. Then they are estimated and committed to by the SCRUM team.

5.10.3 The "Create Tasks" process

The "Create Tasks" process is used to create the tasks required for accomplishing the USER-STORIES of the current sprint.

Figure 5.42 presents the inputs, methods, and outputs of this process and Figure 5.43 presents its data flow diagram. The main inputs, methods, and outputs of this process are detailed in Tables 5.27–5.29, respectively.

The "Create Tasks" Process		
Inputs	**Methods**	**Outputs**
1. Scrum Core Team 2. Approved, Estimated and Committed User Stories	1. Task Planning Meetings 2. Index Cards 3. Decomposition 4. Dependency Determination	1. Task List 2. Updated Approved, Estimated, and Committed User Stories 3. Dependencies

Figure 5.42: The "Create Tasks" process.

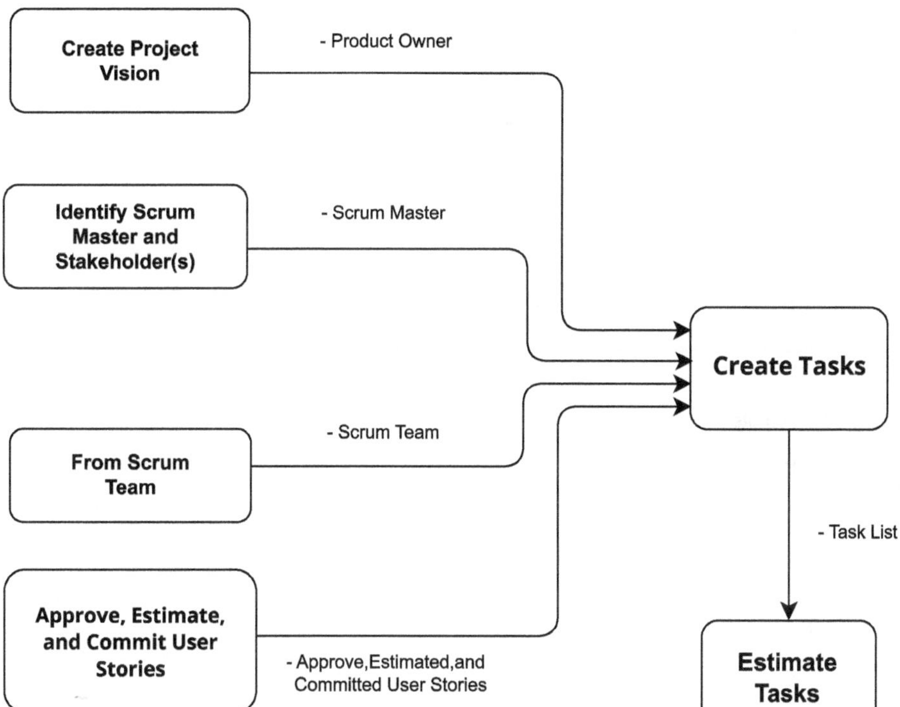

Figure 5.43: The "Create Tasks" process data flow diagram.

Table 5.27: Inputs to the "Create Tasks" process.

Input	Details
Inputs from the SCRUM team	As described earlier in Table 5.9.
Approved, Estimated, and Committed USER-STORIES	As described earlier in Table 5.26.

Table 5.28: Methods used during the "Create Tasks" process.

Method	Details
Task planning meetings	During the task planning meetings, the SCRUM team plans the work for the current sprint. This involves defining and scheduling for the tasks that are required for the USER-STORIES in that sprint.
Decomposition	Decomposition is used to breakdown high-level tasks into lower level ones.
Dependency determination	After selecting the set of USER-STORIES from the Prioritized Product Backlog for the current sprint backlog, the team must figure out the dependencies (e.g., people availability and technical dependencies) among them.

Table 5.29: Outputs of the "Create Tasks" process.

Output	Details
Task list	The task list contains all of the current sprint's tasks committed to by the SCRUM team.
Updated Approved, Estimated, and Committed USER-STORIES	The USER-STORIES may be updated during the Create Tasks process.
Dependencies	As described earlier in Table 5.28.

5.10.4 The "Estimate Tasks" process

The "Estimate Tasks" process is used to estimate the tasks required for accomplishing the USER-STORIES of the current sprint.

Figure 5.44 presents the inputs, methods, and outputs of this process and Figure 5.45 presents its data flow diagram. The main inputs, methods, and outputs of this process are detailed in Tables 5.30–5.32, respectively.

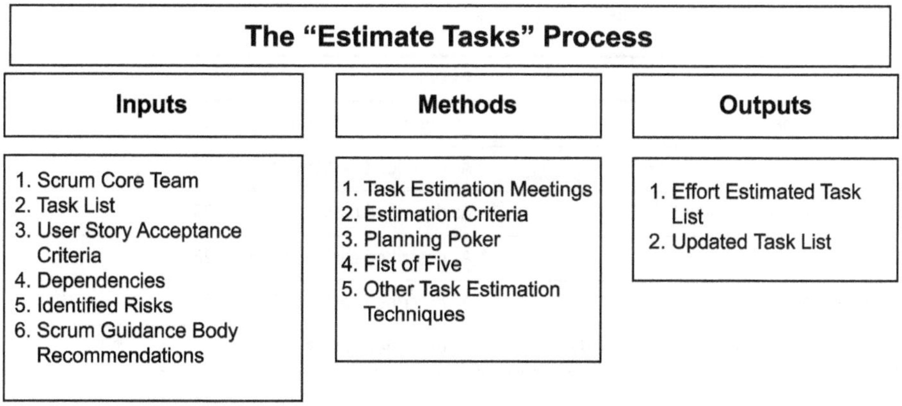

Figure 5.44: The "Estimate Tasks" process.

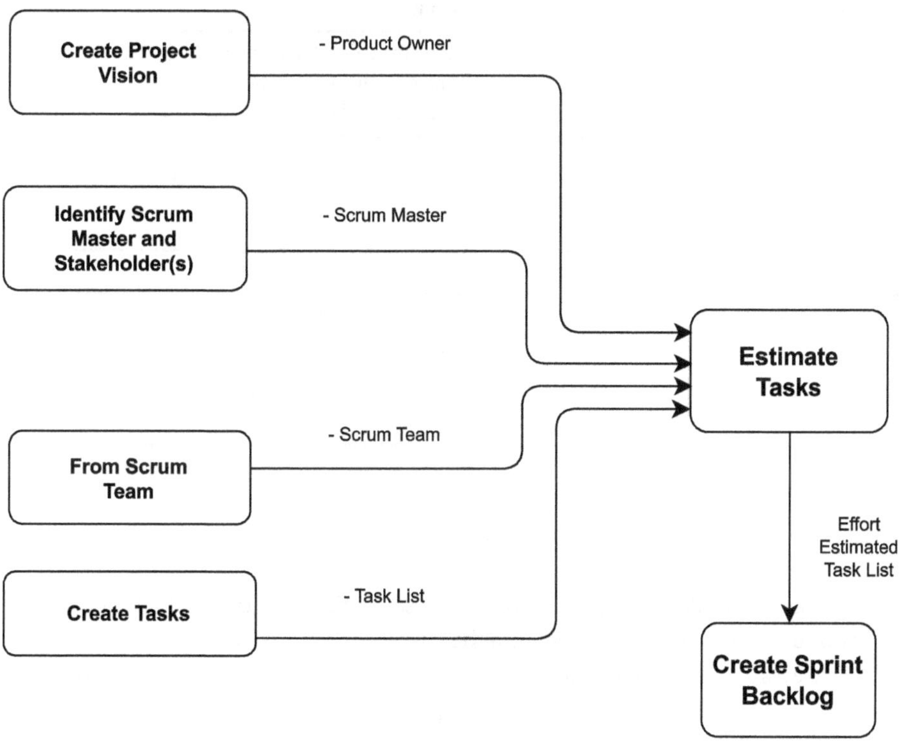

Figure 5.45: The "Estimate Tasks" process data flow diagram.

Table 5.30: Inputs to the "Estimate Tasks" process.

Input	Details
Inputs from the SCRUM team	As described earlier in Table 5.9.
Task list	As described earlier in Table 5.29.
USER-STORIES ACCEPTANCE-CRITERIA	As described earlier in Table 5.23.
Dependencies	As described earlier in Table 5.28.
Identified Risks	As described earlier in Table 5.11.
Inputs from the SCRUM Guidance Body (SGB)	As described earlier in Table 5.24.

Table 5.31: Methods used during the "Estimate Tasks" process.

Method	Details
Task estimation meetings	During the task estimation meetings, the SCRUM team members estimate the efforts required to complete these tasks and to estimate their required resources.

Table 5.32: Outputs of the "Estimate Tasks" process.

Output	Details
Effort Estimated Task List	This is a list of tasks associated with the committed USER-STORIES.
Updated Task List	The task list may be updated during the Estimate Tasks process.

5.10.5 The "Create Sprint Backlog" process

The "Create Sprint Backlog" process is used to select the right USER-STORIES from the prioritized backlog and place them into the current sprint's backlog.

Figure 5.46 presents the inputs, methods, and outputs of this process and Figure 5.47 presents its data flow diagram. The main inputs, methods, and outputs of this process are detailed in Tables 5.33–5.35, respectively.

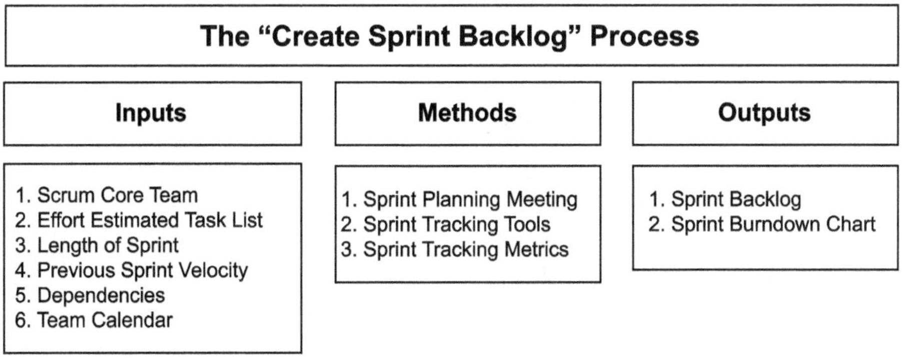

Figure 5.46: The "Create Sprint Backlog" process.

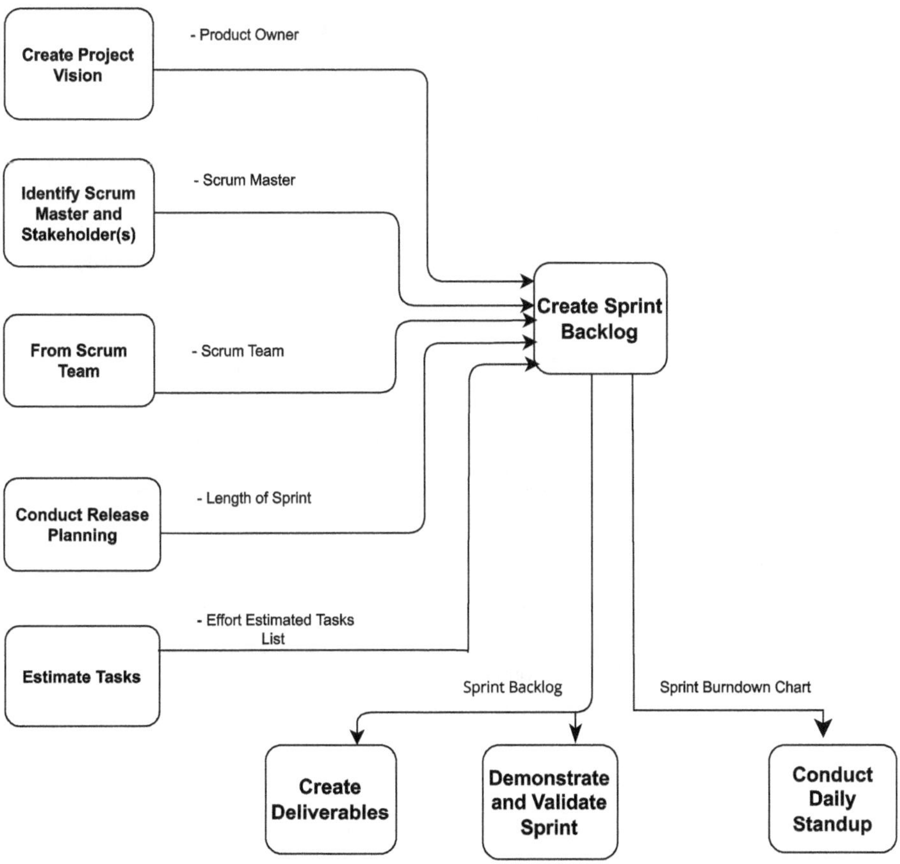

Figure 5.47: The "Create Sprint Backlog" process data flow diagram.

Table 5.33: Inputs to the "Create Sprint Backlog" process.

Input	Details
Inputs from the SCRUM team	As described earlier in Table 5.9.
Effort Estimated Task List	As described earlier in Table 5.32.
Length of sprint	As described earlier in Table 5.20.
Previous sprint velocity	Sprint velocity is the rate at which the SCRUM team can complete a sprint work, and that is an essential factor in identifying the amount of work the SCRUM team can commit to in an upcoming sprint.
Dependencies	As described earlier in Table 5.28.
Team calendar	A team calendar has information with respect to the SCRUM team members' availability including employees' vacation, leaves, events, and holidays.

Table 5.34: Methods used during the "Create Sprint Backlog" process.

Method	Details
Sprint planning meetings	The USER-STORIES that are approved, estimated, and committed during the "Approve, Estimate, and Commit USER-STORIES" process are discussed by the SCRUM team during the sprint planning meetings.
Sprint tracking tools	The sprint tracking tools are meant to track the progress of a sprint and to know where the SCRUM team stands in terms of completing the tasks in the sprint backlog.
Sprint tracking metrics	Metrics used in SCRUM projects include velocity, business value delivered, and number of USER-STORIES.

Table 5.35: Outputs of the "Create Sprint Backlog" process.

Output	Details
Sprint Backlog	A Sprint Backlog is composed of the list of USER-STORIES selected from the prioritized product backlog to be carried out and accomplished during that sprint.
Sprint BURNDOWN-CHART	A Sprint BURNDOWN-CHART is a graph that indicates the amount of work remaining in the current sprint.

5.11 The SCRUM implement phase

In reference to Figure 5.48(a), the SCRUM implement phase involves three processes: Create Deliverables, Conduct Daily Standup, and Groom Prioritized Product Backlog.

Figure 5.48(b) presents the high-level data flow diagram of the SCRUM implement phase and the interaction among its processes.

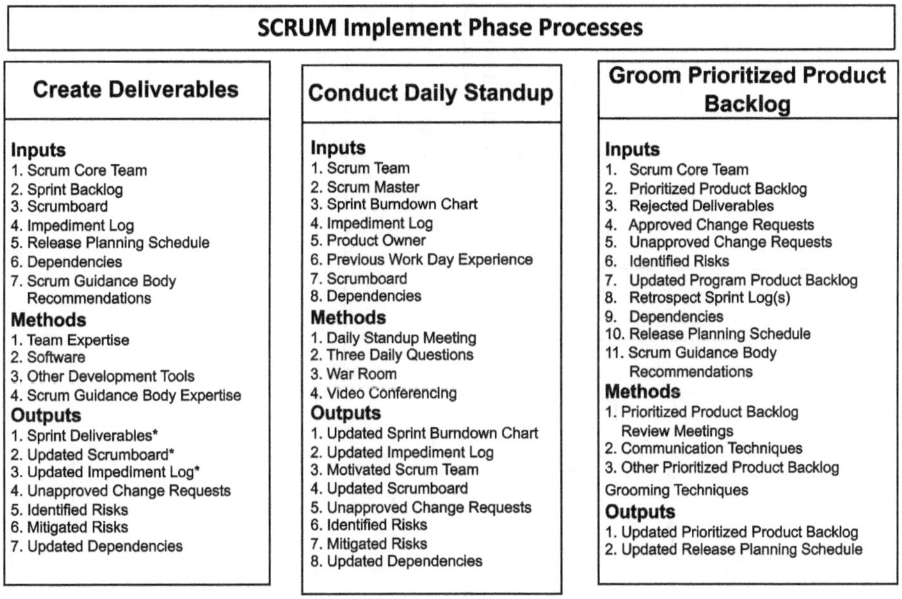

Figure 5.48: (a) SCRUM Implement Phase processes. (b) SCRUM Implement Phase data flow diagram and interactions across its processes.

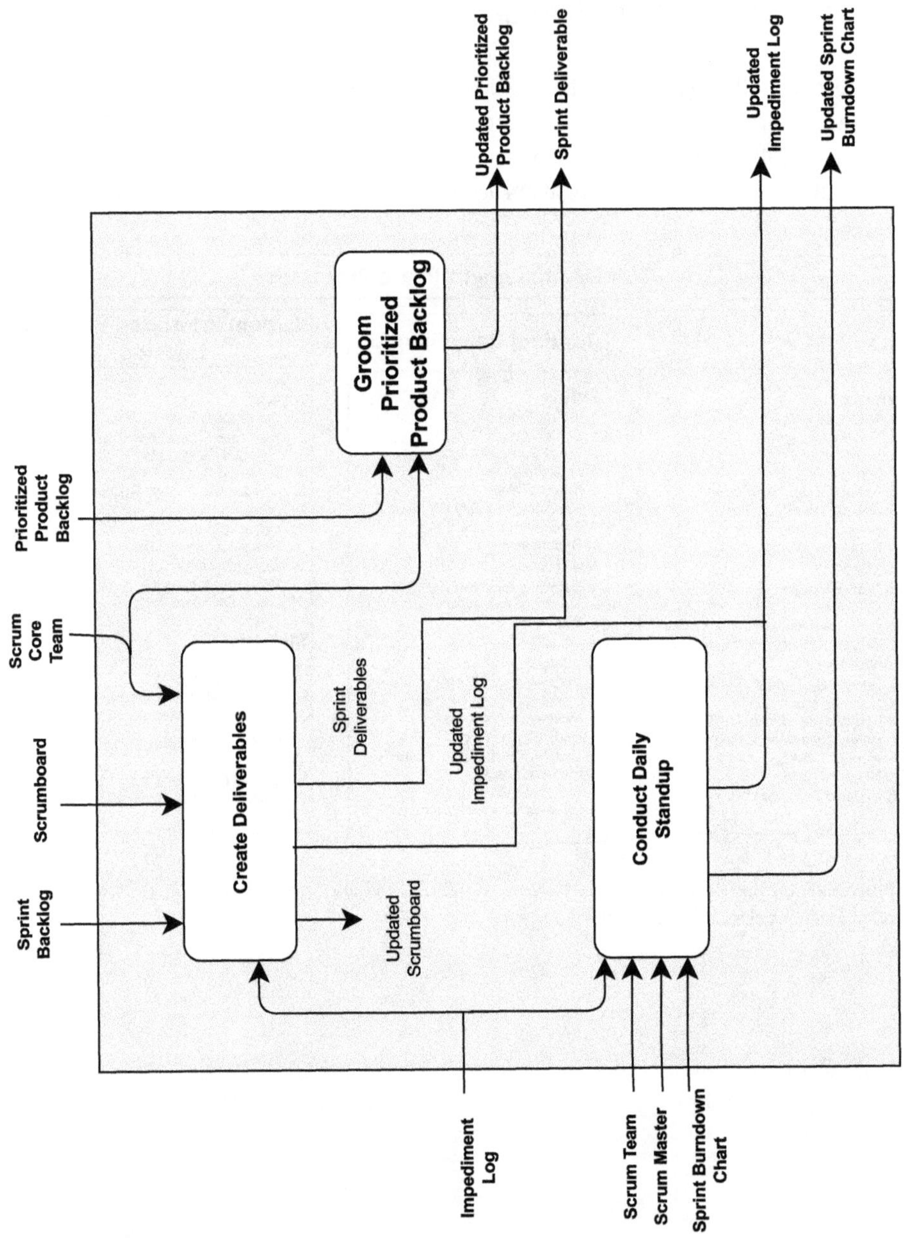

Figure 5.48 (continued)

5.11.1 The “Create Deliverables” process

The “Create Deliverables” process is used to carry out the work in the current sprint and to create its associated product increments.

Figure 5.49 presents the inputs, methods, and outputs of this process and Figure 5.50 presents its data flow diagram. The main inputs, methods, and outputs of this process are detailed in Tables 5.36–5.38, respectively.

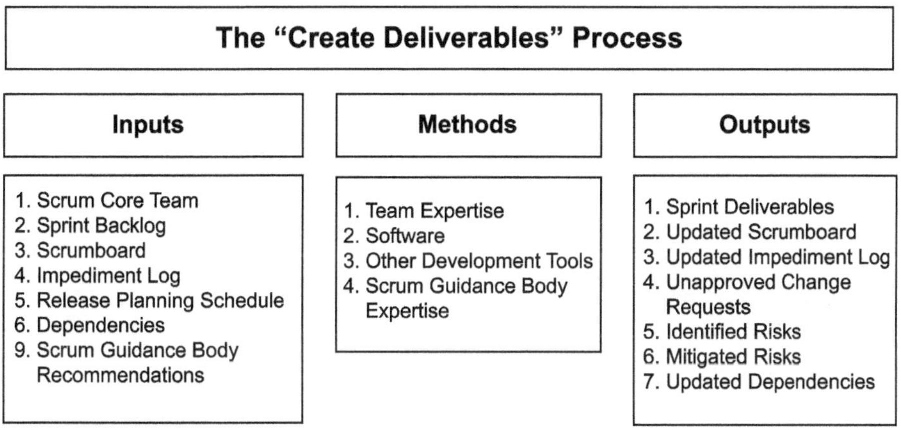

Figure 5.49: The “Create Deliverables” process.

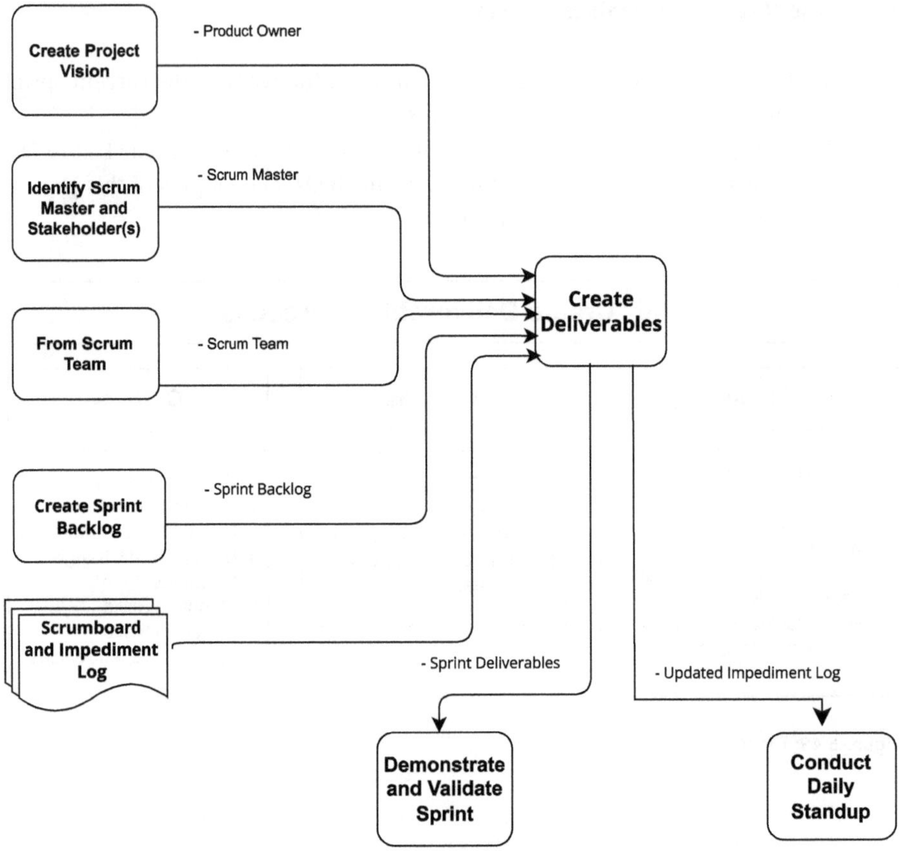

Figure 5.50: The "Create Deliverables" process data flow diagram.

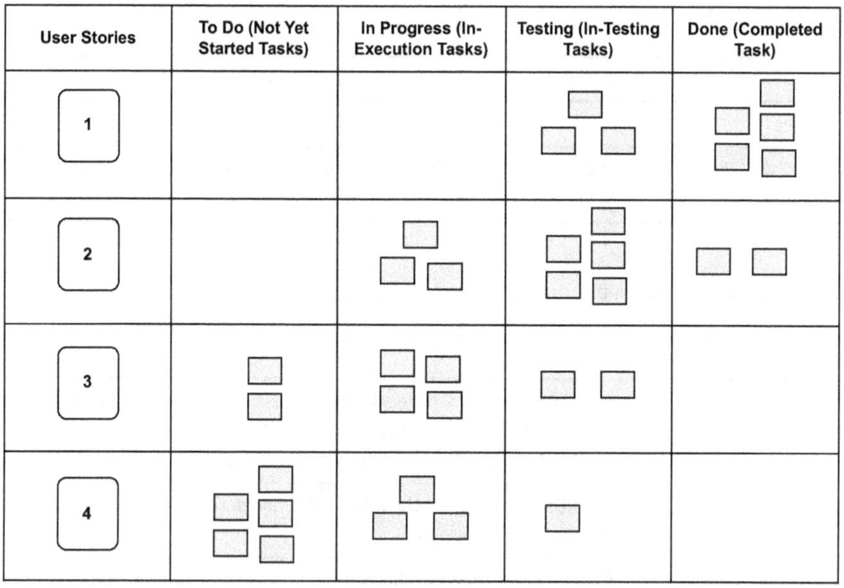

Figure 5.51: A Sample SCRUMBOARD.

Table 5.36: Inputs to the "Create Deliverables" process.

Input	Details
Inputs from the SCRUM team	As described earlier in Table 5.9.
Sprint backlog	As described earlier in Table 5.35.
SCRUMBOARD	In reference to Figure 5.51, a SCRUMBOARD shows the progress of the SCRUM team during each sprint. It contains the following columns to indicate the progress of the sprint estimated tasks: - The "To Do" column includes the tasks that are not yet started. - The "In Progress" column includes the tasks that are started but not yet completed. - The "Testing" column includes the tasks that are completed but in the process of being tested. - The "Done" column includes the tasks that have been completed and SUCCESSFULLY tested.
Impediment log	Impediments impact the SCRUM team's productivity. They must be identified, resolved, and removed in order for the SCRUM team to continue working effectively. Impediments can be internal such as having lack of communication, or external such as having issues with software licenses.
Release planning schedule	As described earlier in Table 5.20.

Table 5.36 (continued)

Input	Details
Dependencies	As described earlier in Table 5.28.
Inputs from the SCRUM Guidance Body (SGB)	The inputs from the SGB to the Create Deliverables process include the BEST-PRACTICES to create deliverables.

Table 5.37: Methods used during the "Create Deliverables" process.

Methods	Details
Team expertise	A SCRUM team must have adequate expertise that enables the team to understand the sprint backlog's USER-STORIES and their associated tasks such that the team can effectively create the targeted deliverables.
Software	A SCRUM team needs scheduling, information collection, and distribution software tools.
SCRUM Guidance Body Expertise	As described earlier in Table 5.10.

Table 5.38: Outputs of the "Create Deliverables" process.

Outputs	Details
Sprint Deliverables	At the end of each sprint, a product increment or deliverable is completed. The deliverable should possess all features and functionality defined in the USER-STORIES of the sprint. Deliverables should be tested SUCCESSFULLY.
Updated SCRUMBOARD	The SCRUMBOARD is updated regularly as the team keeps completing tasks.
Updated Impediment Log	The Impediment Log (e.g., described earlier in Table 5.36) may be updated during the Create Deliverables process.
Unapproved CHANGE-REQUESTS	As described earlier in Table 5.9.
Identified Risks	As described earlier in Table 5.11.
Mitigated Risks	The SCRUM team documents all newly identified risks and the actions taken to mitigate them.
Updated Dependencies	Dependencies among USER-STORIES and among their associated tasks (e.g., described earlier in Table 5.28) may be updated during the Create Deliverables process.

5.11.2 The "Conduct Daily Standup" process

The "Conduct Daily Standup" process is used for the sprint daily standup meetings.

Figure 5.52 presents the inputs, methods, and outputs of this process and Figure 5.53 presents its data flow diagram. The main inputs, methods, and outputs of this process are detailed in Tables 5.39–5.41, respectively.

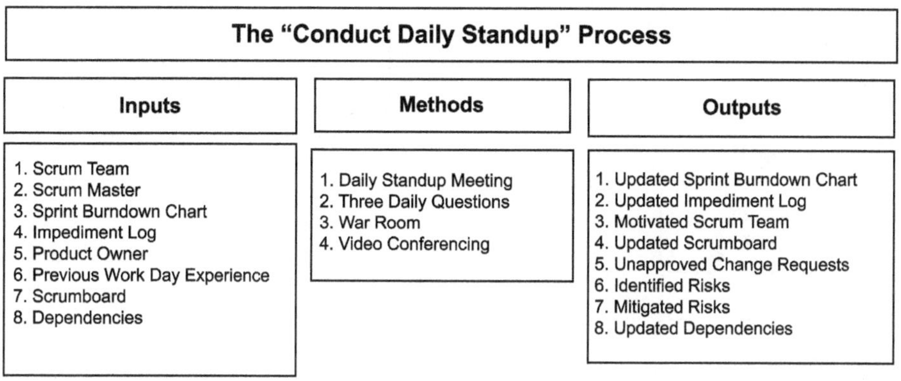

Figure 5.52: The "Conduct Daily Standup" process.

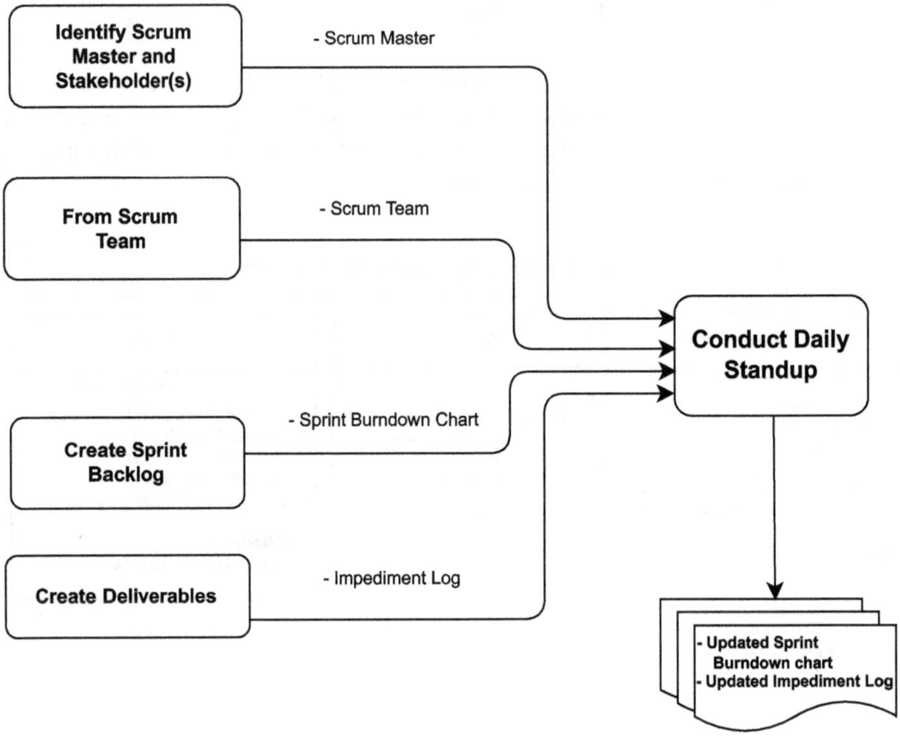

Figure 5.53: The "Conduct Daily Standup" process data flow diagram.

Table 5.39: Inputs to the "Conduct Daily Standup" process.

Input	Details
Inputs from the SCRUM team	As described earlier in Table 5.9.
Inputs from the SCRUM master	As described earlier in Table 5.8 (e.g., Identified SCRUM Master).
Sprint BURNDOWN-CHART	As described earlier in Table 5.35.
Impediment Log	As described earlier in Table 5.36.
Inputs from the Product Owner	As described earlier in Table 5.5 (e.g., identified product owner).
Previous Work Day Experience	The SCRUM team members give status updates to fellow team members in the Daily Standup Meeting. This session is called a standup because members stand throughout the meeting. Team members discuss achievements and experience from the previous work day. This experience is an important input to the Conduct Daily Standup process.
SCRUMBOARD	As described earlier in Table 5.36.
Dependencies	As described earlier in Table 5.28.

Table 5.40: Methods used during the "Conduct Daily Standup" process.

Method	Details
Daily standup meeting and the three daily questions	A Daily Standup Meeting is a 15-min short daily meeting, during which each of the SCRUM team members answers the following daily questions: – What did I complete yesterday? – What will I complete today? – What impediments or obstacles (if any) am I currently facing?
War room	The members of a SCRUM team members are co-located in the same working location that is commonly called the war room.
Video conferencing	When it is not possible to co-locate the entire SCRUM team to be co-located in the same working location, a video conferencing tool can be used to enable face-to-face communication among the team members.

Table 5.41: Outputs of the "Conduct Daily Standup" process.

Outputs	Details
Updated sprint BURNDOWN-CHART	The sprint BURNDOWN-CHART (e.g., described earlier in Table 5.35) may be updated during the Conduct Daily Standup process.
Updated Impediment Log	The Impediment Log (e.g., described earlier in Table 5.36) may be updated during the Conduct Daily Standup Process.
Motivated SCRUM team	As the Daily Standup Meetings enable each member of the SCRUM team to express his/her thoughts and that he/she is already self-organized, the consequent result would be in the form of having motivated members.
Updated SCRUMBOARD	As described earlier in Table 5.38.
Unapproved CHANGE-REQUESTS	As described earlier in Table 5.9.
Identified risks	As described earlier in Table 5.11.
Mitigated risks	As described earlier in Table 5.38.
Updated dependencies	Dependencies among USER-STORIES and among their associated tasks (e.g., described earlier in Table 5.28) may be updated during the Conduct Daily Standup process.

5.11.3 The "Groom Prioritized Product Backlog" process

The "Groom Prioritized Product Backlog" process is used by the Project Product Owner to update the Project Prioritized Product Backlog in accordance with the feedbacks and outcomes of the sprint retrospect meetings.

Figure 5.54 presents the inputs, methods, and outputs of this process and Figure 5.55 presents its data flow diagram. The main inputs, methods, and outputs of this process are detailed in Tables 5.42–5.44, respectively.

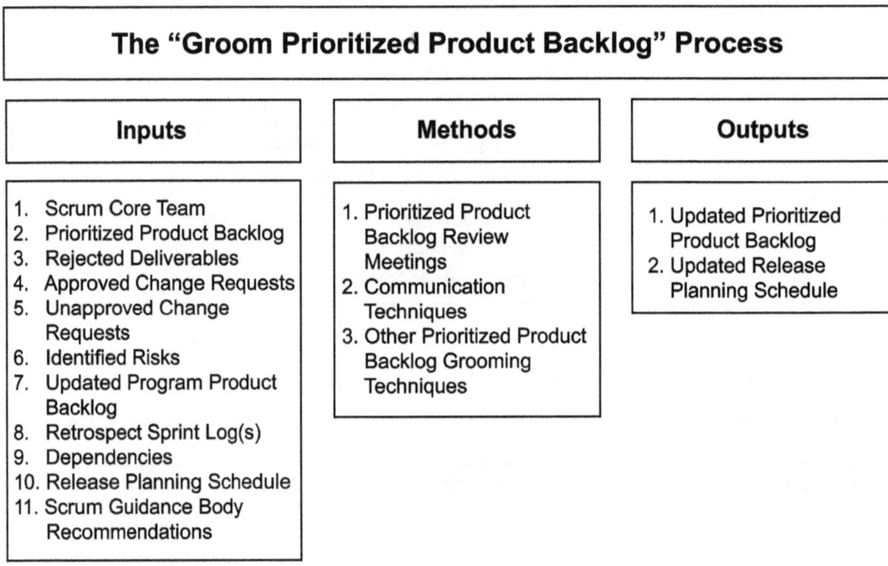

Figure 5.54: The "Groom Prioritized Product Backlog" process.

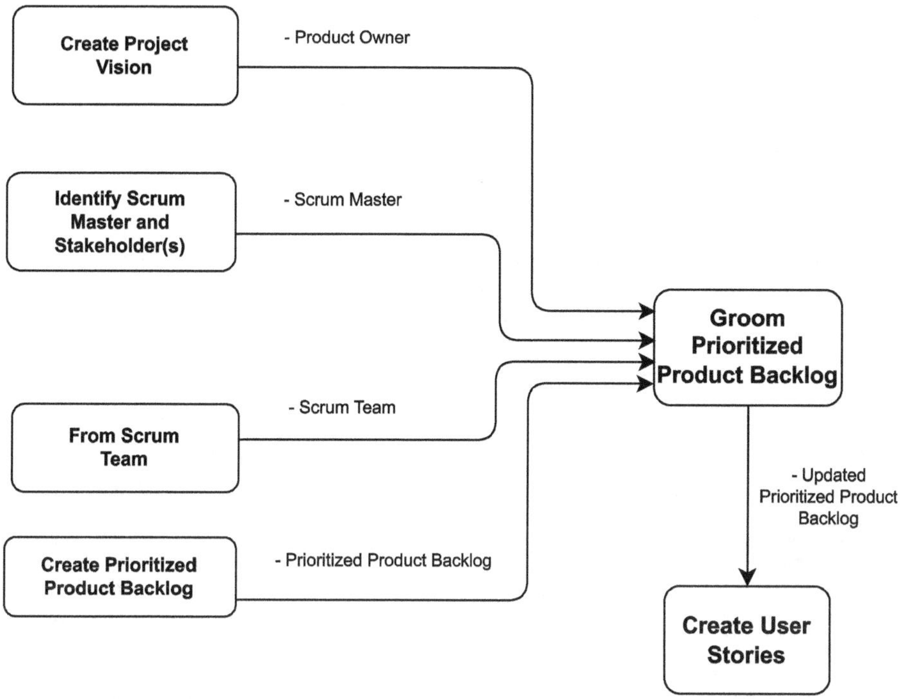

Figure 5.55: The "Groom Prioritized Product Backlog" process data flow diagram.

Table 5.42: Inputs to the "Groom Prioritized Product Backlog" process.

Input	Details
Inputs from the SCRUM team	As described earlier in Table 5.9.
Prioritized Product Backlog	As described earlier in Table 5.17.
Rejected deliverables	Rejected deliverables are those deliverables that do not satisfy the ACCEPTANCE-CRITERIA.
Approved/unapproved CHANGE-REQUESTS	As described earlier in Table 5.9.
Identified risks	As described earlier in Table 5.11.
Updated Program Product Backlog	A Program Product Backlog may be updated as a result of external factors (e.g., changing business scenarios, technology trends, or legal compliance requirements) or internal factors (e.g., modifications in organizational strategy or policies, identified risks, and other factors). Updates in the Program Product Backlog impact the Prioritized Product Backlogs of program's children projects.

Table 5.42 (continued)

Input	Details
Sprint Retrospect Logs	As described earlier in Section 5.4.3.
Dependencies	As described earlier in Table 5.28.
Release planning schedule	As described earlier in Table 5.20.
Inputs from the SCRUM Guidance Body (SGB)	The inputs to the Groom Prioritized Product Backlog from the SCRUM Guidance Body include how to understand and collect requirements from the stakeholders and SCRUM teams and then prioritize them into the Product Backlog.

Table 5.43: Methods used during the "Groom Prioritized Product Backlog" process.

Method	Details
Prioritized product backlog review meetings	The product owner may have meetings with the stakeholders, SCRUM master, and SCRUM team to have adequate information before he/she can update the Prioritized Product Backlog during the Groom Prioritized Product Backlog process.
Communication techniques	SCRUM promotes accurate and effective communications through co-location of the SCRUM team. SCRUM prefers informal and face-to-face interaction communications.

Table 5.44: Outputs of the "Groom Prioritized Product Backlog" process.

Output	Details
Updated Prioritized Product Backlog	The ultimate result of the Groom Prioritized Product Backlog is an updated version of that backlog.
Updated release planning schedule	Another result of the Groom Prioritized Product Backlog is an updated version of the release planning schedule.

5.12 The SCRUM review and retrospect phase

In reference to Figure 5.56(a), the SCRUM Review and Retrospect Phase involves three processes, which are the Convene SCRUM of SCRUMs, Demonstrate and Validate Sprint, and Sprint Retrospect.

Figure 5.56(b) presents the high-level data flow diagram of the SCRUM implement phase and the interaction among its processes.

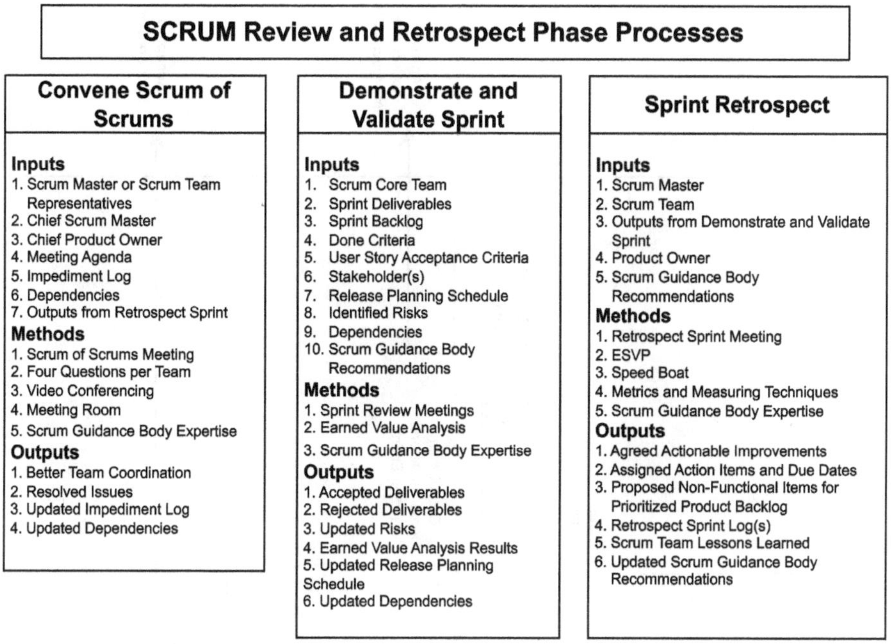

Figure 5.56: (a) SCRUM Review and Retrospect Phase processes. (b) SCRUM Review and Retrospect Phase data flow diagram and interactions across its processes.

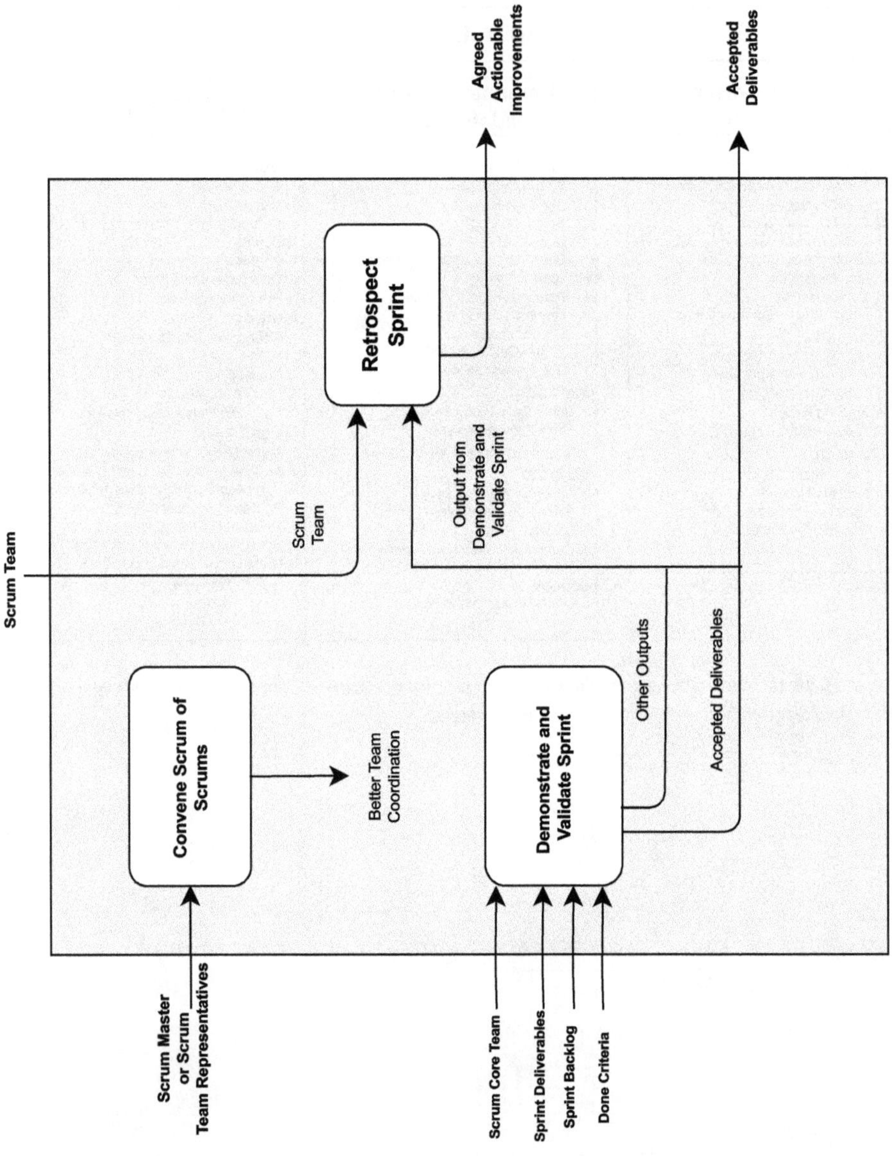

Figure 5.56 (continued)

5.12.1 The "Convene SCRUM of SCRUMs" process

The "Convene SCRUM of SCRUMs" process is used by the project chief product owner to meet with all product owners and SCRUM masters of the project multiple SCRUMs.

Figure 5.57 presents the inputs, methods, and outputs of this process and Figure 5.58 presents its data flow diagram. The main inputs, methods, and outputs of this process are detailed in Tables 5.45–5.47, respectively.

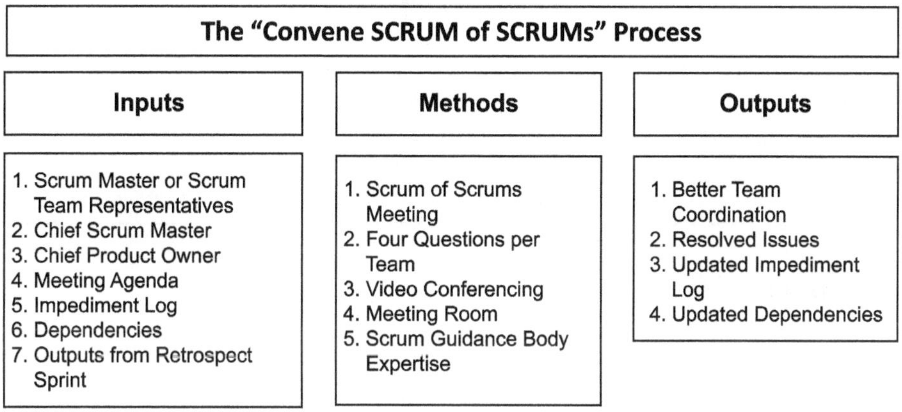

Figure 5.57: The "Convene SCRUM of SCRUMs" process.

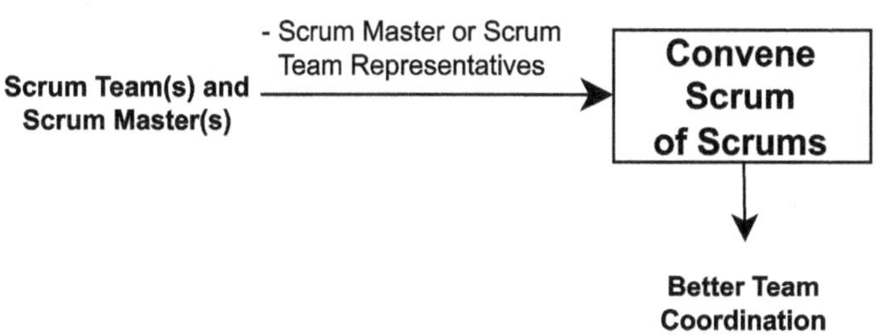

Figure 5.58: The "Convene SCRUM of SCRUMs" process data flow diagram.

Table 5.45: Inputs to the "Convene SCRUM of SCRUMs" process.

Input	Details
SCRUM masters and/or SCRUM teams' other representatives	At a SCRUM of SCRUMs (SoS) meeting, at least the SCRUM master of each SCRUM team should attend. Each person involved in the meeting should be able to proactively identify instances that could cause each other impediments.
Inputs from the chief SCRUM master	As described earlier in Table 5.2.

Table 5.45 (continued)

Input	Details
Inputs from the chief product owners	As described earlier in Table 5.3.
Meeting agenda	An SoS meeting is used to communicate progress between the project SCRUM teams. Prior to a SoS meeting, the chief SCRUM master announces an agenda for that meeting to help the SCRUM teams prepare for that meeting. Any large-scale issue should be communicated at that meeting.
Impediment log	As described earlier in Table 5.36.
Dependencies	As described earlier in Table 5.28.
Outputs from the Sprint Retrospect process	This refers to the feedback and recommendations made during the retrospect meetings at the end of the preceding sprints. The term "sprint retrospect" is described earlier in Section 5.4.8.

Table 5.46: Methods used during the "Convene SCRUM of SCRUMs" process.

Method	Details
SCRUM of SCRUMs meeting	As described earlier in Table 5.45 (e.g., see the meeting agenda item in that table).
Four questions per team	Each SCRUM team representative should update the meeting attendees with respect to his/her SCRUM team's activities and challenges: – The work that has been going on since the last SoS meeting. – The work that has been planned for through the next SoS meeting. – The work that was supposed to be completed in response to needs from other SCRUM teams, but remained uncompleted. – The work that has been planned for, but may impact some or all of the other SCRUM teams.
Video conferencing	When it is not possible to hold a SoS meeting face-to-face, a video conferencing tool may be used to facilitate that meeting.
Meeting room	A SoS meeting should be held in a dedicated conference room.
SCRUM Guidance Body (SGB) Expertise	The inputs from the SGB to the Convene SCRUM of SCRUMs process include information on how to conduct SoS meetings, and how to incorporate their outcomes in the SCRUM teams' works.

Table 5.47: Outputs of the "Convene SCRUM of SCRUMs" process.

Output	Details
Better team coordination	A SCRUM of SCRUMs (SoS) meeting is used to coordinate the project works carried out by the project multiple SCRUM teams.
Resolved issues	One of the ultimate objectives of a SoS meeting is to resolve any SCRUM inter-teams' issues and to ensure that these teams are working coherently.
Updated Impediment Log	The Impediment Log (e.g., described earlier in Table 5.36) may be updated during the Convene SCRUM of SCRUMs process.
Updated dependencies	Dependencies among USER-STORIES, among their associated tasks, and among the SCRUM teams' works (e.g., described earlier in Table 5.28) may be updated during the Convene SCRUM of SCRUMs process.

5.12.2 The "Demonstrate and Validate Sprint" process

The "Demonstrate and Validate Sprint" process is used to have the SCRUM team members to present and demonstrate to the product owner the deliverables of the current sprint to solicit his/her approval before shipping them to the customer.

Figure 5.59 presents the inputs, methods, and outputs of this process and Figure 5.60 presents its data flow diagram. The main inputs, methods, and outputs of this process are detailed in Tables 5.48–5.50, respectively.

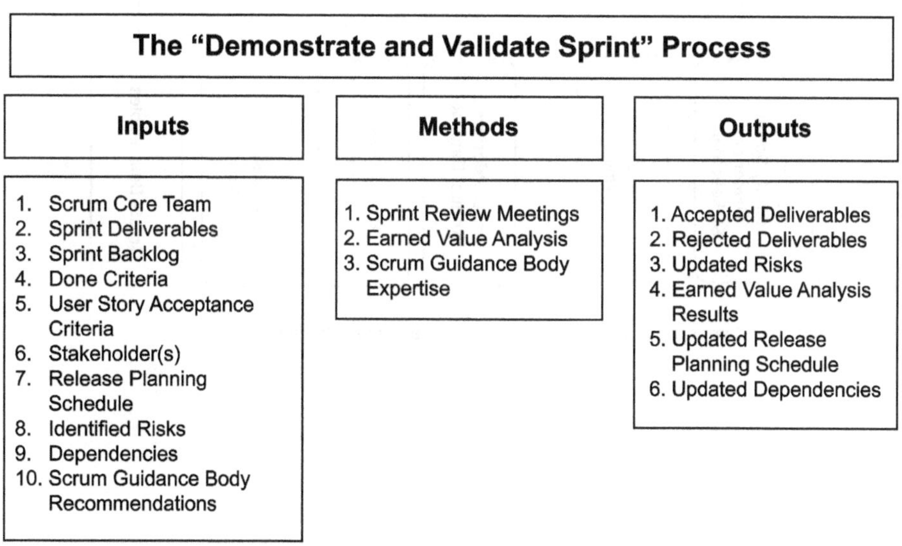

Figure 5.59: The "Demonstrate and Validate Sprint" process.

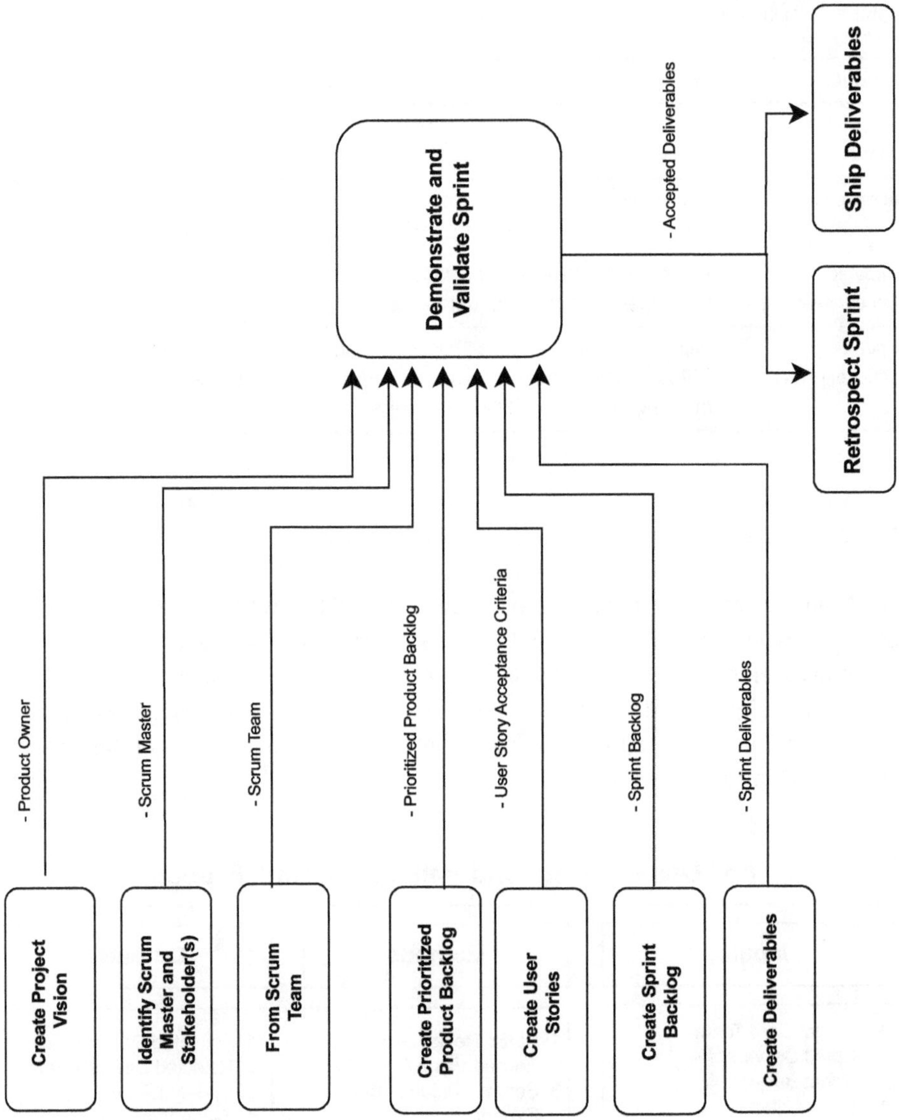

Figure 5.60: The "Demonstrate and Validate Sprint" process data flow diagram.

Table 5.48: Inputs to the "Demonstrate and Validate Sprint" process.

Input	Details
Inputs from the SCRUM team	As described earlier in Table 5.9.
Sprint deliverables	As described earlier in Table 5.38.
Sprint backlog	As described earlier in Table 5.35.
DONE-CRITERIA	As described earlier in Table 5.17.
USER-STORIES ACCEPTANCE-CRITERIA	As described earlier in Table 5.23.
Inputs from the stakeholders	As described earlier in Table 5.9.
Release planning schedule	As described earlier in Table 5.20.
Identified risks	As described earlier in Table 5.11.
Dependencies	As described earlier in Table 5.28.
Inputs from the SCRUM Guidance (SGB) Body	The inputs from the SGB include information on how to Demonstrate and Validate Sprints.

Table 5.49: Methods used during the "Demonstrate and Validate Sprint" process.

Method	Details
Sprint review meeting	A sprint review meeting is used to have the SCRUM team to present and demonstrate the deliverables of the current sprint. The product owner then validates and accepts or rejects these deliverables. If accepted, then he/she makes the necessary arrangements for the SCRUM team to deploy and implement these deliverables at the customer site.
SCRUM Guidance Body (SGB) expertise	An SGB expertise refers to documented BEST-PRACTICES with respect to how to conduct sprint review meetings.

Table 5.50: Outputs of the "Demonstrate and Validate Sprint" process.

Output	Details
Accepted deliverables	This refers to all deliverables that satisfy the USER-STORIES ACCEPTANCE-CRITERIA and accepted by the product owner at the current sprint review meeting. All accepted deliverables are then shipped to the customer and deployed at their site.

Table 5.50 (continued)

Output	Details
Rejected deliverables	This refers to all deliverables that do not satisfy the USER-STORIES ACCEPTANCE-CRITERIA. All rejected deliverables are not shipped to the customer, but rather remain on the SCRUM team's list and could be reworked in an upcoming sprint.
Updated risks	This refers to all risks that may be reassessed at the end of the Demonstrate and Validate Sprint process.
Updated release planning schedule	Another result of the Demonstrate and Validate Sprint process is an updated version of the release planning schedule.
Updated dependencies	Dependencies among USER-STORIES, among their associated tasks, and among the SCRUM teams' works (e.g., described earlier in Table 5.28) may be updated during the Demonstrate and Validate Sprint process.

5.12.3 The "Sprint Retrospect" process

The "Sprint Retrospect" process is used to have the project product owner to meet with the SCRUM master, the SCRUM team, and the concerned stakeholders to discuss what happened during the current sprint, and to assess any feedback from the customer after the current sprint's deliverables have been shipped to the customer and deployed for them. The learned lessons here are used with the upcoming sprints. Also, they can be used to make grooming recommendations to the product owner to update the project prioritized product backlog.

Figure 5.61 presents the inputs, methods, and outputs of this process and Figure 5.62 presents its data flow diagram. The main inputs, methods, and outputs of this process are detailed in Tables 5.51–5.53, respectively.

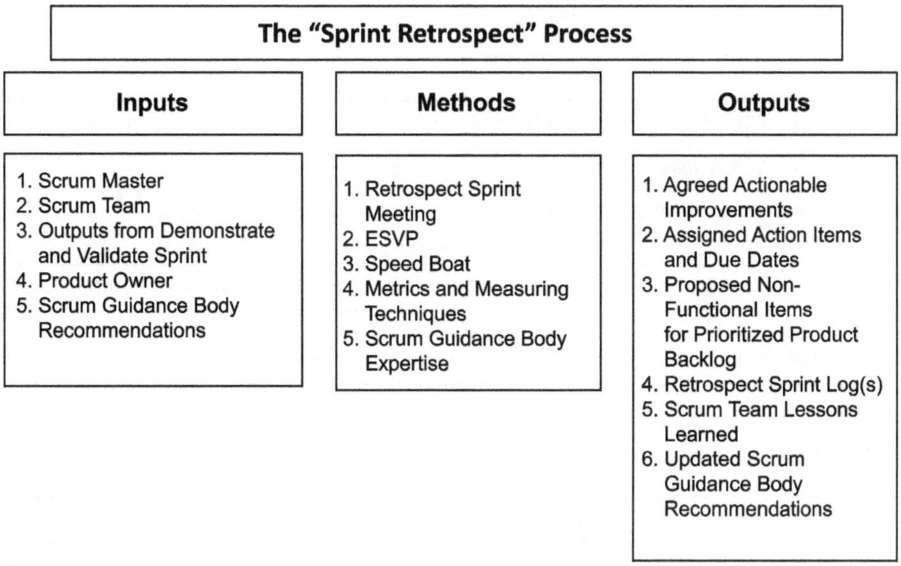

Figure 5.61: The "Sprint Retrospect" process.

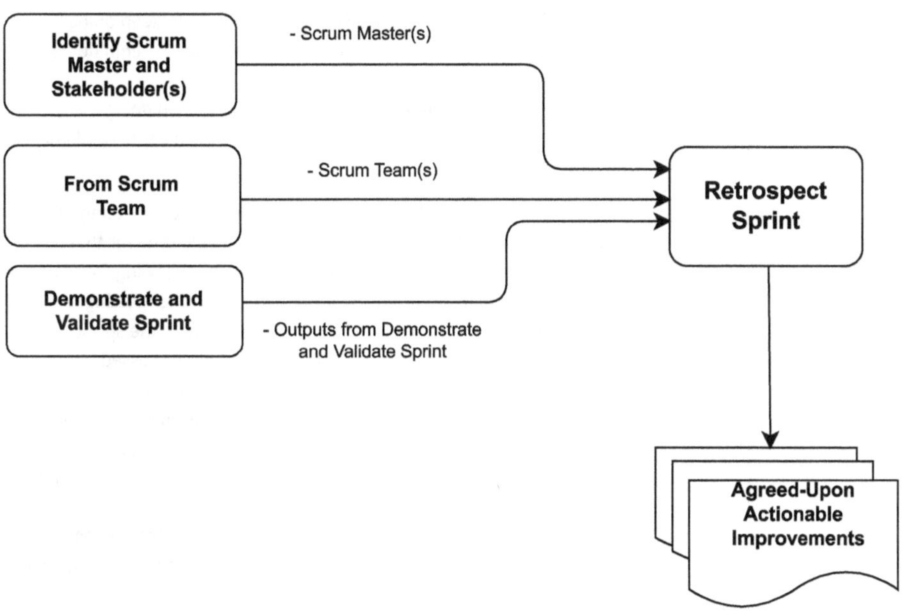

Figure 5.62: The "Sprint Retrospect" process data flow diagram.

Table 5.51: Inputs to the "Sprint Retrospect" process.

Input	Details
Inputs from the SCRUM master	As described earlier in Table 5.8 (e.g., identified SCRUM master).
Inputs from the SCRUM team	As described earlier in Table 5.9.
Outputs from Demonstrate and Validate Sprint	The outputs of the Demonstrate and Validate Sprint process help the process retrospect sprint.
Inputs from the product owner	As described earlier in Table 5.5 (e.g., identified product owner).
Inputs from the SCRUM Guidance Body (SGB)	The inputs from the SCRUM Guidance Body include guidelines for conducting sprint retrospect meetings.

Table 5.52: Methods used during the "Sprint Retrospect" process.

Method	Details
Sprint retrospect meetings	A sprint retrospect meeting is the final step in that sprint and is attended by the SCRUM team, the SCRUM master, the product owner, and the concerned stakeholders. It is meant to discuss what went wrong and what went right, and to identify the following things: – The best practices that the SCRUM team needs to keep doing. – The process improvements that the team needs to begin doing and working on. – The process problems that the team needs to stop doing. Accordingly, a list of agreed-upon actionable improvements would be created.
Explorer/shopper/vacationer/prisoner (ESVP)	This can be done at the beginning of a sprint retrospect meeting to understand the participants and devise a reasonable tone for that meeting. Attendees anonymously select one of the following four options indicating how they feel about their participation in the meeting: – Explorer. An explorer would want to participate in the sprint retrospect meeting and learn about everything discussed. – Shopper. A shopper would want to listen to everything, but choose what he/she would want to take away from the sprint retrospect meeting. – Vacationer. A vacationer would want to relax and take it easy in the sprint retrospect meeting. – Prisoner. A prisoner is attending the sprint retrospect meeting because it is a must to attend, but he would want to be somewhere else.

Table 5.52 (continued)

Method	Details
Metrics and measuring techniques	Examples of metrics used to measure the SCRUM team's performance during a sprint are as follows: – Team velocity. This refers to the number of USER-STORIES completed in a single sprint. – Done SUCCESS rate. This refers to the percentage of USER-STORIES that have been completed versus the ones that were committed to by the SCRUM team. – Estimation effectiveness. This refers to the percentage of deviations between the estimated and actual times spent on the tasks and USER-STORIES. – Review feedback ratings. This refers to the feedback from the stakeholders using quantitative and/or qualitative ratings. – Team morale ratings. This refers to the results from the SCRUM team morale self-assessments.
SCRUM Guidance Body Expertise	The SCRUM Guidance Body helps the Sprint Retrospect process as to how to conduct sprint retrospect meetings.

Table 5.53: Outputs of the "Sprint Retrospect" process.

Outputs	Details
Agreed-upon actionable improvements	The list of agreed-upon actionable improvements is the main output of the Sprint Retrospect process that is meant to address issues and improve processes to enhance the SCRUM team's performance in the upcoming sprints.
Assigned action items and due dates	Once the agreed-upon actionable improvements have been finalized, the SCRUM team needs then to consider appropriate action items to implement these improvements.
Proposed nonfunctional items for Prioritized Product Backlog	The Prioritized Product Backlog is initially developed based on the USER-STORIES and functional requirements. Nonfunctional requirements cannot be fully defined at the beginning, but can become more clear during the sprint review and/or the sprint retrospect meetings. These items should be added to the Prioritized Product Backlog as they are discovered.
Sprint retrospect logs	A sprint retrospect log is a record of opinions, discussions, and actionable items raised in a sprint retrospect meeting.

Table 5.53 (continued)

Outputs	Details
SCRUM team lessons learned	A self-organizing SCRUM team learns from any action and/or event that takes place during a sprint. Such lessons learned can help the SCRUM team improve their performance in the upcoming sprints.
Updated SCRUM Guidance Body recommendations	A sprint retrospect meeting may lead to revising the relevant SCRUM Guidance Body recommendations.

5.13 SCRUM release phase

In reference to Figure 5.63(a), the SCRUM release phase involves two processes: Ship Deliverables and Retrospect Project. Figure 5.63(b) presents the high-level data flow diagram of the SCRUM implement phase and the interaction among its processes.

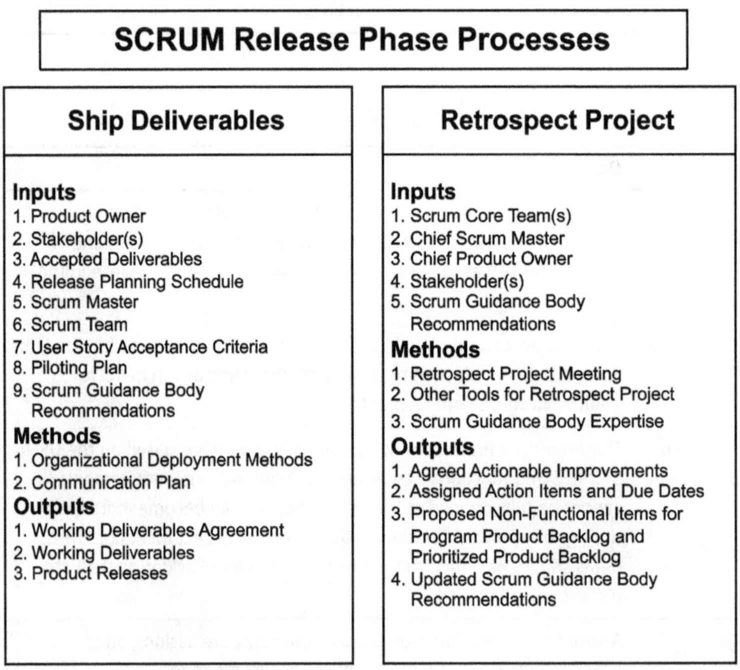

Figure 5.63: (a) SCRUM release phase processes. (b) SCRUM release phase data flow diagram and interactions across its processes.

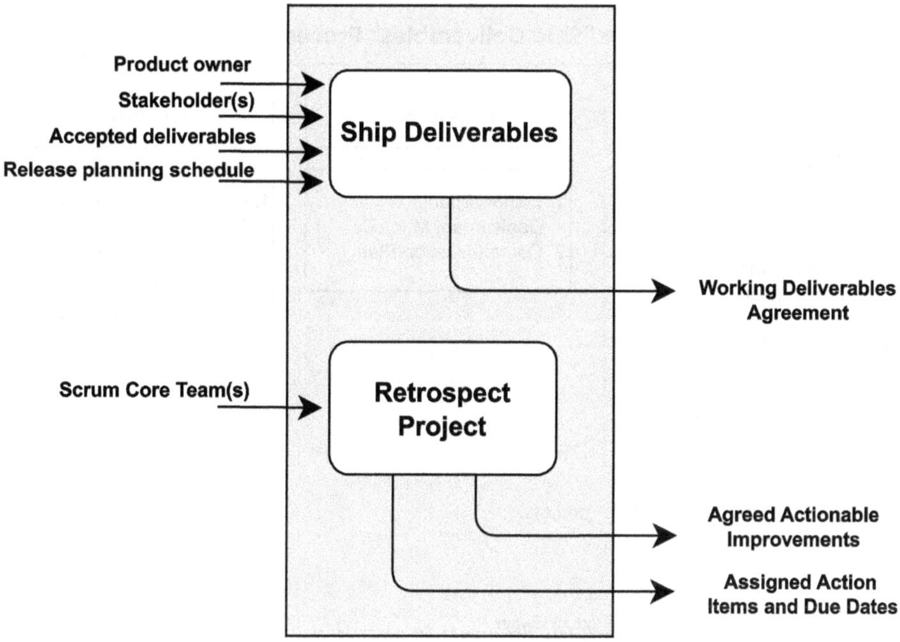

Figure 5.63 (continued)

5.13.1 The "Ship Deliverables" process

The "Ship Deliverables" process enables the project product owner to make the necessary arrangements for the SCRUM team to ship the current sprint's approved deliverables to the customer after approval at the sprint review meeting.

Figure 5.64 presents the inputs, methods, and outputs of this process and Figure 5.65 presents its data flow diagram. The main inputs, methods, and outputs of this process are detailed in Tables 5.54–5.56, respectively.

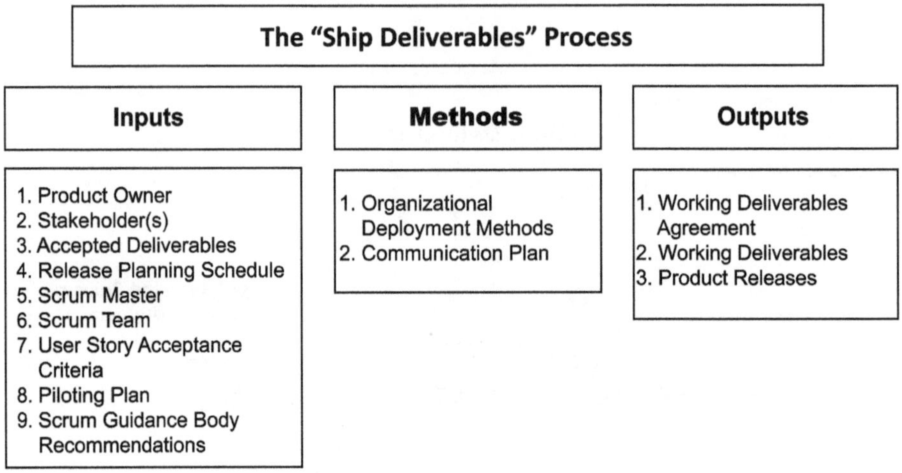

Figure 5.64: The "Ship Deliverables" process.

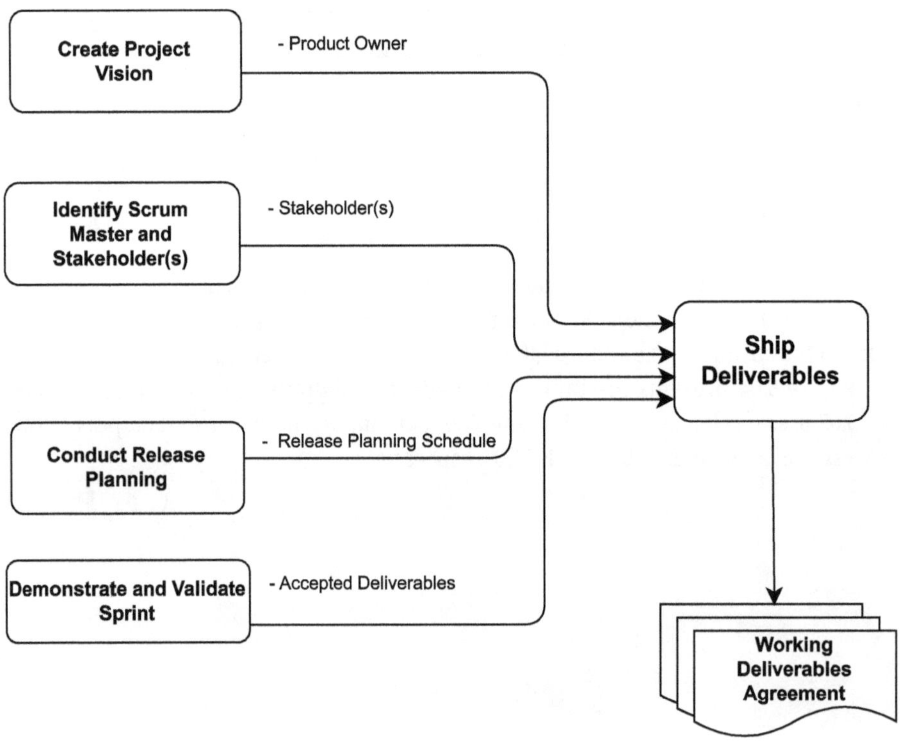

Figure 5.65: The "Ship Deliverables" process data flow diagram.

Table 5.54: Inputs to the "Ship Deliverables" process.

Input	Details
Inputs from the product owner	As described earlier in Table 5.5 (e.g., identified product owner).
Inputs from the stakeholders	As described earlier in Table 5.9.
Accepted deliverables	As described earlier in Table 5.50.
Release planning schedule	As described earlier in Table 5.20.
Inputs from the SCRUM master	As described earlier in Table 5.8 (e.g., identified SCRUM master).
Inputs from the SCRUM team	As described earlier in Table 5.9.
USER-STORIES ACCEPTANCE-CRITERIA	As described earlier in Table 5.23.
Inputs from the SCRUM Guidance Body (SGB)	The SGB can provide recommendations and guidelines regarding the deployment of products.

Table 5.55: Methods used during the "Ship Deliverables" process.

Method	Details
Organizational deployment methods	The deployment methods vary from one organization to another based on their industry, target users, and positioning.
Communication plan	This specifies the records that must be created and maintained throughout the project. A popular communication method is a visual display depicting important information in an easy-to-interpret format, posted in an accessible location, and kept up-to-date with the most current information.

Table 5.56: Outputs of the "Ship Deliverables" process.

Output	Details
Working deliverables agreement	Approved deliverables receive formal approval by the customer.
Working deliverables	This refers to the final shippable deliverable. As new product increments are created, they are continually integrated into the prior ones.
Product releases	A product release should include the following: – Release content. This is the essential information about the deliverables that assist the customer support team. – Release notes. This should include market-relevant shipping CRITERIA for the product to be delivered.

5.13.2 The "Retrospect Project" process

The "Retrospect Project" process enables the project product owner to discuss and learn from what happened throughout the project from the start to the end. All lessons learned here are used for future upcoming projects.

Figure 5.66 presents the inputs, methods, and outputs of this process and Figure 5.67 presents its data flow diagram. The main inputs, methods, and outputs of this process are detailed in Tables 5.57–5.59, respectively.

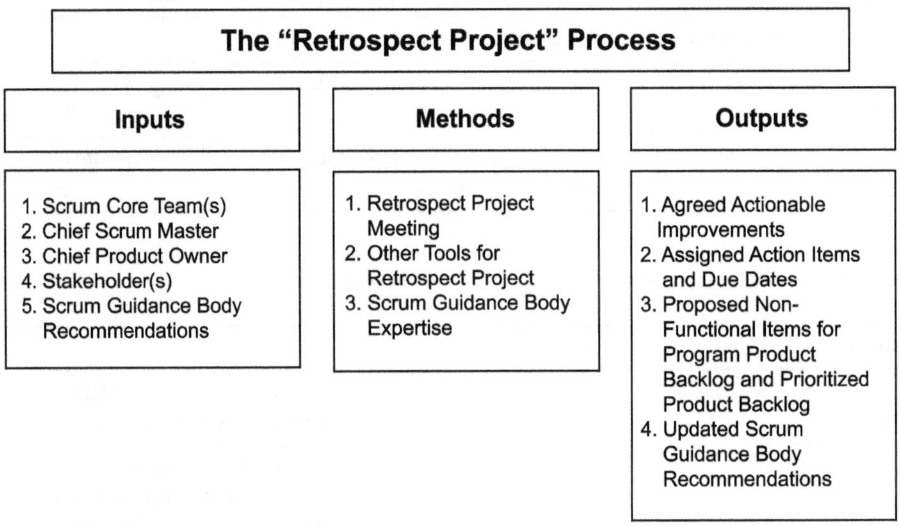

Figure 5.66: The "Retrospect Project" process.

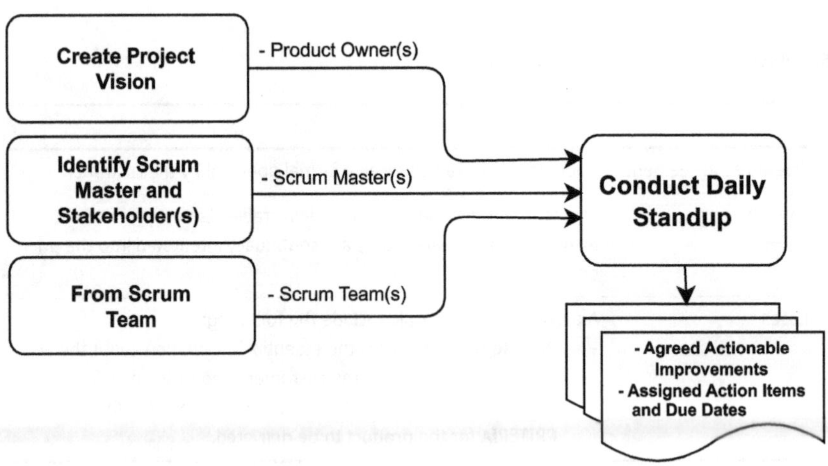

Figure 5.67: The "Retrospect Project" process data flow diagram.

Table 5.57: Inputs to the "Retrospect Project" process.

Input	Details
Inputs from the SCRUM team	As described earlier in Table 5.9.
Inputs from the chief SCRUM master	As described earlier in Table 5.2.
Inputs from the chief product owner	As described earlier in Table 5.3.
Inputs from the stakeholders	As described earlier in Table 5.9.
Inputs from the SCRUM Guidance Body (SGB)	During the Retrospect Project process, the SGB can provide a repository of internal templates that support the future projects and guidance for conducting the retrospect project meeting.

Table 5.58: Methods used during the "Retrospect Project" process.

Method	Details
Retrospect project meeting	A retrospect project meeting is to determine how to improve the team collaboration and effectiveness in future projects.
SCRUM Guidance Body expertise	In the Retrospect Project process, the primary responsibility of the SGB is to ensure that the lessons learned in each project are not lost, but embedded in the organization.

Table 5.59: Outputs of the "Retrospect Project" process.

Outputs	Details
Agreed-upon actionable improvements	As described earlier in Table 5.53.
Assigned action items and due dates	As described earlier in Table 5.53.
Proposed nonfunctional items for Program Product Backlog and Prioritized Product Backlog	When the initial Program Product Backlog and/or Prioritized Product Backlog are developed, they are based on USER-STORIES and required functionalities. Often nonfunctional requirements may not be fully defined in the early stages of the project and can surface during the sprint review, retrospect sprint, or retrospect project meetings. These items should be added to the Program Product Backlog and Prioritized Product Backlog as they are discovered. Some examples of nonfunctional requirements are response times, capacity limitations, and security-related issues.
Updated SCRUM Guidance Body recommendations	The Project Retrospect process may lead to revising the relevant SCRUM Guidance Body recommendations.

Appendix A: the IET competencies

IET competencies (learning outcomes)

Math and science	C1	Apply knowledge of mathematics, statistics, natural science, and engineering principles to the solution of complex problems. Some of the knowledge will be at the forefront of the particular subject of study.
Engineering analysis	C2	Analyze complex problems to reach substantiated conclusions using first principles of mathematics, statistics, natural science, and engineering principles.
	C3	Select and apply appropriate computational and analytical techniques to model complex problems, recognizing the limitations of the techniques employed.
	C4	Select and evaluate technical literature and other sources of information to address complex problems.
Design and innovation	C5	Design solutions for complex problems that meet a combination of societal, user, business, and customer needs as appropriate. This will involve consideration of applicable health and safety, diversity, inclusion, cultural, societal, environmental and commercial matters, codes of practice, and industry standards.
	C6	Apply an integrated or systems approach to the solution of complex problems.
The engineer and society	C7	Evaluate the environmental and societal impact of solutions to complex problems and minimize adverse impacts.
	C8	Identify and analyze ethical concerns and make reasoned ethical choices informed by professional codes of conduct.
	C9	Use a risk management process to identify, evaluate, and mitigate risks (the effects of uncertainty) associated with a particular project or activity.
	C10	Adopt a holistic and proportionate approach to the mitigation of security risks.
	C11	Adopt an inclusive approach to engineering practice and recognize the responsibilities, benefits, and importance of supporting e-quality, diversity, and inclusion.
Engineering practice	C12	Use practical laboratory and workshop skills to investigate complex problems.
	C13	Select and apply appropriate materials, equipment, engineering technologies and processes, recognizing their limitations.
	C14	Discuss the role of quality management systems and continuous improvement in the context of complex problems.
	C15	Apply knowledge of engineering management principles, commercial context, project and change management, and relevant legal matters including intellectual property rights.
	C16	Function effectively as an individual and as a member or leader of a team.
	C17	Communicate effectively on complex engineering matters with technical and nontechnical audiences.
	C18	Plan and record self-learning and development as the foundation for lifelong learning/CPD.

https://doi.org/10.1515/9783111206868-006

Appendix B: Banks of Questions

Appendix B.1: Chapter 2 Bank of Questions

Q1. Project Management is something new emerged during the 1980s of the century.
A) _TRUE_ B) _FALSE_
ANSWER: B

Q2. CPM stands for “Critical Path Method.”
A) _TRUE_ B) _FALSE_
ANSWER: A

Q3. PERT stands for “Program Evaluation and Review Technique.”
A) _TRUE_ B) _FALSE_
ANSWER: A

Q4. During the 1950s, Project Management started to develop mainly from
A) The Civil and Construction Engineering Projects.
B) The Electrical and Computer Engineering Projects.
C) The Mechanical and Aviation Engineering Projects.
D) The Bio-Medical Engineering Projects.
ANSWER: A

Q5. Software Project Management is a discipline of:
A) The Computing and IT.
B) The Software Engineering.
C) The Project Management and that is specific to Software Engineering Management and Software Projects.
D) The Mechanical and Aviation Engineering.
ANSWER: C

Q6. “SMART Objectives” refers to the:
A) Objectives identified at the beginning of any Software Project.
B) Software Engineering Objectives.
C) Project Management Objectives.
D) Objectives that are Specific, Measurable, Achievable/Attainable, Realistic, and Time-Bound.
ANSWER: D

https://doi.org/10.1515/9783111206868-007

Q7. Effective Application of Project Management in Software Projects can lead to:
A) Meeting Business Objectives; Satisfying Shareholders Expectations; Satisfactory Managing Change Requests; Managing and Utilizing Resources; and Handling Problems, Issues, Risks, and Threats.
B) Disturbing the Project Schedule and missing its Deadlines; Disturbing the Project Spendeture and its Allocated Budget; and Failing to meet the Project's predefined Specifications.
C) All Options above are correct.
D) All Options above are Incorrect.
ANSWER: A

Q8. Poor Project Management leads to:
A) Disturbing the Project Schedule and Missing its Deadlines;
B) Disturbing the Project Spendeture and its Allocated Budget; and
C) Failing to meet the Project's predefined Specifications.
D) All of the above options are Correct.
E) None of the above options is Correct.
ANSWER: D

Q9. Responsibility is one of the main concepts involved in the Project Management Code of Ethics and Conduct.
A) _TRUE_ B) _FALSE_
ANSWER: A

Q10. Respect is one of the main concepts involved in the Project Management Code of Ethics and Conduct.
A) _TRUE_ B) _FALSE_
ANSWER: A

Q11. Fairness is one of the main concepts involved in the Project Management Code of Ethics and Conduct.
A) _TRUE_ B) _FALSE_
ANSWER: A

Q12. Honesty is one of the main concepts involved in the Project Management Code of Ethics and Conduct.
A) _TRUE_ B) _FALSE_
ANSWER: A

Q13. A Project can be referred to as a:

A) Temporary Mission that leads to a unique Product, Solution, and/or Service with specific start and end Dates, and within predefined Schedule, Budget, and Specifications.

B) Group of Individuals assembled to carry-on a set of Tasks and Activities related to predefined Objectives and Requirements that should be accomplished within a pre-estimated Time, without exceeding a pre-estimated Budget, and while its Outcomes satisfy a predefined set of Specifications.

C) Both Options above are Correct.

D) Both Options above are Incorrect.

ANSWER: C

Q14. The main Challenge is to achieve the Goals of that Project given the Constraints imposed on the Project by the Customer.

A) _TRUE_ B) _FALSE_

ANSWER: A

Q15. "Utilizing Technological Advancements" is among the main factors that can lead to initiating new Projects.

A) _TRUE_ B) _FALSE_

ANSWER: A

Q16. "Technological Advancements" includes advancements in CPU Speed and Capacity, and Memory Storage and Capacity.

A) _TRUE_ B) _FALSE_

ANSWER: A

Q17. A firm may initiate a new Project to create a new Computer and/or Software System using such advanced Components to build that new System, improve/fix one of their existing Computer and/or Software Systems, and/or implement a new or modify an existing Business and/or Technological Strategy.

A) _TRUE_ B) _FALSE_

ANSWER: A

Q18. "Remaining at the Competitive Edge" is among the main factors that can lead to initiating new Projects.

A) _TRUE_ B) _FALSE_

ANSWER: A

Q19. "Remaining at the Competitive Edge" indicates that if a competing Product's Price is lowered, then an affected firm may initiate a new Project to implement a new or modify an existing Business and/or Technological Strategy such that they lower the Cost of their Production in order to continue having their Product at a Competitive Edge in the Market.
A) _TRUE_ B) _FALSE_
ANSWER: A

Q20. "Responding to Changes in the Market Demand" is among the main factors that can lead to initiating new Projects.
A) _TRUE_ B) _FALSE_
ANSWER: A

Q21. "Responding to Changes in the Market Demand" indicates that if a demand arises for a new Product, then a concerned firm may initiate a new Project to produce that new Product.
A) _TRUE_ B) _FALSE_
ANSWER: A

Q22. "Responding to Customers' Requests" is among the main factors that can lead to initiating new Projects.
A) _TRUE_ B) _FALSE_
ANSWER: A

Q23. "Responding to Customers' Requests" indicates that if a Customer expresses their need for a new Product, or for changing an existing one, then a concerned firm may initiate a new Project in accordance with the Specifications and Requirements requested by that Customer.
A) _TRUE_ B) _FALSE_
ANSWER: A

Q24. A Program is a Group of Projects and probably Subsidiary Programs and Activities managed together by the same entity.
A) _TRUE_ B) _FALSE_
ANSWER: A

Q25. An Example of a Program of Projects would be the Hospital Program of Projects at JUST, which includes all Projects running for that hospital such as the hospital's building maintenance Project, and the hospital's Information Systems and Infrastructure Upgrade Projects.
A) _TRUE_ B) _FALSE_
ANSWER: A

Q26. An Organizational Portfolio is a Group of Projects, Programs, Subsidiary Portfolios, and other Operational Activities, which is managed by a Project Management Office (PMO) leveled at the Top-Management of that Organization.
A) _TRUE_ B) _FALSE_
ANSWER: A

Q27. Program Management and Portfolio Management are different from Project Management in terms of their Objectives, Focus, Activities, life cycles, and Benefits. However, Portfolios, Programs, Projects, and Operations share the same Stakeholders, Human Resources, and Physical Resources.
A) _TRUE_ B) _FALSE_
ANSWER: A

Q28. A Portfolio's Scope of Work composes the Scopes of Work of its Integral Projects, Subsidiary Programs, and Operational Activities.
A) _TRUE_ B) _FALSE_
ANSWER: B

Q29. A Program's Scope of Work composes the Scopes of Work of its Integral Projects, Subsidiary Programs, and Operational Activities.
A) _TRUE_ B) _FALSE_
ANSWER: A

Q30. A Portfolio's Scope of Work composes the Scopes of Work of its Integral Programs, Subsidiary Portfolios, Projects, and Operational Activities. It composes the overall Scope of the concerned Organization, which must be aligned with its Strategic Objectives.
A) _TRUE_ B) _FALSE_
ANSWER: A

Q31. A . . . . Scope of Work composes the Objectives and a high level Description of the Requirements of that Project along with a Description of the Tasks and Activities that will be carried on throughout the life cycle of that Project.
A) Project's B) Program's C) Portfolio's D) Super Project's
ANSWER: A

Q32. A . . . . is used to transform the High-Level Information of the Project Objectives, Scope of Work, and Requirements into a set of Detailed Action Plans for managing, executing, Monitoring, and Controlling that Project throughout its life cycle.
A) Project Planning Method
B) Program Planning Method
C) Portfolio Planning Method
D) Super Project Planning Method
ANSWER: A

Q33. A . . . . is used to guide and monitor the progress of the Plans of its Integral Projects, Subsidiary Programs, and Operational Activities and to identify and watch their Inter-Dependencies.
A) Project Planning Method
B) Program Planning Method
C) Portfolio Planning Method
D) Super Project Planning Method
ANSWER: B

Q34. A . . . . is used to guide and monitor the progress of the Plans of its Integral Programs, Subsidiary Portfolios, and Operational Activities and to identify and watch for their Inter-Dependencies.
A) Project Planning Method
B) Program Planning Method
C) Portfolio Planning Method
D) Super Project Planning Method
ANSWER: C

Q35. The Purpose of Managing a . . . . is to ensure that its Team Members are executing the Project Plan carefully such that the . . . . Objectives and predefined Requirements are accomplished as expected and as they were planned for.
A) Project/Project
B) Project/Program
C) Program/Portfolio
D) Portfolio/Project
ANSWER: A

Q36. The Purpose of Managing a . . . . is to coordinate the Activities of its Integral Projects, Subsidiary Programs, and Operational Activities.
A) Project B) Program C) Portfolio D) Super Project
ANSWER: B

Q37. The Purpose of Managing a . . . . is to coordinate and supervise its Integral Programs, Projects, and Operational Activities.
A) Project B) Program C) Portfolio D) Super Project
ANSWER: C

Q38. In a . . . . managing CHANGE-REQUESTS is used to keep changes Controlled by implementing the necessary Change Processes.
A) Project B) Program C) Portfolio D) Super Project
ANSWER: A

Q39. .... Managers must accept CHANGE-REQUESTS as necessary to facilitate the Delivery of the Components of their ....
A) Project/Projects
B) Program/Programs
C) Portfolio/Portfolios
D) Super Project/Projects.
ANSWER: B

Q40. .... Managers must conduct a series of Quality Assurance and Quality Control Activities throughout the Execution of their ....
A) Project/Projects
B) Program/Programs
C) Portfolio/Portfolios
D) Super Project/Projects.
ANSWER: A

Q41. .... Managers must monitor the Progress of the Components of their Programs. This is to ensure the Satisfaction of the Goals, Schedules, Budget, and Benefits of their ....
A) Project/Projects
B) Program/Programs
C) Portfolio/Portfolios
D) Super Project/Projects.
ANSWER: B

Q42. .... Managers must monitor any Strategic Change across the Components of their ....
A) Project/Projects
B) Program/Programs
C) Portfolio/Portfolios
D) Super Project/Projects.
ANSWER: C

Q43. A . . . . is SUCCESSFUL if and only completed . . . . (as per the predetermined Schedule), ... (Spendeture should not exceed the allocated Budget), ... (Product's Quality is at the expected level), and . . . . (Preserving the Customer's Satisfaction with the delivered Product is at least as expected).
A) Project/ON-TIME/WITHIN-BUDGET/WHILE-MEETING-SPECS/WHILE-ENSURINTG-THE-CUSTOMER-SATISFACTION
B) Program/ON-TIME/WITHIN-BUDGET/WHILE-MEETING-SPECS/WHILE-ENSURINTG-THE-CUSTOMER-SATISFACTION
C) Project/WHILE-MEETING-SPECS/WITHIN-BUDGET/ON-TIME/WHILE-ENSURINTG-THE-CUSTOMER-SATISFACTION
D) Portfolio/ON-TIME/WHILE-MEETING-SPECS/WITHIN-BUDGET/WHILE-ENSURINTG-THE-CUSTOMER-SATISFACTION
ANSWER: A

Q44. A . . . . is SUCCESSFUL if and only if all of its Integral Components are SUCCESSFUL.
A) Project
B) Program
C) Portfolio
D) Both Options of B and C are Correct.
E) All Options above are Correct.
F) All Options above are Incorrect.
ANSWER: D

Q45. Projects are not influenced by the surrounding Environments.
A) _TRUE_ B) _FALSE_
ANSWER: B

Q46. The main Environmental Factors that can influence Projects include:
A) The Enterprise Environmental Factors (EEFs)
B) The Organizational Process Assets Factors (OPAFs)
C) The Organizational Systems-Related Factors
D) The Organizational Governance Frameworks-Related Factors
E) All of the above are Influencing Environmental Factors
F) All of the above are not Influencing Environmental Factors.
ANSWER: E

Q47. Internal to the Organization Enterprise Environmental Factors (EEFs) include the Organizational Culture, Structure, and Governance (e.g., Project Vision, Mission, Values, Culture, Leadership Style, and Code of Ethics); Geographic Distribution of Facilities and Resources (e.g., Project Locations, Virtual Teams, Shared Resources, and Cloud-Computing); Infrastructure; and Resources Capabilities and Availabilities.
A) _TRUE_ B) _FALSE_
ANSWER: A

Q48. External to the Organization Enterprise Environmental Factors (EEFs) include the Marketplace Competition, Recognition, and Trademarks; Legal Restrictions (e.g., Local Regulations related to Security, Data, Business Conduct, and Employment); and Governmental and Industrial Standards.
A) _TRUE_ B) _FALSE_
ANSWER: A

Q49. Organizational Process Assets Factors (OPAFs) are Internal to the Organization itself. They come from the Organization's Portfolios, Programs, Projects, and/or Operations.
A) _TRUE_ B) _FALSE_
ANSWER: A

Q50. The first set of the Organizational Process Assets Factors (OPAFs) includes the Organizational Policies for managing:
A) Human Resources, Health and Safety, Security and Confidentiality, Quality, and Environment
B) Organizational Procedures and Methods for Managing Projects, Estimation and Estimation Metrics, Audits, and Checklists
C) Organizational Templates for Project Management Plans, Project Documents, Reports, Contracts, and Risk Management
D) Organizational Procedures for Conducting Change Control
E) All Options above are Correct
F) All Options above are Incorrect
ANSWER: E

Q51. The second set of the Organizational Process Assets Factors (OPAFs) includes the Organizational Knowledge Repositories containing the Configuration Management Data, Financial Data, Historical Data, Issues Management Data.
A) _TRUE_ B) _FALSE_
ANSWER: A

Q52. Organizational Systems-Related Factors (OSFs) impact the . . . . to act within the Organizational System. The OSFs include the People's Power, Influence, Interests, Competencies, and Political Capabilities.
A) People's Ability
B) Project Team Ability
C) Both Options above are Correct
D) Both Options above are Incorrect
ANSWER: C

Q53. . . . . are used to manage the Authority within the Organization. They include the Organizational Rules, Processes, Policies, Procedures, and Systems. OGFFs influence Projects in how to set and achieve the Organization's Objectives in Alignment with their Organizational Objectives; monitor and assess the Organization's potential Risks; and optimize the Organization's Performance.
A) Organizational Governance Factory-Related Factors (OGFFs)
B) Organizational Governance Firewall-Related Factors (OGFFs)
C) Organizational Governance Firm-Related Factors (OGFFs)
D) Organizational Governance Frameworks-Related Factors (OGFFs)
ANSWER: D

Q54. A . . . . is assigned by his/her Organization to lead a Team that will be responsible for carrying-on a Project to achieve its Objectives and Requirements.
A) Project Manager
B) Functional Manager
C) Operational Manager
D) Portfolio Manager
ANSWER: A

Q55. The . . . . Role. A Functional Manager manages a Functional or a Business Unit.
A) Project Manager
B) Functional Manager
C) Operational Manager
D) Portfolio Manager
ANSWER: B

Q56. The . . . . Role. An Operational Manager is responsible for ensuring that Business Operations go Smooth, Effective, and Efficient.
A) Project Manager
B) Functional Manager
C) Operational Manager
D) Portfolio Manager
ANSWER: C

Q57. A .... must balance the Constraints on his/her Project using the available Resources and to use his/her Soft Interpersonal Skills to communicate with the Stakeholders and/or balance their Goals.

A) Project Manager

B) Functional Manager

C) Program Manager

D) Portfolio Manager

ANSWER: A

Q58. The PMI Talent Triangle composes three sets of Skills that Project Managers must have. These sets are:

A) Technical Project Management Knowledge and Skills, Leadership Knowledge and Skills, and Strategic and Business Management Knowledge and Skills.

B) Innovative Project Management Knowledge and Skills, Innovative Leadership Knowledge and Skills, and Strategic and Business Management Knowledge and Skills.

C) Creative Project Management Knowledge and Skills, Creative Leadership Knowledge and Skills, and Strategic and Business Management Knowledge and Skills.

D) Exceptional Project Management Knowledge and Skills, Creative Leadership Knowledge and Skills, and Strategic and Business Management Knowledge and Skills.

ANSWER: A

Q59. The .... implies that a Software Project Manager must be able to handle the Project Complexities, to analyze the Project Artifacts; to set the Project Goals, Scope, and Implementation Plan; to plan for the Project Budget; to staff for the Project Team; to Control and solve Problems; and to take effective actions.

A) Ability to organize and manage.

B) Ability to organize and lead

C) Ability to be creative and innovative

D) Ability to foresight the future

ANSWER: A

Q60. The .... implies that a Software Project Manager must be able to Handle CHANGE-REQUESTS; to set Directions; to provide Vision; to delegate Authority whenever there is a need for that; to be Positive and Energetic, Flexible, Creative, Patient, and Persistent throughout the Project life cycle; to align People; and to take Meaningful Actions.
A) Ability to cope and lead.
B) Ability to organize and lead
C) Ability to be creative and innovative
D) Ability to foresight the future
ANSWER: A

Q61. The .... implies that a Software Project Manager must be able to effectively communicate, negotiate, and make Decisions; and to build Teams and deal with People.
A) Ability to Communicate and Negotiate Effectively
B) Ability to organize and lead
C) Ability to be creative and innovative
D) Ability to foresight the future
ANSWER: A

Q62. The .... imply that a Software Project Manager must be able to understand his/her own Organization as well as the Customer's Organization; to know enough about the targeted Product and/or Solution; to be well-experienced in the Profession of Project Management and its relevant Technological Tools; and to be comfortable with handling CHANGE-REQUESTS.
A) Ability to Communicate and Negotiate Effectively
B) Ability to organize and lead
C) Hard and Soft Skills
D) Ability to foresight the future
ANSWER: C

Q63. A Project Manager, as a leader must be equipped with a set of Qualities and Skills such as ....
A) Being a Visionary, Optimistic, Positive, Collaborative, Respectful, Courteous, Friendly, Kind, Honest, Trustworthy, Loyal, Ethical, Life-Long Learner, Results/Goals/Actions-oriented, Culturally Sensitive, Courageous, Problem-Solver, Service-oriented, and Decisive.
B) Being able to develop Personal and Professional Networks, manage Relationships, and resolve Conflicts by building Trust and satisfying Concerns.
C) Being a CRITICAL-Thinker and using Analytical-Methods to conclude Decisions.
D) All Options above are Correct.
ANSWER: D

Q64. Radaideh's Classification of the essential Skills for Software Project Managers includes but not limited to:

A) Project Management Skills Class that implies that a Software Project Manager must be well knowledgeable and experienced in the various Project Management Principles, Methods, Processes, and Tools.

B) Software Project Domain-Specific Skills Class that implies that a Software Project Manager must be well experienced in the domain of the Project he/she will be managing.

C) Sociological Skills Class that implies that a Software Project Manager must be well comfortable with the surrounding Humans Sociological Interactions, Behaviors, and Actions.

D) Psychological Skills Class that implies that a Software Project Manager must be well comfortable with the surrounding Humans Psychological Views, and Actions.

E) All Options above are Correct.

ANSWER: E

Q65. Authorities and People from three layers can influence Project Managers. In addition, Project Managers can influence these Authorities and People. These layers are:

A) Layer #1 that includes the Project Team Members, the Peer-Managers, and the Resource-Managers. Layer #2 that includes the Project Sponsors, Governing Bodies, Steering Committee, and the PMO office. Layer #3 that includes the Project Stakeholders, Suppliers, Customers, and Users.

B) Layer #1 that includes the Project Sponsors, Governing Bodies, Steering Committee, and the PMO office. Layer #2 that includes the Project Team Members, the Peer-Managers, and the Resource-Managers. Layer #3 that includes the Project Stakeholders, Suppliers, Customers, and Users.

C) Layer #1 that includes the Project Team Members, the Peer-Managers, and the Resource-Managers. Layer #2 that includes the Project Stakeholders, Suppliers, Customers, and Users. Layer #3 that includes the Project Sponsors, Governing Bodies, Steering Committee, and the PMO office.

D) All Options above are Incorrect.

ANSWER: A

Q66. The Leadership Styles can change in accordance with several Factors such as:

A) The Leader-Characteristics Factor that refers to the Leader's Attitude, Mood, Needs, Values, and Ethics; the Team-Members-Characteristics Factor that refers to the Team-Members' Attitudes, Moods, Needs, Values, and Ethics; the Organizational-Characteristics Factor that refers to the Organization's Purpose, Structure, and Type of Work Performed; and the Environmental-Characteristics Factor that refers to the Work Environment's Social Situation, Economic State, and Political Elements.

B) The Team-Members-Characteristics Factor that refers to the Leader's Attitude, Mood, Needs, Values, and Ethics; the Leader-Characteristics Factor that refers to the Team-Members' Attitudes, Moods, Needs, Values, and Ethics; the Organizational-Characteristics Factor that refers to the Organization's Purpose, Structure, and Type of Work-Performed; and the Environmental-Characteristics Factor that refers to the Work Environment's Social Situation, Economic State, and Political Elements.

C) All Options above are Correct.

D) All Options above are Incorrect.

ANSWER: A

Q67. With the .... Leadership Style, a Project Manager allows his/her Team to make their own Decisions and establish their own Goals.

A) Transactional-Leadership-Style

B) Laissez-Faire-Leadership-Style

C) Servant-Leader-Leadership-Style

D) Transformational-Leadership-Style

ANSWER: B

Q68. With the .... Leadership Style, a Project Manager focuses on the Project Goals, Feedback, and Accomplishments.

A) Transactional-Leadership-Style

B) Laissez-Faire-Leadership-Style

C) Servant-Leader-Leadership-Style

D) Transformational-Leadership-Style

ANSWER: A

Q69. With the .... Leadership Style, a Project Manager commits to serve other People by focusing on their Growth, Learning, Development, and Autonomy.

A) Transactional-Leadership-Style

B) Laissez-Faire-Leadership-Style

C) Servant-Leader-Leadership-Style

D) Transformational-Leadership-Style

ANSWER: C

Q70. With the .... Leadership Style, a Project Manager empowers his/her Team through Idealized Behaviors, and Inspirational Motivation and Encouragement.
A) Transactional-Leadership-Style
B) Laissez-Faire-Leadership-Style
C) Servant-Leader-Leadership-Style
D) Transformational-Leadership-Style
ANSWER: D

Q71. With the .... Leadership Style, a Project Manager inspires his/her Teams by being High-Energetic, Self-Confident, and Enthusiastic.
A) Charismatic-Leadership-Style
B) Laissez-Faire-Leadership-Style
C) Servant-Leader-Leadership-Style
D) Transformational-Leadership-Style
ANSWER: A

Q72. With the .... Style of Leadership, the Project Manager should possess a combination of the Transactional, Transformational, and Charismatic Leadership Styles.
A) Charismatic-Leadership-Style
B) Interactional-Leadership-Style
C) Servant-Leader-Leadership-Style
D) Transformational-Leadership-Style
ANSWER: B

Q73. The .... Leadership Style is not different from Dictatorship.
A) Autocratic
B) Bureaucratic
C) Charismatic
D) Democratic
ANSWER: A

Q74. An .... Project Manager handles Projects with Customers that have .... Work Environments such as Military and Security Services.
A) Autocratic/Autocratic
B) Autocratic/Charismatic
C) Autocratic/Democratic
D) Autocratic/Bureaucratic
ANSWER: A

Q75. An . . . . Customer would not give the Project Manager enough room for Discussion and Negotiation. That Customer simply issues instructions to the Project Manager who in turn obeys and executes these instructions by simply acting . . . . with his/her Team.
A) Autocratic/Autocratically
B) Autocratic/Charismatically
C) Autocratic/Democratically
D) Autocratic/Bureaucratically
ANSWER: A

Q76. The . . . . Leadership Style implies that a Software Project Manager and his/her Team are supposed to follow a set of predefined Processes, Procedures, Guidelines, and/or Regulations.
A) Autocratic
B) Charismatic
C) Democratic
D) Bureaucratic
ANSWER: D

Q77. The . . . . Leadership Style implies that a Software Project Manager acts in accordance with the situation he/she is going through. He/she must spread Hope and Enthusiasm among his/her Team; and encourage his/her Team to move forward.
A) Autocratic
B) Charismatic
C) Democratic
D) Bureaucratic
ANSWER: B

Q78. The . . . . Leadership Style implies that all of the Software Project Team Members are encouraged to participate in the Decision-Making Process, and that their Software Project Manager takes the Final Decision based on the Outcomes of the Team's Discussions and Suggestions.
A) Autocratic
B) Charismatic
C) Democratic
D) Bureaucratic
ANSWER: C

Q79. A .... Leader is a Bureaucratic Leader, but he/she uses Technological Systems to reach the details of the needed Bylaws and Regulations. He/she also uses the Decision Support Systems (DSS).
A) Autocratic
B) Charismatic
C) Technocratic
D) Bureaucratic
ANSWER: C

Q80. A (An) .... Project Manager accepts others for what and who they are.
A) Authentic B) Courteous C) Creative D) Innovative
ANSWER: A

Q81. A (An) .... Project Manager applies appropriate Behaviors.
A) Authentic B) Courteous C) Creative D) Innovative
ANSWER: B

Q82. A (An) .... Project Manager thinks abstractly, sees things differently, and innovates.
A) Authentic B) Courteous C) Creative D) Innovative
ANSWER: C

Q83. A (An) .... Project Manager is sensitive to other Cultures including Values, Norms, and Beliefs.
A) Authentic B) Courteous C) Creative D) Cultural
ANSWER: D

Q84. A (An) .... Project Manager deals with Human Intelligence in multi-Aptitudes.
A) Intellectual B) Courteous C) Creative D) Cultural
ANSWER: A

Q85. A (An) .... Project Manager considers Management Practices.
A) Intellectual B) Managerial C) Creative D) Cultural
ANSWER: B

Q86. A (An) .... Project Manager measures the Political Intelligence and makes things happen.
A) Intellectual B) Managerial C) Political D) Cultural
ANSWER: C

Q87. A (An) .... Project Manager is always willing to serve other People.
A) Intellectual
B) Managerial
C) Political
D) Service-oriented
ANSWER: D

Q88. A (An) .... Project Manager understands and manages People.
A) Social B) Managerial C) Political D) Service-oriented
ANSWER: A

Q89. A Customer develops an RFP document to specify and declare their interest in acquiring a Product, a Service, a Solution, a System, and/or an Asset. The RFP issuing customer must make it clear to the Service-Providers that their Evaluation and Selection Procedures are well structured.
A) _TRUE_ B) _FALSE_
ANSWER: A

Q90. In an RFP document, the Project Information Section is composed of several sub-sections: The Issuing Firm Label Sub-Section that includes the Issuing Firm's Name, Address, Website, and Phone. The Issuing Firm Background Sub-Section that includes the Issuing Firm's History, Organizational Structure, Activities, and Services. The Project Information Sub-Section that includes the Project Title, Issuance Date, Submission Due Date, Questions Time-Window, and Business Domain. The Targeted Categories of Service-Providers Sub-Section that includes the list of Categories and Types of Service-Providers targeted by the RFP. The Intuitive Budget and Budget Constraints Sub-Section that includes the Customer's Suggestions in respect to the Project Budget, and the list of Constraints that the Customer is imposing on the Project Budget. The Focal-Point of Contact Details Sub-Section that includes the Name, Position, Email, Phone, and Office Address of the Project Focal-Point of Contact.
A) _TRUE_ B) _FALSE_
ANSWER: A

Q91. In an RFP document, the Project Description Section presents an Executive Summary of the Project Context.
A) _TRUE_ B) _FALSE_
ANSWER: A

Q92. In an RFP document, the Project Goals and Objectives Section presents the Project Goals and Objectives. They must be well articulated S.M.A.R.T. Objectives (e.g., Specific, Measurable, Achievable, Realistic, and Time-bounded).
A) _TRUE_ B) _FALSE_
ANSWER: A

Q93. In an RFP document, the Project Scope of Work Section includes the Project Formal Statement of Work that the to-be-selected Service-Provider must honor.
A) _TRUE_ B) _FALSE_
ANSWER: A

Q94. In an RFP document, the Project Requirements, Terms and Conditions, and Constraints Section includes: The Business and Technical Requirements that should be satisfied in the Project Outcomes (e.g., Product, Solution, and/or Service). The Project Terms and Conditions, Constraints, Challenges, Risks, and Threats. A Preliminary Architecture and Design of the Targeted Outcome. The Project Targeted Outcomes, Deliverables, and Milestones.
A) _TRUE_ B) _FALSE_
ANSWER: A

Q95. In an RFP document, the Project Privacy, Confidentiality, Non-DisClosure, and Intellectual Property Section presents: (The Rules and Guidelines that must be honored by the Project Parties in respect to the Privacy, Confidentiality, and Non-DisClosure of the Project Information throughout its life cycle. (The Ownership of any Intellectual Properties (IPs) developed during the Project.
A) _TRUE_ B) _FALSE_
ANSWER: A

Q96. In an RFP document, the Proposals Submission Guidelines Section presents a set of clear Guidelines and Requirements that relate to the Proposals Submission Process.
A) _TRUE_ B) _FALSE_
ANSWER: A

Q97. In an RFP document, the Proposals Evaluation CRITERIA Section explains the CRITERIA used to evaluate and assess the to-be-received Proposals.
A) _TRUE_ B) _FALSE_
ANSWER: A

Q98. In an RFP document, the Required Service-Provider Details Section explains to the Service-Provider the details that they must provide in their submitted Proposal. This includes at least the Service-Provider's Size, History, and Domains of Activities; and Profile of Projects, SUCCESS Stories and References, and Focal-Point of Contact (e.g., Name, Email, Phone, and Office Address).
A) _TRUE_ B) _FALSE_
ANSWER: A

Q99. In an RFP document, the Service-Providers Qualifying CRITERIA Section. This section presents the CRITERIA that is to-be-used for qualifying the Service-Providers.
A) _TRUE_ B) _FALSE_
ANSWER: A

Q100. The . . . . FORM is sent by the Customer to the pre-identified Service-Providers to request Information about their Products and Services. Such Information benefits the RFP Development Process.
A) Request for Quotation (RFQ)
B) Request for Information (RFI)
C) Request for Tender (RFT)
D) Invitation to Tender (ITT)
ANSWER: B

Q101. The . . . . FORM is used to only ask for Prices of certain Products (e.g., Equipment, Computers, Laptops, and Software Solutions).
A) Request for Quotation (RFQ)
B) Request for Information (RFI)
C) Request for Tender (RFT)
D) Invitation to Tender (ITT)
ANSWER: A

Q102. The . . . . FORM is similar to the Request for Proposals (RFP), but Governmental Organizations use it.
A) Request for Quotation (RFQ)
B) Request for Information (RFI)
C) Request for Tender (RFT) or Invitation to Tender (ITT)
D) Request for Qualifications or Pre-Qualification Questionnaire (PQQ).
ANSWER: C

Q103. The . . . . FORM is used to build consortiums to jointly carry-on Large-Scale Projects.
A) Request for Association, Partnership, and/or Alliance
B) Request for Information (RFI)
C) Request for Tender (RFT) or Invitation to Tender (ITT)
D) Request for Qualifications or Pre-Qualification Questionnaire (PQQ).
ANSWER: A

Q104. A Service-provider creates a Project Charter based on their Initial Review of the RFP.
A) _TRUE_ B) _FALSE_
ANSWER: A

Q105. A Project Charter documents is composed of several sections such as the Project Vision Section, the Project Objectives Section , the Project Scope of Work Section, the Project Deliverables and Milestones Section, the Project Structure Section, the Project Proposed Implementation Plan Section , the Recommendations Section, the Approval/Disapproval Section, and the Signatory Section.
A) _TRUE_ B) _FALSE_
ANSWER: A

Q106. During the . . . . Phase, the Project Owner identifies and analyzes the Business Requirements; reviews the Organization's current Operations; conducts a High-Level Benefits Analysis; conducts a High-Level Budgeting; identifies the Stakeholders and their Expectations; develops an RFP; and shares the RFP with the Service-Providers.
A) Initiation
B) Planning
C) Execution
D) Monitoring and Control
E) Closure
ANSWER: A

Q107. During the . . . . Phase, Each Service-Provider develops a Project Charter that summarizes the Project Opportunity. This includes the Project Requirements, Terms and Conditions, Cost, Tasks, Deliverables, Milestones, and Schedules; and conducts a SWOT Analysis to identify the Strengths and Weakness that would enable/disable the Service-Provider to carry-on the Project; and to identify the Opportunities and Threats associated with the Project.
A) Initiation
B) Planning
C) Execution
D) Monitoring and Control
E) Closure
ANSWER: A

Q108. The groups of the Project Planning Implicit Questions are . . . .
A) WHAT Questions; WHY Questions; WHEN Questions; WHO Questions; HOW Questions; and WHERE Questions
B) WHAT-IS Questions; WHY-THIS Questions; WHEN Questions; WHO Questions; HOW Questions; and WHERE Questions
C) WHAT Questions; WHY Questions; WHEN Questions; WHO Questions; and HOW Questions
D) WHY Questions; WHEN Questions; WHO Questions; HOW Questions; and WHERE Questions
ANSWER: A

Q109. During the Planning Phase, the Customer reviews and evaluates the Proposals received from the Service-Providers and then creates a short-list of the Service-Providers of the best Proposals (e.g., 3–5 Proposals).
A) _TRUE_ B) _FALSE_
ANSWER: A

Q110. During the Planning Phase, the Customer then invites each of the short-listed Service-Providers to present and discuss their Proposal.
A) _TRUE_ B) _FALSE_
ANSWER: A

Q111. During the Planning Phase, each Service-Provider would then deliver a Presentation on their Proposal to the Customer and negotiate it with them.
A) _TRUE_ B) _FALSE_
ANSWER: A

Q112. During the Planning Phase, the Proposal Negotiation covers (the Project Scope of Work and Requirements, the Work Breakdown Structure (WBS), the Schedule, the Resources Allocation, the Resources Qualifications, the Proposed Solution, the Proposed Development and Management Methods, the Terms and Conditions, and (i) the Proposed Price.

A) _TRUE_ B) _FALSE_

ANSWER: A

Q113. At the end of the Service-Providers' Presentations, the Customer would select the best Proposal and send a Letter of Awarding to their preferred Service-Provider.

A) _TRUE_ B) _FALSE_

ANSWER: A

Q114. At the end of the Planning Phase, the Customer and the selected Service-Provider start the Project Contract Negotiation that involves the same items as in the Proposal Negotiation.

A) _TRUE_ B) _FALSE_

ANSWER: A

Q115. Radaideh's Project Staffing Circles and Options include three circles and three options. In Circle#1 . . . .

A) A Project Manager attempts to hire staff from those who worked with them on previous Projects.

B) If Staffing remains incomplete, then the Project Manager seeks help from his/her colleague Project Managers to hire from those who worked with them on previous Projects.

C) If Staffing is still incomplete, then the Project Manager approaches the Human Resources Department at his/her company to get access to the profiles of all available Technical Staff to fill for as many left Roles as possible.

D) The Project Manager advertises the unfilled Roles searching for hirable independent External Staff Members. However, the Human Resources department should endorse hiring External Staff on full-time bases. Otherwise, the Project Manager will have to consider Option#2.

ANSWER: A

Q116. Radaideh's Project Staffing Circles and Options includes three circles and three options. In Circle#2 . . . .

A) A Project Manager attempts to hire staff from those who worked with them on previous Projects.

B) If Staffing remains incomplete, then the Project Manager seeks help from his/her colleague Project Managers to hire from those who worked with them on previous Projects.

C) If Staffing is still incomplete, then the Project Manager approaches the Human Resources Department at his/her company to get access to the profiles of all available Technical Staff to fill for as many left Roles as possible.

D) The Project Manager advertises the unfilled Roles searching for hirable independent External Staff Members. However, the Human Resources department should endorse hiring External Staff on full-time bases. Otherwise, the Project Manager will have to consider Option#2.

ANSWER: B

Q117. Radaideh's Project Staffing Circles and Options includes three circles and three options. In Circle#3 . . . .

A) A Project Manager attempts to hire staff from those who worked with them on previous Projects.

B) If Staffing remains incomplete, then the Project Manager seeks help from his/her colleague Project Managers to hire from those who worked with them on previous Projects.

C) If Staffing is still incomplete, then the Project Manager approaches the Human Resources Department at his/her company to get access to the profiles of all available Technical Staff to fill for as many left Roles as possible.

D) The Project Manager advertises the unfilled Roles searching for hirable independent External Staff Members. However, the Human Resources department should endorse hiring External Staff on full-time bases. Otherwise, the Project Manager will have to consider Option#2.

ANSWER: C

Q118. Radaideh's Project Staffing Circles and Options include three circles and three options. In Option#1 . . . .

A) A Project Manager attempts to hire staff from those who worked with them on previous Projects.

B) If Staffing remains incomplete, then the Project Manager seeks help from his/her colleague Project Managers to hire from those who worked with them on previous Projects.

C) If Staffing is still incomplete, then the Project Manager approaches the Human Resources Department at his/her company to get access to the profiles of all available Technical Staff to fill for as many left Roles as possible.

D) The Project Manager advertises the unfilled Roles searching for hirable independent External Staff Members. However, the Human Resources department should endorse hiring External Staff on full-time bases. Otherwise, the Project Manager will have to consider Option#2.

ANSWER: D

Q119. Radaideh's Project Staffing Circles and Options include three circles and three options. In Option#2 . . . .

A) A Project Manager attempts to hire staff from those who worked with them on previous Projects.

B) If Staffing remains incomplete, then the Project Manager seeks help from his/her colleague Project Managers to hire from those who worked with them on previous Projects.

C) The Project Manager would offer Contract-Based Employment to those who can't be hired on full-time bases in Option#1.

D) The Project Manager advertises the unfilled Roles searching for hirable independent External Staff Members. However, the Human Resources department should endorse hiring External Staff on full-time bases. Otherwise, the Project Manager will have to consider Option#2.

ANSWER: C

Q120. Radaideh's Project Staffing Circles and Options includes three circles and three options. In Option#3 . . . .

A) A Project Manager attempts to hire staff from those who worked with them on previous Projects.

B) If things do not work fine with Option#2, then the Project Manager approaches External Employment Agencies to get the needed staff.

C) The Project Manager would offer Contract-Based Employment to those who could not be hired on full-time bases in Option#1.

D) The Project Manager advertises the unfilled Roles searching for hirable independent External Staff Members. However, the Human Resources department should endorse hiring External Staff on full-time bases. Otherwise, the Project Manager will have to consider Option#2.

ANSWER: B

Q121. At the beginning of the . . . . Phase, the Project Manager should check the Project Generic Description, the Project Objectives and Anticipated Outcomes, the Project Terms and Conditions, including the Service-Level-Agreement (SLA), the Team Correspondence Process for both of the Horizontal and Vertical Communications, the Team Performance Evaluation Processes, the Contract Management Processes, the sub-Contractors Management Process, the Change Management Process, the Project Accounting Processes, and the Contract Governance and Compliance Process.

A) Initiation

B) Execution

C) Planning

D) Monitoring and Control

ANSWER: B

Q122. During the Monitoring and Control Phase, the ongoing Activities get assessed and measured.

A) _TRUE_ B) _FALSE_

ANSWER: A

Q123. During the Monitoring and Control Phase, a set of Corrective Actions is identified to resolve the Issues observed during the Measurement and Monitoring Activities.

A) _TRUE_ B) _FALSE_

ANSWER: A

Q124. During the Monitoring and Control Phase, a set of Key Performance Indicators System (KPI) is identified for the Project.

A) _TRUE_ B) _FALSE_

ANSWER: A

Q125. During the Monitoring and Control Phase, Functional Testing is conducted to verify the Correctness and Preciseness of the various Functions. Also, Non-Functional Testing is conducted to test the Performance, Availability, Reliability, Scalability, Maintainability, and Testability of the targeted Solution.
A) _TRUE_ B) _FALSE_
ANSWER: A

Q126. The Project .... Phase involves a Post-Implementation Review, Closing the Project Contacts; and Closing the Project.
A) Initiation B) Planning C) Execution D) Closure
ANSWER: D

Q127. A .... Contract can be with a .... or a ....
A) Sales/Sales representative/Sales Agency
B) Purchasing/Purchasing representative/Purchasing Agency
C) Performance/Performance representative/Performance Agency
D) All Options above are Incorrect
ANSWER: A

Q128. By signing a .... Contract, a .... agrees to promote and sell the Products to the Customers, and then he/she will be paid a base Salary plus Sales Commission.
A) Sales Representative/Sales Representative
B) Buying Representative/Buying Representative
C) Both Options above are Correct
D) All Options above are Incorrect
ANSWER: A

Q129. By signing a (an) .... Contract, a (an) .... agrees to promote and sell a predefined list of Products on behalf of the Seller to Customers in a pre-determined region and for predefined Prices and then gets paid a percentage of these Sales.
A) Agent Contract Agreement/Agent
B) Sales Contract Agreement/Seller
C) Both Options above are Correct
D) All Options above are Incorrect
ANSWER: A

Q130. A .... is a Contract between a Customer and a Service-Provider that is promising to sell Products, Solutions, and/or Services in accordance with a set of Terms and Conditions.
A) Sales Contract
B) Purchasing Contract
C) Performance-based Contract
D) All Options above are Incorrect
ANSWER: B

Q131. With a Performance-based Contract, which of the following statements is correct?
A) If the Performance is in Full-Compliance with the Customer's predefined Performance Expectations, then the Service-Provider will get paid the same amount of Money as agreed-upon in the Project Contract.
B) If the Performance is behind the Customer's Expectations, then the Outcomes are rejected by the Customer.
C) If the Performance degree exceeds the Customer's Expectations, then the Service-Provider would be paid some bonus Money on the top of the agreed-upon amount as in the Project Contract.
D) All Options above are Correct
ANSWER: D

Q132. A .... is a Contract that establishes the Terms and Conditions for an Agreement between at least two Parties.
A) Sales Agreement
B) Partnership Agreement
C) Cooperation Agreement
D) Memorandum of Understanding Agreement
ANSWER: B

Q133. A .... is an individual or a Business that is hired by a Prime-Contractor to carry-on part of the work that is defined in their Contract with their Customer.
A) Sub-Contractor
B) Low-class Contractor
C) Important Contractor
D) Negligible Contract
ANSWER: A

Q134. The Incentive to hire a .... is either to reduce the Project Cost, improve its Quality, or mitigate its Risks.
A) Sub-Contractor
B) Low-class Contractor
C) Important Contractor
D) Negligible Contract
ANSWER: A

Q135. A .... signs a Sub-Contract with the Prime-Contractor to supply Material or to execute part of the Prime-Contract. Sub-Contractors of this Category are typically not specialized, but can work across many areas
A) Domestic Sub-Contractor
B) Nominated Sub-Contractor
C) Named Sub-Contractor
D) All Options above are Correct
ANSWER: A

Q136. A .... is directly Contracted. Sub-Contractors of this Category are nominated to join the Project Team because of their specialization related to the part of the Project that they are needed to carry on.
A) Domestic Sub-Contractor
B) Nominated Sub-Contractor
C) Named Sub-Contractor
D) All Options above are Correct
ANSWER: B

Q137. A .... signs a Contract with the Prime-Contractor to supply Materials or to execute part of the work of the Prime-Contract. Sub-Contractors of this Category are listed on the Prime-Contractor's lists of pre-qualified Sub-Contractors.
A) Domestic Sub-Contractor
B) Nominated Sub-Contractor
C) Named Sub-Contractor
D) All Options above are Correct
ANSWER: C

Q138. During the .... Contract Management Phase, the Project Contract is planned for, articulated, and then endorsed and signed by the Parties.
A) Pre-Contract
B) Contract Execution
C) Contract Post-Execution
D) All Options above are Correct
ANSWER: A

Q139. In Terms of Project Management, the . . . . Contract Management Phase is lined up with the Project Management Phases of Initiation, and Planning.
A) Pre-Contract
B) Contract Execution
C) Contract Post-Execution
D) All Options above are Correct
ANSWER: A

Q140. During the . . . . Contract Management Phase, the Project Execution, Monitoring and Control, and Closure Phases' Activities are carried on.
A) Pre-Contract
B) Contract Execution
C) Contract Post-Execution
D) All Options above are Correct
ANSWER: B

Q141. In Terms of Project Management, the . . . . Contract Management Phase is lined up with the Project Management Phases of Execution, Monitoring and Control, and Closure.
A) Pre-Contract
B) Contract Execution
C) Contract Post-Execution
D) All Options above are Correct
ANSWER: B

Q142. During the. . . . . Contract Management Phase, the Contract is checked to ensure its Compliance with its Governing Standards and to ensure having adequate Support to the delivered Products Solutions and/or Services.
A) Pre-Contract
B) Contract Execution
C) Contract Post-Execution
D) All Options above are Correct
ANSWER: C

Q143. In Terms of Project Management, the. . . . . Contract Management Phase is lined up with the Project Management Phases of Monitoring and Control, and Closure.
A) Pre-Contract
B) Contract Execution
C) Contract Post-Execution
D) All Options above are Correct
ANSWER: C

Q144. If a Change Request is received verbally from the Customer, then a verbal or written response should be sent to the Customer regardless whether the Service-Provider accepted it or turned it down. If accepted, then a verbal Agreement or a Contract Amendment must take place to legitimize the Endorsement of that Change Request.

A) _TRUE_ B) _FALSE_

ANSWER: A

Q145. If a Change Request is received in writing from the Customer, then a written response should be sent to the Customer regardless whether the Service-Provider accepted it or turned it down. If accepted, then a written Amendment to the Contract must take place to legitimize the Endorsement of that Change Request.

A) _TRUE_ B) _FALSE_

ANSWER: A

Q146. If a Change Request is received from the Project Team, then such CHANGE-REQUESTS can slightly impact the Project Schedule and/or Specifications.

A) _TRUE_ B) _FALSE_

ANSWER: A

Q147. If a Change Request is received from the Project Accountant, then such CHANGE-REQUESTS can slightly impact the Project Budget distribution across the Project various Components. They are handled internally without involving the Customer.

A) _TRUE_ B) _FALSE_

ANSWER: A

Q148. The .... step is meant to identify all potential Risks and Threats and their anticipated Impacts along with the Probabilities of their potential Occurrences.

A) Risks Identification

B) Risks Containment

C) Both Options above are Correct

D) Both Options above are Incorrect

ANSWER: A

Q149. The .... step is to classify the identified potential Risks into three Categories, which are the Risks with Low-Impacts Category, Risks with Moderate-Impacts Category, and Risks with High-Impacts Category.

A) Risks Identification

B) Risks Classification

C) Both Options above are Correct

D) Both Options above are Incorrect

ANSWER: B

Q150. In the case of. . . . . all what is needed is proactively identify these Risks. However, if it happens that a Low-Impact Risk occurs during the Project life cycle, then the Project Manager should reactively devise a Plan and immediately implement it to counterpart that Risk.
A) Low-Impact Risks
B) Moderate-Impact Risks
C) High-Impact Risks
D) All Options above are Incorrect
ANSWER: A

Q151. In the case of . . . . all what is needed is proactively identify and plan for these Risks. However, if it happens that a Moderate-Impact Risk occurs during the Project life cycle, then the Project Manager should reactively implement the pre-prepared Plan to counterpart that Risk.
A) Low-Impact Risks
B) Moderate-Impact Risks
C) High-Impact Risks
D) All Options above are Incorrect
ANSWER: B

Q152. In the case of . . . . all what is needed is proactively identify, plan for encountering these Risks, and then immediately implement these Plans.
A) Low-Impact Risks
B) Moderate-Impact Risks
C) High-Impact Risks
D) All Options above are Incorrect
ANSWER: C

Q153. The . . . . Approach is a variation from the Phased-Based Project Management (PBPM) Approach. Although it goes through the same five Phases of the PBPM Approach, it is meant to produce Deliverables in less time and with less waste by RE-CYCLING Components from previous Products, Solutions, and/or Services.
A) Lean-Based Project Management (LBPM)
B) Lean-Based Project Planning (LBPP)
C) Lean-Based Project Execution (LBPE)
D) Lean-Based Project Manager
ANSWER: A

Q154. The . . . . Approach is not suitable for Large-Size Projects with Fast-Changing Requirements that are not clearly predefined. It is not suitable as well for Projects that have much Risks, Uncertainties, and/or Dependencies. Several IIBPM Methods have evolved to handle such Complexities. It is important to notice that the IIBPM Methods are considered part of the Agile Methods.
A) Iterative-Incremental-Based Project Management (IIBPM)
B) Iterative-Incremental-Based Scrum (IIBS)
C) Iterative-Incremental-Based Project Planning (IIBPP)
D) Iterative-Incremental-Based Project Manager
ANSWER: A

Q155. The . . . . Approach is designed to deal with the Uncertainties in Managing Projects such as having limited Availability of the needed Human Resources and/or Physical Resources. Constrained Tasks and Activities are given Higher Priorities during the Project Planning, Resource Allocation, and Execution.
A) Critical-Chain-Based Project Management (CCBPM)
B) Critical-Chain-Based Project Planning (CCBPP)
C) Critical-Chain-Based Project Execution (CCBPE)
D) Critical-Chain-Based Project Manager
ANSWER: A

Q156. The . . . . Approach is a Structured Project Management Approach that identifies the Project targeted Products and then produces the targeted Outcome Products, Solutions, and/or Services using the materialized and informational Inputs at a predefined Production Volume for each of these Outcomes.
A) Product-and-Production-Based Project Management (PPBPM)
B) Product-and-Production-Based Project Planning (PPBPP)
C) Product-and-Production-Based Project Execution (PPBPE)
D) Product-and-Production-Based Project Manager
ANSWER: A

Q157. The Projects Success ON-TIME Pillar refers to . . . .
A) Completing the Project in accordance with its predefined Schedule.
B) Completing the Project without exceeding the pre-allocated Budget.
C) Completing the Project with its Outcome Product's Specifications meeting the Project predefined Specifications.
D) Having the Customer to accept the Project Deliverables. In other words, the Deliverables should pass the Customer Testing and Acceptance-Criteria.
ANSWER: A

Q158. The Projects Success WITHIN-BUDGET Pillar refers to . . . .
A) Completing the Project in accordance with its predefined Schedule.
B) Completing the Project without exceeding the pre-allocated Budget.
C) Completing the Project with its Outcome Product's Specifications meeting the Project predefined Specifications.
D) Having the Customer to accept the Project Deliverables. In other words, the Deliverables should pass the Customer Testing and Acceptance-Criteria.
ANSWER: B

Q159. The Projects Success MEET-SPECS Pillar refers to . . . .
A) Completing the Project in accordance with its predefined Schedule.
B) Completing the Project without exceeding the pre-allocated Budget.
C) Completing the Project with its Outcome Product's Specifications meeting the Project predefined Specifications.
D) Having the Customer to accept the Project Deliverables. In other words, the Deliverables should pass the Customer Testing and Acceptance-Criteria.
ANSWER: C

Q160. The Projects Success SATISFY-CUSTOMER Pillar refers to . . . .
A) Completing the Project in accordance with its predefined Schedule.
B) Completing the Project without exceeding the pre-allocated Budget.
C) Completing the Project with its Outcome Product's Specifications meeting the Project predefined Specifications.
D) Having the Customer to accept the Project Deliverables. In other words, the Deliverables should pass the Customer Testing and Acceptance-Criteria.
ANSWER: D

Q161. Among the 4Ps for Project Success, the PLAN refers to . . . .
A) The Project Core Action Plan that composes an estimated Work Breakdown Structure (WBS), an estimated Schedule with careful considerations of the Tasks Inter-Dependencies, a carefully estimated Allocation of Resources, and a set of carefully defined Deliverables and Milestones.
B) The set of Project Management Processes needed besides the Project Core Action Plan to create the Project Detailed Action Plan.
C) Allocating the right staff members as part of the Project Team.
D) The necessary Level of Power and Authority that the Project Manager and the Project Team should have in order to enable them to carry-on their duties with less Issues.
ANSWER: A

Q162. Among the 4Ps for Project Success, the PROCESSES refers to . . . .

A) The Project Core Action Plan that composes an estimated Work Breakdown Structure (WBS), an estimated Schedule with careful considerations of the Tasks Inter-Dependencies, a carefully estimated Allocation of Resources, and a set of carefully defined Deliverables and Milestones.

B) The set of Project Management Processes needed besides the Project Core Action Plan to create the Project Detailed Action Plan.

C) Allocating the right staff members as part of the Project Team.

D) The necessary Level of Power and Authority that the Project Manager and the Project Team should have in order to enable them to carry on their duties with less Issues.

ANSWER: B

Q163. Among the 4Ps for Project Success, the PEOPLE refers to . . . .

A) The Project Core Action Plan that composes an estimated Work Breakdown Structure (WBS), an estimated Schedule with careful considerations of the Tasks Inter-Dependencies, a carefully estimated Allocation of Resources, and a set of carefully defined Deliverables and Milestones.

B) The set of Project Management Processes needed besides the Project Core Action Plan to create the Project Detailed Action Plan.

C) Allocating the right staff members as part of the Project Team.

D) The necessary Level of Power and Authority that the Project Manager and the Project Team should have in order to enable them to carry on their duties with less Issues.

ANSWER: C

Q164. Among the 4Ps for Project Success, the POWER refers to . . . .

A) The Project Core Action Plan that composes an estimated Work Breakdown Structure (WBS), an estimated Schedule with careful considerations of the Tasks Inter-Dependencies, a carefully estimated Allocation of Resources, and a set of carefully defined Deliverables and Milestones.

B) The set of Project Management Processes needed besides the Project Core Action Plan to create the Project Detailed Action Plan.

C) Allocating the right staff members as part of the Project Team.

D) The necessary Level of Power and Authority that the Project Manager and the Project Team should have in order to enable them to carry on their duties with less Issues.

ANSWER: D

Q165. Among the Success Factors for a Project is defining the Goals, the Scope of Work and Requirements for that Project.

A) _TRUE_ B) _FALSE_

ANSWER: A

Q166. Among the Success Factors for a Project is involving the End-Users as early as possible in the Project life cycle. The Goals, the Scope of Work, and Requirements for that Project.

A) _TRUE_ B) _FALSE_

ANSWER: A

Q167. For a Project Manager to be ...., he/she must be a professional Project Manager, have adequate knowledge of the subject matter of the Project's Domain, have the ability to handle any Sociological and Psychological issues that may arise during the Project's life cycle, have adequate Technological Skills, and have the ability to manager and lead.

A) Competent B) Experienced C) Capable D) Innovative

ANSWER: A

Q168. Among the Success Factors for a Project is having the Work Breakdown Structure (WBS) in place, the Schedule in place, the Resource Allocations completed, the Deliverables and Milestones well defined, the needed Processes defined and in place, the Project Team is well qualified, the Test Cases are comprehensive, and so forth.

A) _TRUE_ B) _FALSE_

ANSWER: A

Q169. The Process of moving from Software Development and Project Management Best Practices → to → Software Development and Project Management Standardization goes through the following steps:

A) Generic Practices → Best Practices → De facto → Standards

B) Best Practices → De facto → Standards

C) Best Practices → Standards

D) Generic Practices → De facto → Standards

ANSWER: A

Appendix B.2: Chapter 3 Bank of Questions

Q1. According to PMI, the Project Integration Management is one of the Project Management Knowledge Areas that incorporates the following Processes:

A) Develop Project Charter, Develop Project Management Plan, Direct and Manage Project Work, Manage Project Knowledge, Monitor and Control Project Work, Perform Integrated Change Control, and Close Project or Phase.

B) Plan Scope Management, Collect Requirements, Define Scope, Create WBS, Validate Scope, and Control Scope.

C) Plan Schedule Management, Define Activities, Sequence Activities, Estimate Activity Durations, Develop Schedule, and Control Schedule.

D) Plan Cost Management, Estimate Costs, Determine Budget, and Control Costs.

ANSWER: A

Q2. According to PMI, the Project Scope Management is one of the Project Management Knowledge Areas that incorporates the following Processes:

A) Develop Project Charter, Develop Project Management Plan, Direct and Manage Project Work, Manage Project Knowledge, Monitor and Control Project Work, Perform Integrated Change Control, and Close Project or Phase.

B) Plan Scope Management, Collect Requirements, Define Scope, Create WBS, Validate Scope, and Control Scope.

C) Plan Schedule Management, Define Activities, Sequence Activities, Estimate Activity Durations, Develop Schedule, and Control Schedule.

D) Plan Cost Management, Estimate Costs, Determine Budget, and Control Costs.

ANSWER: B

Q3. According to PMI, the Project Schedule Management is one of the Project Management Knowledge Areas that incorporates the following Processes:

A) Develop Project Charter, Develop Project Management Plan, Direct and Manage Project Work, Manage Project Knowledge, Monitor and Control Project Work, Perform Integrated Change Control, and Close Project or Phase.

B) Plan Scope Management, Collect Requirements, Define Scope, Create WBS, Validate Scope, and Control Scope.

C) Plan Schedule Management, Define Activities, Sequence Activities, Estimate Activity Durations, Develop Schedule, and Control Schedule.

D) Plan Cost Management, Estimate Costs, Determine Budget, and Control Costs.

ANSWER: C

Q4. According to PMI, the Project Cost Management is one of the Project Management Knowledge Areas that incorporates the following Processes:

A) Develop Project Charter, Develop Project Management Plan, Direct and Manage Project Work, Manage Project Knowledge, Monitor and Control Project Work, Perform Integrated Change Control, and Close Project or Phase.

B) Plan Scope Management, Collect Requirements, Define Scope, Create WBS, Validate Scope, and Control Scope.

C) Plan Schedule Management, Define Activities, Sequence Activities, Estimate Activity Durations, Develop Schedule, and Control Schedule.

D) Plan Cost Management, Estimate Costs, Determine Budget, and Control Costs.

ANSWER: D

Q5. According to PMI, the Project Quality Management is one of the Project Management Knowledge Areas that incorporates the following Processes:

A) Plan Quality Management, Manage Quality, and Control Quality.

B) Plan Resource Management, Estimate Activity Resources, Acquire Resources, Develop Team, Manage Team, and Control Resources.

C) Plan Communications Management, Manage Communications and Monitor Communications.

D) Plan Risk Management, Identify Risks, Perform Qualitative Risk Analysis, Perform Quantitative Risk Analysis, Plan Risk Responses, Implement Risk Responses, and Monitor Risks.

ANSWER: A

Q6. According to PMI, the Project Resource Management is one of the Project Management Knowledge Areas that incorporates the following Processes:

A) Plan Quality Management, Manage Quality, and Control Quality.

B) Plan Resource Management, Estimate Activity Resources, Acquire Resources, Develop Team, Manage Team, and Control Resources.

C) Plan Communications Management, Manage Communications and Monitor Communications.

D) Plan Risk Management, Identify Risks, Perform Qualitative Risk Analysis, Perform Quantitative Risk Analysis, Plan Risk Responses, Implement Risk Responses, and Monitor Risks.

ANSWER: B

Q7. According to PMI, the Project Communications Management is one of the Project Management Knowledge Areas that incorporates the following Processes:
A) Plan Quality Management, Manage Quality, and Control Quality.
B) Plan Resource Management, Estimate Activity Resources, Acquire Resources, Develop Team, Manage Team, and Control Resources.
C) Plan Communications Management, Manage Communications, and Monitor Communications.
D) Plan Risk Management, Identify Risks, Perform Qualitative Risk Analysis, Perform Quantitative Risk Analysis, Plan Risk Responses, Implement Risk Responses, and Monitor Risks.
ANSWER: C

Q8. According to PMI, the Project Risk Management is one of the Project Management Knowledge Areas that incorporates the following Processes:
A) Plan Quality Management, Manage Quality, and Control Quality.
B) Plan Resource Management, Estimate Activity Resources, Acquire Resources, Develop Team, Manage Team, and Control Resources.
C) Plan Communications Management, Manage Communications, and Monitor Communications.
D) Plan Risk Management, Identify Risks, Perform Qualitative Risk Analysis, Perform Quantitative Risk Analysis, Plan Risk Responses, Implement Risk Responses, and Monitor Risks.
ANSWER: D

Q9. According to PMI, the Project Procurement Management is one of the Project Management Knowledge Areas that incorporates the following Processes:
A) Plan Procurement Management, Conduct Procurements, and Control Procurements.
B) Identify Stakeholders, Plan Stakeholder Engagement, Manage Stakeholder Engagement, and Monitor Stakeholder Engagement.
C) Plan Communications Management, Manage Communications, and Monitor Communications.
D) Plan Risk Management, Identify Risks, Perform Qualitative Risk Analysis, Perform Quantitative Risk Analysis, Plan Risk Responses, Implement Risk Responses, and Monitor Risks.
ANSWER: A

Q10. According to PMI, the Project Stakeholder Management is one of the Project Management Knowledge Areas that incorporates the following Processes:

A) Plan Procurement Management, Conduct Procurements, and Control Procurements.

B) Identify Stakeholders, Plan Stakeholder Engagement, Manage Stakeholder Engagement, and Monitor Stakeholder Engagement.

C) Plan Communications Management, Manage Communications, and Monitor Communications.

D) Plan Risk Management, Identify Risks, Perform Qualitative Risk Analysis, Perform Quantitative Risk Analysis, Plan Risk Responses, Implement Risk Responses, and Monitor Risks.

ANSWER: B

Q11. According to PMI, the main Inputs to the "Develop Project Charter" Process are the following:

A) Business Case, Benefits Management Plan, Any earlier, but relevant Contracts and Agreements, EEFs, and OPAFs.

B) Expert judgment, Data gathering approaches, Interpersonal and team skills, Meetings.

C) Business Case, Benefits Management Plan, Assumption Log, and EEFs and OPAFs.

D) All Options above are Correct.

ANSWER: A

Q12. According to PMI, the main Methods used during the "Develop Project Charter" Process are the following:

A) Brainstorming, Interviews, Facilitation, Conflict Management, Meeting Management, and so forth.

B) Meetings to discuss all issues that concern the Project Charter Development.

C) Both Options above are Correct.

D) All Options above are Incorrect.

ANSWER: C

Q13. According to PMI, the main Outputs of the "Develop Project Charter" Process are the following:

A) A Project Charter and an Assumption Log.

B) A Business Document, an Intuitive Project Management Plan, and a Project Charter.

C) Both Options above are Correct.

D) All Options above are Incorrect.

ANSWER: A

Q14. According to PMI, the Outputs of the “Develop Project Charter” Process impact the following Processes:

A) The Close Project/Phase, the Collect Requirements, the Plan Schedule Management, the Plan Cost Management, the Plan Quality Management, the Plan Resource Management, the Plan Communications Management, the Plan Risk Management, the Plan Procurement Management, the Plan Stakeholder Engagement, the Plan Scope Management, the Develop Project Management Plan, the Plan Scope Management, the Define Scope, and the Identify Risk.

B) Only the Plan Schedule Management, the Plan Cost Management, the Plan Quality Management, the Plan Resource Management, the Plan Communications Management, the Plan Risk Management, the Plan Procurement Management, the Plan Stakeholder Engagement, the Plan Scope Management, and the Plan Scope Management.

C) Both Options above are Correct.

D) All Options above are Incorrect.

ANSWER: A

Q15. According to PMI, the main Business Documents that are used as Inputs to the “Develop Project Charter” Process are the following:

A) The main Business Documents used as Inputs to this Process are the Business Case, Market Demand Report, Customer Requests, Organizational Needs, Technological Advances, Legal Requirements, Ecological Impacts, and Social Needs.

B) The Plan Schedule Management, the Plan Cost Management, the Plan Quality Management, the Plan Resource Management, the Plan Communications Management, the Plan Risk Management, the Plan Procurement Management, the Plan Stakeholder Engagement, the Plan Scope Management, and the Plan Scope Management.

C) Both Options above are Correct.

D) All Options above are Incorrect.

ANSWER: A

Q16. According to PMI, the main Agreements that are used as Inputs to the "Develop Project Charter" Process are the following:
A) The Project Contract including all of its Amendments, Memorandum of Understanding (MoU), Alliance Agreements, Association Agreements, and Sub-Contracting Agreements.
B) The Plan Schedule Management, the Plan Cost Management, the Plan Quality Management, the Plan Resource Management, the Plan Communications Management, the Plan Risk Management, the Plan Procurement Management, the Plan Stakeholder Engagement, the Plan Scope Management, and the Plan Scope Management.
C) Both Options above are Correct.
D) All Options above are Incorrect.
ANSWER: A

Q17. According to PMI, a Contract is an Agreement that obligates the Service-Provider to provide the Customer with the agreed-upon Products, Solutions, and/or Services; obligates the Customer to compensate the Service-Provider for their provision; and legalizes the Contractual Relationship between the Service-Provider and the Customer.
A) _TRUE_ B) _FALSE_
ANSWER: A

Q18. The main Components of an Agreement include the following:
A) Procurement Statement of Work, Deliverables, Schedule, Milestones, Performance Reporting, Pricing and Payment Terms, Inspection, and Quality.
B) The Acceptance-Criteria, Warranty and Support, Incentives and Penalties, Insurance and Performance Bonds, and Sub-Contractors Approvals.
C) The Terms and Conditions, CHANGE-REQUESTS Handling, Termination Clause, and Dispute Resolution Methods.
D) All Options above are Correct.
ANSWER: D

Q19. The main EEFs used as Inputs to the Develop Project Charter Process include the following:
A) The Governmental Standards, Industrial Standards, Legal and Regulatory Requirements, Marketplace Conditions, Organizational Culture, Political Climate, Organizational Governance Framework, and Stakeholder Requirements.
B) Only the Governmental Standards, Legal and Regulatory Requirements, Organizational Culture, Political Climate, and Stakeholder Requirements.
C) Both Options above are Correct.
D) All Options above are Incorrect.
ANSWER: A

Q20. The main Data Gathering Methods used during the Develop Project Charter Process include the following:
A) The Brainstorming, Checklists, Focus Groups, Interviews, and Questionnaires.
B) Only the Brainstorming and some other tools.
C) Both Options above are Correct.
D) All Options above are Incorrect.
ANSWER: A

Q21. The Assumption Log presents the --------------------------:
A) High-Level Strategic Assumptions and Constraints throughout the Project Life Cycle.
B) The Operational Brainstorming and some other tools.
C) Both Options above are Correct.
D) All Options above are Incorrect.
ANSWER: A

Q22. The main items that compose the Project Charter are the Project Description, Purpose, Objectives, SUCCESS-CRITERIA, Requirements Description, Deliverables, Terms and Conditions, Risks, Milestones, Schedule, Stakeholders, Financial Requirements, APPROVAL-CRITERIA, CLOSURE-CRITERIA, Project Manager, and Project Sponsor.
A) _TRUE_ B) _FALSE_
ANSWER: A

Q23. The "Develop Project Management Plan" Process is meant to define, prepare, coordinate, and consolidate the Project Management Plan.
A) _TRUE_ B) _FALSE_
ANSWER: A

Q24. The Outputs of the "Develop Project Charter" and some other Processes (e.g., mainly related to the Subsidiary Plans) are used as Inputs to the "Develop Project Management Plan" Process.
A) _TRUE_ B) _FALSE_
ANSWER: A

Q25. The main Enterprise Environmental Factors (EEFs) that influence the "Develop Project Management Plan" Process include the following:
A) Governmental and Industry Standards; Legal and Regulatory Requirements; Project Management Body of Knowledge; and Focus Areas and Focus Groups.
B) Operational Environment; Safety; Risks; Software Development Methods; and Organizational Structure, Culture, Management Practices.
C) Organizational Governance Framework that controls, directs, and coordinates People, Policies, and Processes; and Organizational Infrastructure.
D) All Options above are Correct.
ANSWER: D

Q26. The main Organizational Process Assets Factors (OPAFs) that influence the "Develop Project Management Plan" Process include the following:
A) Organizational Standard Policies, Processes, and Procedures; Project Management Plan Templates; and Change Control Procedures.
B) Monitoring and Reporting Methods; Risks Control Procedures; and Communication Requirements.
C) Information from similar past Projects.
D) All Options above are Correct.
ANSWER: D

Q27. At a meeting during the "Develop Project Management Plan" Process, the following topics are discussed as determined:
A) The Project Management Approach.
B) The Project Work Execution.
C) The Project Monitoring and Control.
D) All of the above Options are Correct.
ANSWER: D

Q28. The Project Management Plan describes how the Project will be Executed, Monitored, Controlled, and Closed.
A) _TRUE_ B) _FALSE_
ANSWER: A

Q29. The main Components of a Project Management Plans are a set of Subsidiary Plans and a set of Baseline Documents.
A) _TRUE_ B) _FALSE_
ANSWER: A

Q30. The Outputs of the "Develop Project Management Plan" Process include a set of subsidiary Project Plans such as a Scope Management Plan, Requirements Management Plan, Schedule Management Plan, Cost Management Plan, Quality Management Plan, Resource Management Plan, Communications Management Plan, Risk Management Plan, Procurement Management Plan, Stakeholders Engagement Plan, Change Management Plan, and Configuration Management Plan.
A) _TRUE_ B) _FALSE_
ANSWER: A

Q31. The Outputs of the "Develop Project Management Plan" Process include a set of Subsidiary Project Baseline Documents such as a Scope Baseline, Schedule Baseline, Cost Baseline, and Performance Measurement Baseline.
A) _TRUE_ B) _FALSE_
ANSWER: A

Q32. The complete list of the project documents updated or produced during the "Develop Project Management Plan" Process are the Activity Attributes, Activity List, Assumption Log, Basis of Estimates, Change Log, Cost Estimates, Cost Forecasts, Duration Estimates, Issue Log, Lessons Learned Register, Milestone List, Physical Resource Assignments, Project Calendars, Project Communications, Project Schedule, Project Schedule Network Diagram, Project Scope Statement, Project Team Assignments, Quality Control Measurements, Quality Metrics, Quality Report, Requirements Documentation, Requirements Traceability Matrix, Resource Breakdown Structure, Resource Calendars, Resource Requirements, Risk Register, Risk Report, Schedule Forecasts, Stakeholder Register, Team Charter, and Test and Evaluation Documents.

A) _TRUE_ B) _FALSE_

ANSWER: A

Q33. The "Direct and Manage Project Work" Process is followed to lead the work as defined in the Project Management Plan and then implement the approved CHANGE-REQUESTS to achieve the Project Objectives.

A) _TRUE_ B) _FALSE_

ANSWER: A

Q34. The Inputs to the "Direct and Manage Project Work" Process include the following:

A) The Project Management Plan along with its subsidiary Plans and Documents.
B) The Project Management Plan without its subsidiary Plans and Documents, but with the Approved Change Requests.
C) The Project Management Plan without any Enterprise Environmental Factors.
D) The Project Management Plan along with only its relevant Enterprise Environmental Factors.

ANSWER: A

Q35. The Outputs of the "Direct and Manage Project Work" Process include the following:

A) The Project Deliverables along with the Project Work Performance Data/Information/Reports.
B) The Project Deliverables along with an updated version of the Issue Log document.
C) The Project Deliverables along with an updated version of the Project Management Plan and updated versions of many of the Project Documents.
D) All Options above are Correct.

ANSWER: D

Q36. The "Manage Project Knowledge" Process is to use and extend existing Knowledge to achieve the Project Objectives and to contribute to the Organizational Learning.
A) _TRUE_ B) _FALSE_
ANSWER: A

Q37. The main Inputs to the "Manage Project Knowledge" Process include the following:
A) The Project Management Plan along with many of its subsidiary Project Documents.
B) The Project Deliverables along with many of its relevant Enterprise Environmental Factors.
C) The Project Management Plan along with many of its relevant Organizational Assets Factors.
D) All Options above are Correct.
ANSWER: D

Q38. The main Outputs of the "Manage Project Knowledge" Process include the following:
A) An updated version of the Lessons Learned Register Document.
B) An updated version of the Project Management Plan.
C) Updated versions of many of the subsidiary Project Documents.
D) All Options above are Correct.
ANSWER: D

Q39. The "Monitor and Control Project Work" Process is to track, review, and report the Project ongoing progress and that enables the Stakeholders to understand the Project Status, recognize any Corrective Actions taken to resolve Performance Issues, and realize the Project Cost and Schedule Forecasts.
A) _TRUE_ B) _FALSE_
ANSWER: A

Q40. The main Inputs to the "Monitor and Control Project Work" Process include the following:
A) The Project Management Plan along with many of its subsidiary Plans and Project Documents.
B) The Project Management Plan along with the Project Work Performance Data/Information/Reports and the Project Agreements.
C) The Project Management Plan along with many of its relevant Enterprise Environmental Factors and Organizational Process Assets Factors.
D) All Options above are Correct.
ANSWER: D

Q41. The Data Analysis used during the "Monitor and Control Project Work" Process includes the Alternatives Analysis, Cost-Benefit Analysis, Earned-Value Analysis, Root-Cause Analysis, Trend Analysis, and Variance Analysis.
A) _TRUE_ B) _FALSE_
ANSWER: A

Q42. The main Decision Making Method used during the "Monitor and Control Project Work" Process is "Voting."
A) _TRUE_ B) _FALSE_
ANSWER: A

Q43. The main Outputs of the "Monitor and Control Project Work" Process include the following:
A) The Project Performance Data/Information/Reports along with a set of Change Requests.
B) An updated version of the Project Management Plan.
C) Updated versions of many of the Project Management Plan subsidiary Plans and/or Project Documents.
D) All Options above are Correct.
ANSWER: D

Q44. The "Perform Integrated Change Control" Process is to manage and review all CHANGE-REQUESTS and approve and implement some of them.
A) _TRUE_ B) _FALSE_
ANSWER: A

Q45. The main Inputs to the "Perform Integrated Change Control" Process include the following:
A) The Project Management Plan along with many of its subsidiary Plans and Project Documents.
B) The Project Management Plan along with the Project Performance Data/Information/Reports and a set of Change Requests.
C) The Project Management Plan along with many of its relevant Enterprise Environmental Factors and Organizational Process Asset Factors.
D) All Options above are Correct.
ANSWER: D

Q46. The main Data Analysis Methods used during the "Perform Integrated Change Control" Process are the Alternatives Analysis and Cost-Benefit Analysis.
A) _TRUE_ B) _FALSE_
ANSWER: A

Q47. The main Decision-Making Methods used during the "Perform Integrated Change Control" Process are the Voting, Automatic Decision Making, and Multi-CRITERIA Decision Analysis.
A) _TRUE_ B) _FALSE_
ANSWER: A

Q48. The Change Control Tools used during the "Perform Integrated Change Control" Process include Identifying the Configuration Team; Recording and Reporting the Items Statuses; Performing the Configuration Items, Verifications and Audits; and Identifying, Documenting, Deciding on, and Tracking Changes.
A) _TRUE_ B) _FALSE_
ANSWER: A

Q49. The main Outputs of the "Perform Integrated Change Control" Process are the Project Work Data/Information/Reports, the Approved Change Requests, an updated version of the Project Management Plan, and updated versions of the many of the Project Documents.
A) _TRUE_ B) _FALSE_
ANSWER: A

Q50. The "Close Project or Phase" Process is to:
A) Finalize all of the Project Activities such as ensuring that all Documents and Deliverables are up-to-date and that all Issues are resolved; Ensure that all Costs are charged to the Project; confirming the delivery and the Formal Acceptance of Deliverables by the Customer; Measure the Stakeholders' Satisfaction; and Reassign Human Resources, dealing with excess Project material and reallocating the Project Facilities and Equipment.
B) Prepare the Project Reports in accordance with the Organizational Policies; Manage Knowledge sharing and transfer; Identify Lessons Learned; and Transfer the Project Products, Solutions, and/or Services to the next Phase or to Production/Operations.
C) Collect Suggestions for improving the Organizational Policies and Procedures; Archive the Project Information for future use by the Organization; and Close the Project Accounts.
D) All Options above are Correct.
ANSWER: D

Q51. The main Inputs to the "Close Project or Phase" Process include the following:
A) The Project Charter; and the Project Management Plan and its subsidiary Plans and Project Documents.
B) The set of Accepted Deliverables, the relevant Business Documents, and the Procurement Documentation.
C) The Project Agreements along with the set of relevant Organizational Process Assets Factors.
D) All Options above are Correct.
ANSWER: D

Q52. The main Data Analysis Methods used during the "Close Project or Phase" Process are the Documents Analysis, Regression Analysis, Trend Analysis, and Variance Analysis.
A) _TRUE_ B) _FALSE_
ANSWER: A

Q53. The main Outputs of the "Close Project or Phase" Process are the Final Product/Solution/Service transition to the Customer; updated versions of many of the Project Documents; the Project Final Report; and a set of relevant Organizational Process Assets.
A) _TRUE_ B) _FALSE_
ANSWER: A

Q54. The Project Scope Management Knowledge Area involves six processes, which are the "Plan Scope Management" Process, the "Collect Requirements" Process, the "Define Scope" Process, the "Create WBS" Process, the "Validate Scope" Process, and the "Control Scope" Process.
A) _TRUE_ B) _FALSE_
ANSWER: A

Q55. The term "SCOPE" refers to:
A) The Project SCOPE.
B) The Product SCOPE.
C) Both Options above are Correct.
D) Both Options above are Incorrect.
ANSWER: D

Q56. The "Plan Scope Management" Process is to create a Scope Management Plan that illustrates how to define, validate, and control the SCOPES of the Project and its anticipated Product(s). In addition, it shows how to manage the Project and the Product Scopes throughout the Project life cycle.
A) _TRUE_ B) _FALSE_
ANSWER: A

Q57. The main Inputs to the "Plan Scope Management" Process are the Project Charter, the Project Management Plan, a set of relevant Organizational Process Assets, and a set of relevant Enterprise Environmental Factors.
A) _TRUE_ B) _FALSE_
ANSWER: A

Q58. The main Data Analysis Method used during the "Plan Scope Management" Process is the Alternatives Analysis.
A) _TRUE_ B) _FALSE_
ANSWER: A

Q59. The Scope Management Plan is an Integral Component of the Project Management Plan that describes how to define, monitor, control, and validate the Project and Product Scopes. It includes Processes for preparing the Project Scope Statement, creating the Work Breakdown Structure (WBS), approving and maintaining the Scope Baseline, and obtaining/accepting the Deliverables.
A) _TRUE_ B) _FALSE_
ANSWER: A

Q60. The Requirements Management Plan is an Integral Component of the Project Management Plan that describes how to analyze, document, and manage the Project and Product Requirements. It includes Processes for planning, tracking, and reporting Activities; analyzing their Impacts; tracing, tracking, and reporting them; and for prioritizing the Requirements.
A) _TRUE_ B) _FALSE_
ANSWER: A

Q61. The "Collect Requirements" Process is to extract and organize the Requirements from the Formal Documents such as the RFP.
A) _TRUE_ B) _FALSE_
ANSWER: A

Q62. The main Inputs to the "Collect Requirements" Process include the following:
A) The Project Charter and the Project Management Plan along with its subsidiary Plans and Project Documents.
B) The Project Business Documents along with all of the Project Agreements.
C) A set of relevant Organizational Process Assets, a set of relevant Enterprise Environmental Factors.
D) All Options above are Correct.
ANSWER: D

Q63. The main Data Gathering Methods used during the "Collect Requirements" Process are the Brainstorming, Interviews, Focus Groups, Questionnaires/Surveys, and Benchmarking.
A) _TRUE_ B) _FALSE_
ANSWER: A

Q64. The main Documents analyzed during the "Collect Requirements" Process are the Project Agreements, Business Plans, Business Rules Repositories, Interface Document, Issue Log, Policies, Procedures, and Use-Cases.
A) _TRUE_ B) _FALSE_
ANSWER: A

Q65. The Prototypes are to demonstrate parts of the targeted System at early stages of the Project.
A) _TRUE_ B) _FALSE_
ANSWER: A

Q66. The Requirements Documentation includes the following:
A) Business Requirements that describe the Business Issues or Opportunities, and why the Project has been carried-on); Stakeholder Requirements that describe the Stakeholders Needs; and Solution Requirements that describe the targeted Solution's Functional and Non-Functional Requirements.
B) Transition and Readiness Requirements that describe the Data conversion and training Requirements; Project Requirements that describe the Actions and Processes that should be satisfied by the Project; and Quality Requirements that describe the CRITERIA for validating the SUCCESSFUL completion of a Project.
C) Both Options above are Correct.
D) All Options above are Incorrect.
ANSWER: D

Q67. The Requirements Traceability Matrix includes the Business Needs, Opportunities, Project Objectives, Scope, Deliverables, Solution Design, Development Method, Testing Strategy, and Testing Scenario.
A) _TRUE_ B) _FALSE_
ANSWER: A

Q68. The "Define Scope" Process is to develop a Project Scope of Work during the Project Initiation and the Project Planning Phases.
A) _TRUE_ B) _FALSE_
ANSWER: A

Q69. The main Inputs to the "Define Scope" Process include the following:
A) The Project Charter and the Project Management Plan along with its subsidiary Plans and Project document,
B) The Project Charter along with all of the Project Agreements.
C) A set of relevant Organizational Process Assets and a set of relevant Enterprise Environmental Factors.
D) All Options above are Correct.
ANSWER: D

Q70. The main Product Analysis Methods used during the "Define Scope" Process are the Product Breakdown, Requirements Analysis, Systems Analysis and Engineering, and Value Analysis and Engineering.
A) _TRUE_ B) _FALSE_
ANSWER: A

Q71. The "Create WBS" Process subdivides the Project Work into small Components that are easy to manage.
A) _TRUE_ B) _FALSE_
ANSWER: A

Q72. The Work Breakdown Structure (WBS) is a Hierarchical Breakdown and Decomposition of the work to be carried-on by the Project Team to accomplish the Project Objectives and to create its Deliverables.
A) _TRUE_ B) _FALSE_
ANSWER: A

Q73. The main Inputs to the "Create WBS" Process include the following:
A) The Project Management Plan along with the Project Scope Statement and the Project Requirements Documentation.
B) The Project Management Plan along with a set of relevant Organizational Process Assets.
C) The Project Management Plan along with a set of relevant Enterprise Environmental Factors.
D) All Options above are Correct.
ANSWER: D

Q74. The Decomposition Method used during the "Create WBS" Process is to divide the Project Work into small and manageable Components.
A) _TRUE_ B) _FALSE_
ANSWER: A

Q75. The Scope Baseline includes the approved Scope Statement; WBS; and WBS Dictionary including an Identifier, Work Description, Constraints, Schedule Milestones, Cost Estimates, and so forth.
A) _TRUE_ B) _FALSE_
ANSWER: A

Q76. The "Validate Scope" Process Formalizes the Project Deliverables Acceptance-Criteria.
A) _TRUE_ B) _FALSE_
ANSWER: A

Q77. The main Inputs to the "Validate Scope" Process include the Project Management Plan and its subsidiaries, a set of the Project Documents, the set of Verified Deliverables, and the Work Performance Data/Information/Reports.
A) _TRUE_ B) _FALSE_
ANSWER: A

Q78. The Inspection Method includes measuring, Examining, and validating the Project Work and Deliverables in order to satisfy the Requirements and pass the Product Acceptance-Criteria.
A) _TRUE_ B) _FALSE_
ANSWER: A

Q79. This includes all Deliverables that are accepted by the . . . .
A) Project Sponsor
B) Project Manager
C) Project Stakeholders
D) Project Team
ANSWER: A

Q80. The "Control Scope" Process is to monitor the Project status and the Project and Product Scopes, and manage Changes to the Scope Baseline. The "Control Scope" Process ensures that all CHANGE-REQUESTS and their Corrective Actions are processed through the "Perform Integrated Change Control" Process.
A) _TRUE_ B) _FALSE_
ANSWER: A

Q81. The main Inputs to the "Control Scope" Process include the Project Management Plan along with its subsidiary Plans and Project Documents; a set of relevant Organizational Process Assets; and the Work Performance Data/Information/Reports.
A) _TRUE_ B) _FALSE_
ANSWER: A

Q82. The main Data Analysis Methods used during the "Control Scope" Process are the Variance Analysis and Trend Analysis.
A) _TRUE_ B) _FALSE_
ANSWER: A

Q83. The main Outputs of the "Control Scope" Process are:
A) An updated version of the Project Management Plan. Particularly its subsidiary Scope Management Plan, Schedule Baseline, Cost Baseline, Performance Measurements Baseline, Lessons Learned Register, Requirements Documentations, and Requirements Traceability Matrix.
B) The Project Work Performance Data/Information/Reports and the set of Change Requests.
C) Both Options above are Correct.
D) All Options above are Incorrect.
ANSWER: C

Q84. The Project Scope Management Knowledge Area includes six processes that are necessary to complete the Project on time per its predefined Schedule.
A) _TRUE_ B) _FALSE_
ANSWER: A

Q85. The Processes of the Project Scope Management Knowledge Area are the "Plan Schedule Management" Process, the "Define Activities" Process, the "Sequence Activities" Process, the "Estimate Activity Durations" Process, the "Develop Schedule" Process, and the "Control Schedule" Process.
A) _TRUE_ B) _FALSE_
ANSWER: B

Q86. The main Inputs to the "Plan Schedule Management" Process are the Project Charter, the Project Management Plan along with some of its subsidiaries, a set of relevant EEFs, and a set of relevant OPAFs.
A) _TRUE_ B) _FALSE_
ANSWER: A

Q87. The main Data Analysis Method used during the "Plan Schedule Management" Process is the Alternatives Analysis.
A) _TRUE_ B) _FALSE_
ANSWER: A

Q88. The Schedule Management Plan is an Integral Component of the Project Management Plan that establishes the CRITERIA and the Activities that are required to develop, monitor, and control the Project Schedule.
A) _TRUE_ B) _FALSE_
ANSWER: A

Q89. The Schedule Management Plan is an Integral Component of the Project Management Plan that establishes the Project Schedule Model Development, Release and Iteration Length, Accuracy Level, Measurement Units, Organizational Procedures Links, Schedule Model Maintenance, Control Thresholds, Performance Measurement Rules, and Reporting Formats.
A) _TRUE_ B) _FALSE_
ANSWER: A

Q90. The "Define Activities" Process is to identify and document what is required to produce the Project Deliverables.
A) _TRUE_ B) _FALSE_
ANSWER: A

Q91. The main subsidiaries of the Project Management Plan used as Inputs to the "Define Activities" Process are the Schedule Management Plan, and the Scope Baseline.
A) _TRUE_ B) _FALSE_
ANSWER: A

Q92. The main Enterprise Environmental Factors used as Inputs to the "Define Activities" Process are the Organizational Culture and Structure, Availability of Commercial Databases, and Availability of a Project Management Information System (PMIS).
A) _TRUE_ B) _FALSE_
ANSWER: A

Q93. The main Organizational Process Assets used as Inputs to the "Define Activities" Process are the Lessons Learned Repositories and the Standard Processes.
A) _TRUE_ B) _FALSE_
ANSWER: A

Q94. WBS Decomposition is a Method used to divide and subdivide the Project Scope and Deliverables into smaller, but more manageable parts.
A) _TRUE_ B) _FALSE_
ANSWER: A

Q95. The Activities Log includes Activities scheduled for the Project, while the Activity Attributes includes the Attributes of each Activity such as the Activity ID, Duration, and so forth.
A) _TRUE_ B) _FALSE_
ANSWER: A

Q96. The "Sequence Activities" Process identifies and documents the Project inter-Activities Relationships.
A) _TRUE_ B) _FALSE_
ANSWER: A

Q97. The main subsidiaries of the Project Management Plan used as Inputs to the "Sequence Activities" Process are the Schedule Management Plan, the Scope Baseline, the Activity Attributes, the Activity List, the Assumption Log, and the Milestone List.
A) _TRUE_ B) _FALSE_
ANSWER: A

Q98. The main Enterprise Environmental Factors used as Inputs to the "Sequence Activities" Process are the Governmental and Industry Standards, the Project Management Information System (PMIS), and the Availability of Scheduling Tools. While the main Organizational Process Assets used as Inputs to this Process are the Portfolio and Program Plans, the Project Dependencies and Relationships; the existing Planning-Related Policies, Procedures, and Guidelines; and the Lessons Learned Repository.
A) _TRUE_ B) _FALSE_
ANSWER: A

Q99. The Precedence Diagramming Method (PDM) is to construct a Schedule Model, in which Activities are represented by Nodes and inter-linked graphically by Logical Relationships to show the Sequence, in which the Activities are performed.
A) _TRUE_ B) _FALSE_
ANSWER: A

Q100. The PDM Relationships include the following:
A) The Finish-to-Start (FS), with which a Successor Activity cannot start before its Predecessor Activity is completed; and the Finish-to-Finish (FF), with which both of a Successor Activity and its Predecessor Activity must be completed at the same time.
B) The Start-to-Start (SS), with which a Successor Activity and its Predecessor Activity must be started at the same time; and the Start-to-Finish (SF), with which a Successor Activity cannot be completed before its Predecessor Activity is started.
C) Both Options above are Correct.
D) All Options above are Incorrect.
ANSWER: C

Q101. The Dependency Determination and Integration (DDI) includes the Dependencies among the Project Activities that impose some Restrictions on the Activities Sequencing in the Schedule.
A) _TRUE_ B) _FALSE_
ANSWER: A

Q102. The Leads and Lags implies that when a Successor Activity does not need an input from its Predecessor when it starts, but rather it will need that input after some amount of time (*e.g., lead time*), then the Successor Activity can proceed in parallel with its Predecessor for that lead time.
A) _TRUE_ B) _FALSE_
ANSWER: A

Q103. The "Estimate Activity Durations" Process estimates the work periods needed to complete the Project Activities.
A) _TRUE_ B) _FALSE_
ANSWER: A

Q104. The main subsidiaries of the Project Management Plan used as Inputs to the "Estimate Activity Durations" are the Schedule Management Plan, the Scope Baseline, the Activities Attributes, the Activity List, the Assumption Log, the Lessons Learned Register, the Milestone List, the Project Team Assignments, the Resource Requirements, the Risk Register, the Resource Calendars, and the Resource Breakdown Structure.
A) _TRUE_ B) _FALSE_
ANSWER: A

Q105. The main Enterprise Environmental Factors used as Inputs to the "Estimate Activity Durations" Process are the Duration Estimating Databases, Productivity Metrics, and Team Members Locations. While the main Organizational Process Assets used as inputs to this Process are the Historical Duration Information, Scheduling Method, Estimating Policies, Project Calendars, and Lessons Learned Repository.
A) _TRUE_ B) _FALSE_
ANSWER: A

Q106. The "Develop Schedule" Process is to analyze the Activities sequences, Durations, Resource Requirements, and Schedule Constraints to create a Project Schedule Model.
A) _TRUE_ B) _FALSE_
ANSWER: A

Q107. The main subsidiaries of the Project Management Plan used as Inputs to the "Develop Schedule" Process include the following:
A) The Schedule Management Plan and the Scope Baseline.
B) The Activities Attributes, the Activity List, the Assumption Log, the Basis of Estimates and their Durations, and the Lessons Learned Register.
C) The Milestone List, the Project Schedule Network Diagrams, the Project Team Assignments, the Resource Calendars, the Resource Requirements, and the Risk Register.
D) All Options above are Correct.
ANSWER: D

Q108. The main Enterprise Environmental Factors used as Inputs to the "Develop Schedule" Process are the Governmental and Industry Standards, and Communication Channels. While the main Organizational Process Assets used as Inputs to this Process are the Scheduling Method including the Policies that govern the Schedule and its Maintenance, and the Project Calendars.
A) _TRUE_ B) _FALSE_
ANSWER: A

Q109. The main Outputs of the "Develop Schedule" Process include the following:
A) The Schedule Baseline, the Schedule Data, and the Project Schedule.
B) The Project Calendars and the set of Change Requests.
C) Updated versions of the Schedule Management Plan, the Activity Attributes, the Assumption Log, the Duration Estimates, the Lessons Learned Register, the Resource Requirements, and the Risk Register.
D) All Options above are Correct.
ANSWER: D

Q110. The "Control Schedule" Process is to monitor the Project status, update its Schedule, and manage Changes to the Schedule Baseline.
A) _TRUE_ B) _FALSE_
ANSWER: A

Q111. The main subsidiaries of the Project Management Plan used as Inputs to the "Control Schedule" Process include the following:
A) The Schedule Management Plan, Schedule Baseline, the Scope Baseline, and the Performance Measurement Baseline.
B) The Lessons Learned Register, Project Calendars, Project Schedule, Resource Calendars, and Schedule Data.
C) Both Options above are Correct.
D) All Options above are Incorrect.
ANSWER: C

Q112. Among the Inputs to the "Control Schedule" Process are the Project Work Performance Data/Information/Reports and a set of relevant Organizational Process Assets.
A) _TRUE_ B) _FALSE_
ANSWER: A

Q113. The main Organizational Process Assets used as Inputs to the "Control Schedule" Process are the existing Schedule Control-Related Policies, Procedures, and Guidelines and available Schedule Controlling, Monitoring, and Reporting Tools.
A) _TRUE_ B) _FALSE_
ANSWER: A

Q114. The main Data Analysis Methods used during the "Control Schedule" Process include the following:
A) The Earned-Value Analysis, the Iteration BURNDOWN-CHART, and the Performance Reviews.
B) The Trend Analysis, the Variance Analysis, and the What-If Scenario Analysis.
C) Both Options above are Correct.
D) All Options above are Incorrect.
ANSWER: C

Q115. The main Outputs of the "Control Schedule" Process include the following:
A) The Work Performance Data/Information/Reports, the Schedule Forecasts, and the set of Change Requests.
B) The Project Management Plan subsidiaries including the Schedule Management Plan, the Schedule Baseline, the Cost Baseline, the Performance Measurement Baseline, the Assumption Log, the Basis of Estimates, the Lessons Learned Register, the Project Schedule, the Resource Calendars, the Risk Register, and the Schedule Data.
C) Both Options above are Correct.
D) All Options above are Incorrect.
ANSWER: C

Q116. Project Cost Management includes four Processes that are necessary to plan, estimate, manage, monitor, and control the Project Cost and Budget. These Processes are the "Plan Cost Management" Process, the "Estimate Costs" Process, the "Determine Budget" Process, and the "Control Costs" Process.
A) _TRUE_ B) _FALSE_
ANSWER: A

Q117. The Project Costs are estimated, budgeted, managed, monitored, and controlled in accordance with the "Plan Cost Management" Process.
A) _TRUE_ B) _FALSE_
ANSWER: A

Q118. The main Inputs to the "Plan Cost Management" Process include the Project Charter, the Project Management Plan along with some of its subsidiaries, a set of relevant EEFs, and a set of relevant OPAFs.
A) _TRUE_ B) _FALSE_
ANSWER: A

Q119. The main Output of the "Plan Cost Management" Process is the Cost Management Plan that establishes the Cost Units of Measure, Cost Levels of Precision and Accuracy, Organizational Procedures Links, Cost Control Thresholds, Rules of Performance Measurement, and the Reporting Forms.
A) _TRUE_ B) _FALSE_
ANSWER: A

Q120. The "Estimate Costs" Process is to estimate the Cost of Resources required to carry on and complete the Project Work.
A) _TRUE_ B) _FALSE_
ANSWER: A

Q121. The main Inputs to the "Estimate Costs" Process include the following:
A) The Project Charter along with the Project Management Plan and some of its subsidiaries.
B) A set of relevant Enterprise Environmental Factors and a set of relevant Organizational Process assets.
C) The Cost Management Plan, the Quality Management Plan, the Project Scope Statement, the Project Work Breakdown Structure. The WBS Directory, the Lessons Learned Register, the Project Schedule, the Resource Requirements, and the Risk Register.
D) All Options above are Correct.
ANSWER: D

Q122. The main EEFs used as Inputs to the "Estimate costs" Process are the Marketplace Conditions, Published Commercial Information, and Exchange Rates and Inflation. While the main OPAFs used as Inputs to this Process are the Cost Estimation Policies and Templates and Lessons Learned Register.
A) _TRUE_ B) _FALSE_
ANSWER: A

Q123. The main Outputs of the "Estimate Costs" Process are the Cost Estimates; the Basis of Estimates; and updated versions of the Assumption Log, the Lessons Learned Register, and the Risk Register Project Documents.

A) _TRUE_ B) _FALSE_

ANSWER: A

Q124. The "Determine Budget" Process is to aggregate the estimated Costs of Activities and establish a Cost Baseline and determine the Project Budget, which includes all the funds required to execute the Project.

A) _TRUE_ B) _FALSE_

ANSWER: A

Q125. The main Inputs to the "Determine Budget" Process include the following:

A) The Project Charter; the Project Management Plan subsidiaries including the Cost Management Plan, the Resource Management Plan, the Scope Baseline, the Basis of Estimates, the Cost Estimates, the Project Schedule, and the Risk Register.
B) A set of the Project Documents, the Business Case, the Benefits Management Plan, and the Project Agreements.
C) A set of relevant EEFs including the Marketplace Conditions, Published Commercial Information, and Exchange Rates and Inflation.
D) All Options above are Correct.

ANSWER: D

Q126. The OPAFs used as Inputs to the "Determine Budget" Process are the Cost Budgeting-Related Policies, Procedures, and Guidelines; Lessons Learned Repository; Available Cost Budgeting Tools; and Used Reporting Methods.

A) _TRUE_ B) _FALSE_

ANSWER: A

Q127. The "Control Costs" Process is to monitor the Project status, update the Project Costs, and manage changes to the Cost Baseline.

A) _TRUE_ B) _FALSE_

ANSWER: A

Q128. The “Costs Control” Process involves:

A) Managing all CHANGE-REQUESTS in timely manners and ensuring that Cost Expenditures do not exceed the authorized Funding by period, by WBS Component, and/or by Activity;

B) Monitoring the Cost Performance to isolate any Variances from the Approved Cost Baseline and monitoring the Work Performance against Funds spent.

C) Preventing unapproved CHANGE-REQUESTS from being included in the reported Cost or Resource Usage.

D) All Options above are Correct.

ANSWER: D

Q129. The main subsidiaries of the Project Management Plan used as Inputs to the “Control Costs” Process are the Cost Management Plan, the Cost Baseline, the Performance Measurement Baseline, and the Lessons Learned Register.

A) _TRUE_ B) _FALSE_

ANSWER: A

Q130. The main Data Analysis Methods used during the “Control Costs” Process are the following:

A) The Earned-Value Analysis including the Planned-for Value, Earned Value, and Actual Costs.

B) The Variance Analysis including the Schedule Variance, Cost Variance, Schedule Performance Index, and Cost Performance Index.

C) The Trend Analysis including Charts and Forecasting; and the Reserve Analysis.

D) All Options above are Correct.

ANSWER: D

Q131. The main Outputs of the “Control Costs” Process include the following:

A) The Project Work Performance Data/Information/Reports, the Project Cost Forecasts, and the set of Change Requests.

B) The Project Management Plan and its subsidiaries including the Cost Management Plan, the Cost Baseline, the Performance Measurement Baseline, the Assumption Log, the Basis of Estimates, the Cost Estimates, the Lessons Learned Register, and the Risk Register.

C) Both Options above are Correct.

D) All Options above are Incorrect.

ANSWER: C

Q132. The Project Quality Management Knowledge Area includes three processes that are necessary to incorporate the Organization's Quality Policy that is used to plan, manage, and control the Project and the Product Quality Requirements. These processes are the "Plan Quality Management" Process, the "Manage Quality" Process, and the "Control Quality" Process.
A) _TRUE_ B) _FALSE_
ANSWER: A

Q133. The "Plan Quality Management" Process is to identify the Project and Product Quality Requirements and/or Standards and to document the Project Compliance with these Quality Requirements and/or Standards.
A) _TRUE_ B) _FALSE_
ANSWER: A

Q134. The subsidiaries of the Project Management Plan used as Inputs to the "Plan Quality Management" Process include the following:
A) The Requirements Management Plan, the Risk Management Plan, the Stakeholder Management Plan, and the Scope Baseline.
B) The Assumption Log, the Requirements Documentation, the Requirements Traceability Matrix, the Risk Register, and the Stakeholder Register.
C) Both Options above are Correct.
D) All Options above are Incorrect.
ANSWER: C

Q135. The main Enterprise Environment Factors used as Inputs to the "Plan Quality Management" Process are the relevant Governmental Regulations, Rules, Standards, and Guidelines; the Geographical Distribution of Facilities; the Organizational Structure and Culture; and the Marketplace Conditions. While the Organizational Process Assets are the Organizational Quality Management Policies, Procedures, and Guidelines; the Quality Templates; and the Lessons Learned Register.
A) _TRUE_ B) _FALSE_
ANSWER: A

Q136. The main components of the Quality Management Plan are the Project Quality Standards, Deliverables, Quality Roles and Responsibilities, and Quality Control and Management Activities throughout the Project life cycle.
A) _TRUE_ B) _FALSE_
ANSWER: A

Q137. The Quality Metrics are to verify the Compliance of the Project and the Product to the Quality Requirements and Standards.
A) _TRUE_ B) _FALSE_
ANSWER: A

Q138. Among the Outputs of the "Plan Quality Management" Process are updated versions of the Project Management Plan and its subsidiaries that include the following:
A) The Risk Management Plan and the Scope Baseline.
B) The Lessons Learned Register, the Requirements Traceability Matrix, the Risk Register, and the Stakeholder Register.
C) Both Options above are Correct.
D) All Options above are Incorrect.
ANSWER: C

Q139. The "Manage Quality" Process is to translate the Quality Management Plan into a set of Quality Activities to incorporate the Organization's Quality Policies into the Project.
A) _TRUE_ B) _FALSE_
ANSWER: A

Q140. The main Inputs to the "Manage Quality" Process include the following:
A) The Project Management Plan along with some of its subsidiary the Quality Management Plan.
B) The Lessons Learned Register and the Quality Control Measurements.
C) The Quality Metrics and the Risk Register.
D) All Options above are Correct.
ANSWER: D

Q141. The main Organizational Process Assets used as Inputs to the "Manage Quality" Process are the Organizational Quality Management Policies, Procedures, and Guidelines; Quality Templates; Quality Results from similar Projects and Lessons Learned Repository.
A) _TRUE_ B) _FALSE_
ANSWER: A

Q142. The purpose of conducting Audits during the "Manage Quality" Process is to:
A) Identify the BEST-PRACTICES, Nonconformities, Gaps, and Shortcomings.
B) Share BEST-PRACTICES, and offer assistance to help improve the Team Productivity.
C) Highlight Audits Contributions in the Lessons Learned Repository.
D) All Options above are Correct.
ANSWER: D

Q143. The Quality Reports include all of the Project and the Product Quality Monitoring and Control Conclusions. While the Test and Evaluation Documents present all of the Project Test and Evaluation Results.
A) _TRUE_ B) _FALSE_
ANSWER: A

Q144. The subsidiaries of the Project Management Plan that may be updated during the "Manage Quality" Process are the Quality Management Plan, the Scope Baseline, the Schedule Baseline, the Cost Baseline, the Issue Log, the Lessons Learned Register, and the Risk Register.
A) _TRUE_ B) _FALSE_
ANSWER: A

Q145. The "Control Quality" Process is to monitor and record the Outcomes of the Quality Management Activities conducted to assess the Project Outcomes' Performance and to determine if the Project Outcomes comply with the Requirements, Regulations, Specifications, and Standards.
A) _TRUE_ B) _FALSE_
ANSWER: A

Q146. The main subsidiaries of the Project Management Plan used as Inputs to the "Control Quality" Process are the Quality Management Plan, the Lessons Learned Register, the Quality Control Measurements, the Quality Metrics, and the Test and Evaluation Documentation.
A) _TRUE_ B) _FALSE_
ANSWER: A

Q147. Among the Inputs to the "Quality Control" Process are the Approved Change Requests, the Deliverables, and the Project Work Performance Data/Information/Reports.
A) _TRUE_ B) _FALSE_
ANSWER: A

Q148. The main Enterprise Environmental Factors used as Inputs to the "Quality Control" Process are the Project Quality Management Systems and the Governmental Regulations, Rules, Standards, and Guidelines. While the main Organizational Process Assets used as Inputs to this Process are the Organizational Quality Management Standards/Policies/ Procedures/Guidelines, the Quality Templates, the Quality Results from similar previous Projects, and the Defect Reporting Procedures.
A) _TRUE_ B) _FALSE_
ANSWER: A

Q149. The main Outputs of the "Quality Control" Process are the Quality Control Measurements, the Verified Deliverables, the Project Work Performance Data/Information/Reports, the Change Requests, an updated version of the Quality Management Plan, updated versions of several of the Project Documents.
A) _TRUE_ B) _FALSE_
ANSWER: A

Q150. The Project Resource Management Knowledge Area includes six processes that are necessary to (i) identify, acquire, and manage Resources; and (ii) ensure that the right Resources will be available at the right time and place. These Processes are the "Plan Resource Management" Process, the "Estimate Activity Resource" Process, the "Acquire Resources" Process, the "Develop Team" Process, the "Manage Team" Process, and the "Control Resources" Process.
A) _TRUE_ B) _FALSE_
ANSWER: A

Q151. The "Plan Resource Management" Process helps the Project Manager in estimating, acquiring, managing, and utilizing the Project Human and Physical Resources.
A) _TRUE_ B) _FALSE_
ANSWER: A

Q152. The main Inputs to the "Plan Resource Management" Process include the following:
A) The Project Charter and the Project Management Plan along with some of its subsidiary including the Quality Management Plan and the Scope Baseline.
B) The Project Charter and the Project Management Plan along with some of its subsidiary including the Project Schedule, the Requirements Documentations, the Risk Register, and the Stakeholder Register.
C) A set of relevant Enterprise Environmental Factors including the Organizational Culture and Structure, Resources and Facilities Geographic Distribution, Resources Competencies, and Marketplace Conditions; and a set of Organizational Process Assets including the Human Resource Policies and Procedures, the Salary Policies, the Safety and Security Policies, and Information of similar Projects.
D) All Options above are Correct.
ANSWER: D

Q153. A Resource Management Plan composes the following components:
A) Methods for deriving the Project Organization Charts; Methods for identifying and acquiring Resources; and Methods for building the Project Team.
B) Methods for defining the Project Team's Roles and Responsibilities; and Methods for training, developing, controlling, recognizing, and managing the Project Team.
C) Both Options above are Correct.
D) All Options above are Incorrect.
ANSWER: C

Q154. A Role is the position such as Programmer; a Responsibility is the duties assigned to a given Role; a Competency is the set of Skills that enable a person to take over his/her assigned Role; and an Authority is the rights to use Resources, make Decisions, sign approvals, and so forth.
A) _TRUE_ B) _FALSE_
ANSWER: A

Q155. A Team Charter includes the Project Team Values; Team Communications Guidelines, Decision-Making Process, Conflict Resolution Process, Meeting Guidelines, and Project Team Agreements.
A) _TRUE_ B) _FALSE_
ANSWER: A

Q156. The "Estimate Activity Resources" Process is to estimate the Project Team Resources, and the type and quantities of Materials, Equipment, and Supplies necessary to carry on the Project Work.
A) _TRUE_ B) _FALSE_
ANSWER: A

Q157. The main Inputs to the "Estimate Activity Resources" Process include the following:
A) The Project Management Plan along with its subsidiary Plans, which are the resource management plan, and the scope baseline.
B) The Project Management Plan along with its subsidiary Project Documents, which are the activity attributes, activity list, assumption log, cost estimates, resource calendars, and risk register.
C) Both Options above are Correct.
D) All Options above are Incorrect.
ANSWER: C

Q158. The Project Management Plan along with its subsidiaries; a set of relevant Enterprise Environmental Factors that includes the Resource Location, Resource Availability, Resource Skills, Organizational Culture, Estimated Data, and Marketplace Conditions; and a set of Organizational Process Assets that includes the Staffing Policies and Procedures, Handling Suppliers Policies and Procedures, and Historical Information of similar Projects.
A) _TRUE_ B) _FALSE_
ANSWER: A

Q159. The Basis of Estimates includes the Methods for Estimates Development, Assumptions made for Estimates, Constraints, Estimates ranges, and so forth.
A) _TRUE_ B) _FALSE_
ANSWER: A

Q160. The “Acquire Resources” Process is to obtain the Project Team, Facilities, Equipment, Materials, and Supplies that are necessary to complete the Project Work.
A) _TRUE_ B) _FALSE_
ANSWER: A

Q161. The main Inputs to the “Acquire Resources” Process include the following:
A) The Project Management Plan along with its subsidiary Plans, which are the Resource Management Plan, and the Cost Estimates; and a set of relevant Enterprise Environmental Factors that includes the Resources Availabilities, Locations, Skills, and experience Levels; the Organizational Structure; and the Market Place.
B) The Project Management Plan along with its subsidiary Project Documents, which are the Project Schedule, Resource Calendar, Resource Requirements, and Stakeholder Register; and a set of relevant Organizational Process Assets that includes the Staffing Policies and Procedures.
C) Both Options above are Correct.
D) All Options above are Incorrect.
ANSWER: C

Q162. A Virtual Team is a group of People with a shared Goal to fulfill their Roles with little or no time spent on meeting face to face. Virtual Teams makes it possible to establish a Team without having its members to be physically in the same place.
A) _TRUE_ B) _FALSE_
ANSWER: A

Q163. A Resource Calendar depicts the working days, shifts, Business hours, weekends, public holidays, and so forth.
A) _TRUE_ B) _FALSE_
ANSWER: A

Q164. The “Develop Team” Process is to improve the Project Team Members’ competencies, Interactions, and their work Environment. This Process improves Teamwork, enhances interpersonal skills and competencies, motivates employees, and improves the Project Performance.
A) _TRUE_ B) _FALSE_
ANSWER: A

Q165. The main Inputs to the “Develop Team” Process includes the Project Management Plan along with its subsidiary Plans and Project Documents; a set of relevant Enterprise Environmental Factors; and a set of relevant Organizational Process Assets.
A) _TRUE_ B) _FALSE_
ANSWER: A

Q166. Co-Location involves placing most of the Project Team Members in the same Physical location to enhance their ability to perform as a collaborative Team.
A) _TRUE_ B) _FALSE_
ANSWER: A

Q167. The main Methods used during the "Develop Team" Process to transfer Information to the Project Stakeholders are Conversations, Meetings, Databases, Documents, Websites, Social Media, and so forth.
A) _TRUE_ B) _FALSE_
ANSWER: A

Q168. The Communication Technology SELECTION-CRITERIA depends on many Factors such the Urgency of the Need for the Information, Availability and Reliability of Technology, Ease of Use Technology, Project Environment, and Sensitivity and Confidentiality of Information.
A) _TRUE_ B) _FALSE_
ANSWER: A

Q169. Recognition, Training, and Team Assessments are among the Methods used during the "Develop Team" Process.
A) _TRUE_ B) _FALSE_
ANSWER: A

Q170. The man Outputs of the "Develop Team" Process include the following:
A) The Team Performance Assessments, the set of change requests, and an updated version of the project management plan along with its subsidiary plans and project documents.
B) An updated set of relevant Enterprise Environmental Factors including the employee development plan records and the skills assessments.
C) An updated set of relevant Organizational Process Assets including the training requirements and the personnel assessments.
D) All Options above are Correct.
ANSWER: D

Q171. The "Manage Team" Process is to track the Performance of the Team Members, to provide them with feedback, to resolve their issues, and to manage their changes.
A) _TRUE_ B) _FALSE_
ANSWER: A

Q172. The main Inputs to the “Manage Team” Process are the Resource Management Plan; the Issue Log; the Lessons Learned Register; the Project Team Assignments; the Project Work Performance Data/Information/Reports; the Team Charter; the Enterprise Environmental Factor of the Human Resource Management Policies; and the Organizational Process Asset of the Certificates of Appreciation.
A) _TRUE_ B) _FALSE_
ANSWER: A

Q173. The “Control Resources” Process ensures that the Project Physical Resources are available in accordance with their Resource Management Plan. In addition, it monitors the planned for versus actual utilization of Resources and take Corrective Action as necessary.
A) _TRUE_ B) _FALSE_
ANSWER: A

Q174. The main Inputs to the “Control Resources” Process are the Project Management Plan along with its subsidiary Resource Management Plan, and its subsidiary Project Documents of the Issue Log, the Lessons Learned Register, the Physical Resource Equipments, the Project Schedule, the Resource Breakdown Structure, the Resource Requirements, and the Risk Register. In addition, the Inputs to this Process include the Project Work Performance Data/Information/Reports; the Project Agreements; and a set of relevant Organizational Process Assets including the Resource Control and Assignment Policies, Issues Escalation Procedure, and Lessons Learned from similar Projects.
A) _TRUE_ B) _FALSE_
ANSWER: A

Q175. The main Data Analysis Methods used during the “Control Resources” Process are the Alternatives Analysis, Cost-Benefit Analysis, Performance Reviews, and Trend Analysis.
A) _TRUE_ B) _FALSE_
ANSWER: A

Q176. The Problem-Solving Method used during the “Control Resources” Process goes through six steps to solve a problem. These steps are the Problem Identification, the Problem Definition, the Problem Investigation, the Problem Analyzation, the Problem Solving, and the Solution Checking and Validating.
A) _TRUE_ B) _FALSE_
ANSWER: A

Q177. The main Outputs of the "Control Resources" Process include the following:
A) The Project Work Performance Data/Information/Reports and a set of Change Requests.
B) Updated versions of the Resource Management Plan, the Schedule Baseline, and the Cost Baseline.
C) Updated versions of the Assumption Log, the Issue Log, the Lessons Learned Register, the Physical Resource Assignments, and the Risk Register.
D) All Options above are Correct.
ANSWER: D

Q178. The Project Communications Management Knowledge Area includes three processes that are necessary to ensure that the project informational needs are satisfied through a set of activities designed for effective information exchange.
A) _TRUE_ B) _FALSE_
ANSWER: A

Q179. The Processes of the Project Communications Management Knowledge Area are the "Plan Communications Management" Process, the "Manage Communications" Process, and the "Monitor Communications" Process.
A) _TRUE_ B) _FALSE_
ANSWER: A

Q180. The exchanged Information includes ideas, instructions, and/or emotions; and that can be in Writing or Spoken, formal or informal, through gestures or media.
A) _TRUE_ B) _FALSE_
ANSWER: A

Q181. The 5Cs Method is to reduce misunderstanding in Communications. It stands for:
A) **Correct** grammar and spelling/**Concise** expression and elimination of excess words/**Clear** purpose and expression directed to the Needs of the reader/**Coherent** logical flow of ideas/**Controlling** flow of words and ideas.
B) **Collect** grammar and spelling/**Concise** expression and elimination of excess words/**Clear** purpose and expression directed to the Needs of the reader/**Coherent** logical flow of ideas/**Controlling** flow of words and ideas.
C) **Correct** grammar and spelling/**Consolidate** expression and elimination of excess words/**Clear** purpose and expression directed to the Needs of the reader/**Coherent** logical flow of ideas/**Controlling** flow of words and ideas.
D) **Correct** grammar and spelling/**Concise** expression and elimination of excess words/**Clear** purpose and expression directed to the Needs of the reader/**Compromise** logical flow of ideas/**Controlling** flow of words and ideas.
ANSWER: A

Q182. The "Plan Communications Management" Process is to plan for the Project Communications.

A) _TRUE_ B) _FALSE_

ANSWER: A

Q183. The main Inputs to the "Plan Communications Management" Process include the following:

A) The Project Charter; and the Project Management Plan along with its subsidiary Resource Management Plan, the Stakeholder Engagement Plan, the Requirements Documentation, and the Stakeholder Register.

B) A set of relevant Enterprise Environmental Factors including the Organizational Culture, the Organizational Politics, the Organizational Governance, the Human Resources Policies, the Stakeholder Risk Thresholds, and the Geographical Distribution of Resources.

C) A set of relevant Organizational Process Assets including the Organizational Policies and Procedures for Social Media, Ethics, Security, Risks, CHANGE-REQUESTS, and Data Management; the Organizational Communications Requirements; and the Communication Data from similar Projects.

D) All Options above are Correct.

ANSWER: D

Q184. The main Outputs of the "Plan Communications Management" Process are the Communications Management Plan; an updated version of the Stakeholders Engagement Plan; and updated versions of the Project Schedule, and the Stakeholder Register.

A) _TRUE_ B) _FALSE_

ANSWER: A

Q185. The main Communication Models used during the "Plan Communications Management" Process are (i) the Basic Sender/Receiver Communication Model, with which Information is first encoded then transmitted, and once arrived at the receiver side, Information gets decoded and used; and (ii) Interactive Communication Model, with which once Information is received, the receiver acknowledges the reception of Information, then decode that.

A) _TRUE_ B) _FALSE_

ANSWER: A

Q186. The main Communication Methods used during the "Plan Communications Management" Process are the Interactive Communication, the Push Communication, the Pull Communication, the Interpersonal Communication, the Small Group Communication, the Public Communication, the Mass Communication, and the Networks and Social Computing Communication.

A) _TRUE_ B) _FALSE_

ANSWER: A

Q187. The "Manage Communications" Process is used to ensure effective and timely collection, creation, distribution, storage, retrieval, management, monitoring, and disposition of the Project Information.
A) _TRUE_ B) _FALSE_
ANSWER: A

Q188. The main Inputs to the "Manage Communications" Process include the resource management plan, the communications management plan, the stakeholders engagement plan, the change log, the issue log, the lessons learned register, the quality report, the risk report, the stakeholder register, the project work performance reports, a set of relevant enterprise environmental factors, and a set of relevant organizational process assets.
A) _TRUE_ B) _FALSE_
ANSWER: A

Q189. The main Enterprise Environmental Factors used as Inputs to the "Manage Communications" Process are the Organizational Culture, the Organizational Political Climate, the Organizational Governance Framework, the Human Resources Policies, the Stakeholder Risk Thresholds, and the Communication Channels. While The main Organizational Process Assets used as Inputs to this Process are the Corporate Policies and Procedures for Social Media, Ethics, Security, Risks, CHANGE-REQUESTS, and Data Management; the Organizational Communications Requirements; the Guidelines for Information Development, Exchange, Storage, and Retrieval; the Data from similar Projects; and the Lessons Learned Repository.
A) _TRUE_ B) _FALSE_
ANSWER: A

Q190. The main Outputs of the "Manage Communications" Process include the Project Communications including the Performance Reports, the Deliverable Status, the Schedule Progress, the Cost Incurred, and the Presentations; and updated versions of the Communications Management Plan, the Stakeholders Engagement Plan, the Issues Log, the Lessons Learned Register, the Project Schedule, the Risk Register, and the Stakeholder Register.
A) _TRUE_ B) _FALSE_
ANSWER: A

Q191. The "Monitor Communications" Process is to ensure that the project and its stakeholders' informational needs are satisfied.
A) _TRUE_ B) _FALSE_
ANSWER: A

Q192. The main Inputs to the "Monitor Communications" Process include the following:
- A) The Resource Management Plan, the Communications Plan, the Stakeholders Plan, the Issue Log, the Lessons Learned Register, and the Project Communications.
- B) The Project Work Performance Data/Information/Reports.
- C) A set of relevant Enterprise Environmental Factors including the Organizational Culture, the Organizational Political Climate, the Organizational Governance Framework, the Human Resources Policies, the Stakeholder Risk Thresholds, and the Communication Channels.
- D) A set of relevant Organizational Process Assets including the Corporate Policies and Procedures for Social Media, Ethics, Security, Risks, Change, and Data Management; the Organizational Communications Requirements; the Guidelines for Information Development, Exchange, Storage, and Retrieval; Data from similar Projects; and the Lessons Learned Repository.
- E) All Options above are Correct.

ANSWER: E

Q193. The main Outputs of the "Monitor Communications" Process include the following:
- A) The Project Work Performance Data/Information/Reports; and the set of Change Requests.
- B) Updated versions of the Communications Management Plan, the Stakeholder Engagement Plan, the Issue Log, the Lessons Learned Register, and the Stakeholder Register.
- C) Both Options above are Correct.
- D) All Options above are Incorrect.

ANSWER: C

Q194. The Project Risk Management Knowledge Area includes seven processes that are necessary to:
- A) Conduct Project Risks Management Planning.
- B) Identify and analyze these Risks.
- C) Plan, implement, and monitor responses to these Risks.
- D) All Options above are Correct.

ANSWER: D

Q195. The Project Risk Management Processes are the "Plan Risk Management" Process, the "Identify Risk" Process, the "Performance Qualitative Risk Analysis" Process, the "Performance Quantitative Risk Analysis" Process, the "Plan Risk Responses" Process, the "Implement Risk Responses" Process, and the "Monitor Risks" Process.

A) _TRUE_ B) _FALSE_

ANSWER: A

Q196. The "Plan Risk Management" Process is to define how to conduct the Project Risk Management Activities and should begin when the Project starts and should end early in the Project.

A) _TRUE_ B) _FALSE_

ANSWER: A

Q197. The Inputs to the "Plan Risk Management" Process include the following:

A) The Project Charter; the Risk Management Plan and the Stakeholder Register.

B) A set of relevant Organizational Process Assets including the Organizational Risk Policy, the Risk Categories, the Risk Statement Formats, and the Risk Management Plan Template.

C) The Enterprise Environmental Factor of the Risk Thresholds.

D) All Options above are Correct.

ANSWER: D

Q198. The Data Representation Method of the Stakeholders Analysis is to determine Risks that are associated with the Project Stakeholders.

A) _TRUE_ B) _FALSE_

ANSWER: A

Q199. The Risk Management Plan includes the following elements:

A) The **Risk Management Strategy** element that describes the general Approach to Managing Projects Risks; the **Risk Management Method** element that defines the Approaches, Tools, and Data Sources used to perform the Project Risk Management; and the **Roles and Responsibilities** element that defines the Risk Management Team Members and clarifies their Responsibilities.

B) The **Funding** element that identifies the funds needed to perform the Project Risk Management; the **Timing** element that schedules the Project Risk Management Processes throughout the Project life cycle; and the **Risk Categories and Risks Breakdown Structure RBS** Element.

C) The **Definitions of Risk Probability and Impacts** element; the **Probability and Impact Matrix** element; and the **Reporting Formats** element.

D) All Options above are Correct.

ANSWER: D

Q200. The "Identify Risks" Process is to identify and document the Project Risks and their characteristics.

A) _TRUE_ B) _FALSE_

ANSWER: A

Q201. The main Inputs to the "Identify Risks" Process include the Requirements Management Plan, the Schedule Management Plan, the Cost Management Plan, the Quality Management Plan, the Resource Management Plan, the Risk Management Plan, the Scope Baseline, the Schedule Baseline, and the Cost Baseline. In addition, the Inputs to this Process include the Assumption Log, the Cost Estimates, the Duration Estimates, the Issue Log, the Lesson Learned Register, the Requirements Documentation, the Resource Requirements, and the Stakeholder Register.
A) _TRUE_ B) _FALSE_
ANSWER: A

Q202. The Project Agreements, the Procurement Documentation, and relevant EEFs and OPAFs are used as Inputs to the "Identify Risks" Process.
A) _TRUE_ B) _FALSE_
ANSWER: A

Q203. The main Data Gathering Methods used during the "Identify Risks" Process are the brainstorming, the checklists, and the interviews.
A) _TRUE_ B) _FALSE_
ANSWER: A

Q204. The main Data Analysis Methods used during the "Identify Risks" Process are the Root-Cause Analysis, Assumption-and-Constraint Analysis, SWOT Analysis, and Documents Analysis.
A) _TRUE_ B) _FALSE_
ANSWER: A

Q205. The Risk Register contains the Results from the Perform Qualitative Risk Analysis Process, the Plan Risk Responses Process, the Implement Risk Responses Process, and the Monitor Risks Process. At the end of the Identify Risks Process, the Risk Register can contain things like the list of identified Risks, the Risk Owner, and the List of Risk Responses.
A) _TRUE_ B) _FALSE_
ANSWER: A

Q206. The Project Documents updated during the "Identify Risks" Process are the Assumption Log, the Issue Log, and the Lessons Learned Register.
A) _TRUE_ B) _FALSE_
ANSWER: A

Q207. The "Perform Qualitative Risk Analysis" Process is to prioritize the Project Risks by assessing mainly their probability and impact.
A) _TRUE_ B) _FALSE_
ANSWER: A

Q208. The main Inputs to the "Perform Qualitative Risk Analysis" Process include the Risk Management Plan, the Assumption Log, the Risk Register, and the Stakeholder Register.
A) _TRUE_ B) _FALSE_
ANSWER: A

Q209. The "Perform Quantitative Risk Analysis" Process is to analyze the combined effect of the Project Risks.
A) _TRUE_ B) _FALSE_
ANSWER: A

Q210. The main Inputs to the "Perform Quantitative Risk Analysis" Process are the Risk Management Plan, the Scope Baseline, the Schedule Baseline, the Cost Baseline, the Assumption Log, the Basis of Estimates, the Cost Estimates, the Cost Forecasts, the Duration Estimates, the Milestone List, the Resource Requirements, the Risk Register, and the Schedule Forecasts.
A) _TRUE_ B) _FALSE_
ANSWER: A

Q211. The main Data Analysis Methods used during the "Perform Quantitative Risk Analysis" Process are the Simulation Method, the Sensitivity Analysis, and the Decision Tree Analysis.
A) _TRUE_ B) _FALSE_
ANSWER: A

Q212. The main Documents that may be updated during the "Perform Quantitative Risk Analysis" Process are the Assessment of the Overall Project Risk Exposure, the Detailed Probabilistic Analysis of the Project, the Trends in Quantitative Risk Analysis, and the Recommended Risk Responses.
A) _TRUE_ B) _FALSE_
ANSWER: A

Q213. The "Plan Risk Responses" Process is to develop options, select Strategies, and agree on Actions to address and treat the project risks.
A) _TRUE_ B) _FALSE_
ANSWER: A

Q214. The main Inputs to the "Plan Risk Responses" Process include the following:
A) The Resource Management Plan, the Risk Management Plan, the Cost Baseline, the Lessons Learned Register, the Project Schedule, the Project Team Assignments, the Resource Calendars, the Risk Register, the Risk Report, and the Stakeholder Register.
B) The Enterprise Environmental Factor of the Stakeholders Risk Thresholds; and the Organizational Process Assets of the Risk Management Plans Templates, the Risk Register, the Risk Report, and the Lessons Learned from similar Projects.
C) Both Options above are Correct.
D) All Options above are Incorrect.
ANSWER: C

Q215. The main Outputs of the "Plan Risk Responses" Process include the following:
A) The Schedule Management Plan, the Cost Management Plan, the Quality Management Plan, the Resource Management Plan, the Procurement Management Plan, the Scope Baseline, the Schedule Baseline, and the Cost Baseline.
B) The Assumption Log, the Cost Forecasts, the Lessons Learned Register, the Project Schedule, the Project Team Assignments, the Risk Register, and the Risk Report.
C) Both Options above are Correct.
D) All Options above are Incorrect.
ANSWER: C

Q216. The "Implement Risk Responses" Process is to implement a set of pre-determined Risk Responses.
A) _TRUE_ B) _FALSE_
ANSWER: A

Q217. The main Inputs to the "Implement Risk Responses" Process include:
A) The Risk Management Plan, the Lesson Learned Register, the Risk Register, and the Risk Report.
B) The Organizational Process Asset of the Lessons Learned Repository.
C) Both Options above are Correct.
D) All Options above are Incorrect.
ANSWER: C

Q218. The main Outputs of the "Implement Risk Responses" Process are the Change Requests; and updated versions of the Issue Log, the Lessons Learned Register, the Project Team Assignments, the Risk Register, and the Risk Report.
A) _TRUE_ B) _FALSE_
ANSWER: A

Q219. The “Monitor Risks” Process is to monitor the Implementation of the predefined Risk Responses and track them, identify, and analyze new Risks.
A) _TRUE_ B) _FALSE_
ANSWER: A

Q220. The main Inputs to the “Monitor Risks” Process are the Risk Management Plan, the Issue Log, the Lessons Learned Register, the Risk Register, the Risk Report, and the Project Work Performance Data/Information/Reports.
A) _TRUE_ B) _FALSE_
ANSWER: A

Q221. The main Data Analysis Methods used during the “Monitor Risks” Process are the Technical Performance Analysis, and the Reserve Analysis.
A) _TRUE_ B) _FALSE_
ANSWER: A

Q222. The main Outputs of the “Monitor Risks” Process are the Change Requests, the Project Work Performance Data/Information/Reports, the Project Management Plan along with all of its subsidiary Plans, and the Assumption Log, the Issue Log, the Lessons Learned Register, the Risk Register, and the Risk Report.
A) _TRUE_ B) _FALSE_
ANSWER: A

Q223. The Project Procurement Management Knowledge Area includes three processes that are necessary to purchase Products, Solutions, and/or Services needed from outside the Project. These Processes are the “Project Procurement Management” Process, the “Conduct Procurements” Process, and the “Control Procurements” Process.
A) _TRUE_ B) _FALSE_
ANSWER: A

Q224. The “Plan Procurement Management” Process is to document the Procurement Decisions and identify possible sellers.
A) _TRUE_ B) _FALSE_
ANSWER: A

Q225. The main Inputs to the "Plan Procurement Management" Process include the following:

A) The Project Charter, the Scope Management Plan, the Quality Management Plan, the Resource Management Plan, the Scope Baseline, the Milestone List, the Project Team Assignments, the Requirements Documentation, the Requirements Traceability Matrix, the Resource Requirements, the Risk Register, and the Stakeholder Register.

B) The Business Documents of the Business Case, and the Benefits Management Plan.

C) The Enterprise Environmental Factors of the Marketplace Conditions, the Potential Suppliers, the Products and Services Availability in the Market, the Contract Management Systems, and the Financial Accounting Payments Systems.

D) All Options above are Correct.

ANSWER: D

Q226. The main Data Gathering Method used during the "Plan Procurement Management Process is the market research method.

A) _TRUE_ B) _FALSE_

ANSWER: A

Q227. The main Data Analysis Method used during the "Plan Procurement Management" Process is the make-or-buy method.

A) _TRUE_ B) _FALSE_

ANSWER: A

Q228. The main Source-Selection Analysis Methods used during the "Plan Procurement Management" Process are the Least-Cost Analysis, the Qualifications-only Analysis, the Quality-Based Analysis, the Cost-Based Analysis, the Sole-Source Analysis, and the Fixed Budget Analysis.

A) _TRUE_ B) _FALSE_

ANSWER: A

Q229. The main Outputs of the "Plan Procurement Management" Process are the Procurement Management Plan, the Procurement Strategy, the Bid Documents, the Procurement Statement of Work, the Source Selection Criteria, the Make-or -Buy Decisions, and the Independent Cost Estimates.

A) _TRUE_ B) _FALSE_

ANSWER: A

Q230. The "Conduct Procurements" Process is to solicit and obtain proposals from service providers, before selecting one of them, and negotiating and signing a contract with the selected service provider.

A) _TRUE_ B) _FALSE_

ANSWER: A

Q231. The main Inputs to the "Conduct Procurements" Process include the following:
A) The Scope Management Plan, the Requirements Management Plan, the Communications Management Plan, the Risk Management Plan, the Procurement Management Plan, the Configuration Management Plan, the Cost Baseline, the Lessons Learned Register, the Project Schedule, the Requirements Documentation, the Risk Register, and the Stakeholder Register.
B) The Procurement Documents including the Bid Documents, the Procurement Statement of Work, the Independent Cost Estimates, and the Source Selection Criteria.
C) The Service Providers Proposals in response to the Procurement Request for Proposals (RFP).
D) All Options above are Correct.
ANSWER: D

Q232. A Bidder Conference is to conduct a meeting between the Customer and a Prospective Service-Provider prior to their Proposals Submission.
A) _TRUE_ B) _FALSE_
ANSWER: A

Q233. The List of the Shortlisted Service Providers includes the initially selected Seller/Bidders/Service Providers based on the outcomes of the evaluation of their proposals.
A) _TRUE_ B) _FALSE_
ANSWER: A

Q234. The "Control Procurements" Process is to manage the Procurement Relationships, monitor the Contract Performance and make changes and corrections as appropriate and close out Contracts.
A) _TRUE_ B) _FALSE_
ANSWER: A

Q235. The Inputs to the "Control Procurements" Process include the following:
A) The Requirements Management Plan, the Risk Management Plan, the Procurement Management Plan, the Change Management Plan, the Schedule Baseline, the Assumption Log, the Lessons Learned Register, the Milestone List, the Quality Reports, the Requirements Documentation, the Requirements Traceability Matrix, the Risk Register, and the Stakeholder Register.
B) The Project Agreements; the Change Requests, the Project Work Performance Data/Information/Reports; and the Procurement Documentation including the Statement of Work, the Payment Information, the Contractor Work Performance Information, the Project Plans/Drawings, and the Correspondence Details.
C) Both Options above are Correct.
D) All Options above are Incorrect.
ANSWER: C

Q236. The Outputs of the "Control Procurements" Process include the following:

A) The Closed Procurements, the Change Requests, and the Project work Performance Data/Information/Reports.

B) Updated versions of the Procurements Documents including the Project Contract/Schedules, the Approved and Unapproved Change Requests, the Technical Documents, and so forth.

C) Updated versions of the Risk Management Plan, the Procurement Management Plan, the Schedule Baseline, the Cost Baseline, the Lessons Learned Register, the Resource Requirements, the Requirements Traceability Matrix, the Risk Register, and the Stakeholder Register.

D) All Options above are Correct.

ANSWER: D

Q237. The Project Stakeholder Management Knowledge Area includes four processes that are necessary to (i) identify the People that impact or be impacted by the Project, (ii) analyze the Stakeholder Expectations and their Impact on the Project, and (iii) develop appropriate Management Strategies to effectively engage Stakeholders in Project.

A) _TRUE_ B) _FALSE_

ANSWER: A

Q238. The "Identify Stakeholders" Process is to identify the Project Stakeholders and their interests, involvement, inter-dependencies, and influence.

A) _TRUE_ B) _FALSE_

ANSWER: A

Q239. The main Inputs to the "Identify Stakeholders" Process include the Project Charter, the Business Case Document, the Benefits Management Plan, the Communications Management Plan, the Stakeholders Engagement Plan, the Change Log, the Issue Log, the Requirements Documentation, and the Project Agreements.

A) _TRUE_ B) _FALSE_

ANSWER: A

Q240. The main Data Gathering Methods used during the "Identify Stakeholders" Process are the Questionnaires, Surveys, and Brainstorming.

A) _TRUE_ B) _FALSE_

ANSWER: A

Q241. The main Data Analysis Methods used during the "Identify Stakeholders" Process are the Stakeholder Analysis, and the Documents Analysis.

A) _TRUE_ B) _FALSE_

ANSWER: A

Q242. The Stakeholder Register contains information about the Project Stakeholders such as their Identification Information, Assessment Information, and Stakeholder Classification.
A) _TRUE_ B) _FALSE_
ANSWER: A

Q243. The "Plan Stakeholder Engagement" Process is to involve the Project Stakeholders based on their Needs, Expectations, interests, and potential Impact on the Project.
A) _TRUE_ B) _FALSE_
ANSWER: A

Q244. The main Inputs to the "Plan Stakeholder Engagement" Process are the Project Charter, the Resource Management Plan, the Communications Management Plan, the Risk Management Plan, the Assumption Log, the Change Log, the Issue Log, the Project Schedule, the Risk Register, and the Stakeholder Register.
A) _TRUE_ B) _FALSE_
ANSWER: A

Q245. The "Manage Stakeholder Engagement" Process is to communicate with Stakeholders to meet their expectations, to address their Issues, and to foster appropriate stakeholder involvement.
A) _TRUE_ B) _FALSE_
ANSWER: A

Q246. The main Inputs to the "Manage Stakeholder Engagement" Process are the Communications Management Plan, the Risk Management Plan, the Stakeholders Engagement Plan, the Change Management Plan, the Change Log, the Issue Log, the Lessons Learned Register, and the Stakeholder Register.
A) _TRUE_ B) _FALSE_
ANSWER: A

Q247. The "Monitor Stakeholder Engagement" Process is to monitor the Stakeholders' Relationships and to tailor Strategies for engaging Stakeholders through the modification of Engagement Strategies and Plans.
A) _TRUE_ B) _FALSE_
ANSWER: A

Q248. The main Inputs to the "Monitor Stakeholder Engagement" are the Resource Management Plan, the Communications Management Plan, the Stakeholders Engagement Plan, the Issue Log, the Lessons Learned Register, the Project Communications, the Risk Register, the Stakeholder Register, and the Project Work Performance Data/Information/Reports.
A) _TRUE_ B) _FALSE_
ANSWER: A

Q249. The main Outputs of the "Monitor Stakeholder Engagement" Process are the Project Work Performance Data/Information/Reports, the Change Requests, the Resource Management Plan, the Communications Management Plan, the Stakeholders Engagement Plan, the Issue Log, the Lesson Learned Register, the Risk Register, and the Stakeholder Register.
A) _TRUE_ B) _FALSE_
ANSWER: A

Appendix B.3: Chapter 4 Bank of Questions

Q1. What is Software?

A) Software solutions are for particular customers or for general customers in the markets.

B) Software is a Computer Program that comes along with specific documentation.

C) Software is a Java Program that comes along with specific documentation.

D) All Options above are Correct Answers.

ANSWER: B

Q2. What are the Attributes of Good Software?

A) Software is a Computer Program that comes along with specific good documentation.

B) Software is a Java Program that comes along with specific good documentation.

C) Good Software solutions provide the required functionalities and performance levels. Also, they are expected to be maintainable, reliable, secured, efficient, and easy-to-use.

D) All Options above are Correct Answers.

ANSWER: C

Q3. What are the Fundamental Software Engineering Activities?

A) The Fundamental Software Engineering Activities are the Software Specification, Software Development, Software Validation, and Software Evolution.

B) The Fundamental Software Engineering Activities are the Software Initiation, Software Development, Software Validation, and Software Evolution.

C) The Fundamental Software Engineering Activities are the Software Specification, Software Development, Software Deployment, Software Validation, and Software Evolution.

D) All Options above are Correct Answers.

ANSWER: A

Q4. What are the Differences between Software Engineering and Computer Science?

A) Computer Science focuses on Computer-related theories and fundamentals, while Software Engineering focuses on the practicalities of developing and delivering useful Software solutions.

B) The Fundamental Software Engineering Activities are the Software Specification, Software Development, Software Validation, and Software Evolution.

C) Both Options above are Correct Answers.

D) All Options above are Incorrect.

ANSWER: C

Q5. What are the Differences between Software Engineering and Systems Engineering?

A) Engineering focuses on the practicalities of developing and delivering useful Software solutions.

B) Software Engineering is part of the Systems Engineering, which is concerned with all Aspects of Computer-Software-Based Systems Development including Hardware, Software, and Process Engineering.

C) Both Options above are Correct Answers.

D) All Options above are Incorrect.

ANSWER: B

Q6. What are the Challenges Facing Software Engineering?

A) The main challenges that Software Engineering is facing are the need to cope with the increasing diversity, the demands for reduced delivery times, and the need to develop trustworthy Software solutions.

B) The main challenges that Software Engineering is facing include the fact that it is part of the Systems Engineering, which is concerned with all aspects of Computer-Software-based Systems Development including Hardware, Software, and Process Engineering.

C) Both Options above are Correct Answers.

D) All Options above are Incorrect.

ANSWER: A

Q7. How much is the Cost of Software Engineering?

A) About 60% of Software Costs are mainly Development Costs and the left 40% are Testing Costs.

B) Most of the Software Costs are mainly Development Costs, not Testing Costs.

C) Both Options above are Correct Answers.

D) All Options above are Incorrect.

ANSWER: C

Q8. What are the Best Software Engineering Techniques?

A) While all Software Projects have to be professionally managed and developed, different methods are good for different types of Systems.

B) Different methodologies are good for managing different types of Software Projects.

C) Both Options above are Correct Answers.

D) All Options above are Incorrect.

ANSWER: C

Q9. What is the Computer Scientist Role?

A) A Computer Scientist works in multi-Application Software and IT Domains. He/she collects and analyzes the Requirements of his/her Client and design, implement, test, and deploy their targeted Solutions.

B) A Computer Scientist develops Solutions that solve specific Problems for his/her Clients. He/she uses Computers and Programming Languages along with relevant Tools, and Methods to complete his/her Tasks and Activities within predefined Time-Boundaries.

C) A Computer Scientist proves Theorems about Algorithms, designs Computer Programming Languages, and defines Knowledge Representation Schemes. He/she can take his/her time to complete his/her Tasks.

D) All Options above are Incorrect.

ANSWER: C

Q10. What is the Software Engineer Role?

A) A Software Engineer works in multi-Application Software and IT Domains. He/she collects and analyzes the Requirements of his/her Client and design, implement, test, and deploy their targeted Solutions.

B) A Software Engineer develops Solutions that solve specific Problems for his/her Clients. He/she uses Computers and Programming Languages along with relevant Tools and Methods to complete his/her Tasks and Activities within predefined Time-Boundaries.

C) A Software Engineer proves Theorems about Algorithms, designs Computer Programming Languages, and defines Knowledge Representation Schemes. He/she can take his/her time to complete his/her Tasks.

D) All Options above are Incorrect.

ANSWER: A

Q11. What is the Engineer Role?

A) An Engineer works in multi-Application Software and IT Domains. He/she collects and analyzes the Requirements of his/her Client and design, implement, test, and deploy their targeted Solutions.

B) An Engineer develops Solutions that solve specific Problems for his/her Clients. He/she uses Computers and Programming Languages along with relevant Tools and Methods to complete his/her Tasks and Activities within predefined Time-Boundaries.

C) An Engineer proves Theorems about Algorithms, designs Computer Programming Languages, and defines Knowledge Representation Schemes. He/she can take his/her time to complete his/her Tasks.

D) All Options above are Incorrect.

ANSWER: B

Q12. The latest version of the ISO/IEC TR 19759 standard for Software Engineering (Guide to the Software Engineering Body of Knowledge SWEBOK) was issued in year 2005.
A) _TRUE_ B) _FALSE_
ANSWER: B

Q13. A Software Requirement is an Essential Functional or Non-Functional Feature that must be:
A) Identified, analyzed, and counted-in during the Software Design and Implementation; and
B) Tested, qualified, and brought into an operational mode as part of the overall Software Solution.
C) Both Options above are Correct.
D) All Options above are Incorrect.
ANSWER: C

Q14. The Software Requirements Knowledge Area is concerned with the Elicitation, Analysis, Specification, Verification, and Validation of Software Requirements.
A) _TRUE_ B) _FALSE_
ANSWER: A

Q15. The Product Requirements refer to what is needed in the targeted Product, including the Constraints imposed on the Product by the Customer.
A) _TRUE_ B) _FALSE_
ANSWER: A

Q16. The Functional Requirements include the Functionality of the targeted Software Solution.
A) _TRUE_ B) _FALSE_
ANSWER: A

Q17. The Non-Functional Requirements include the Quality Factors that constrain the Solution. These Quality Factors include Performance, Privacy, Security, Safety, Scalability, Maintainability, Serviceability, Supportability, Availability, Interoperability, Scalability, and Reliability. Each Quality Factor must have a set of predefined Quality Metrics that can be used to assess and evaluate the concerned Solution.
A) _TRUE_ B) _FALSE_
ANSWER: A

Q18. A System is a composition of Components such as Hardware, Software, Firmware, People, Information, Techniques, Tools, Facilities, and Services.
A) _TRUE_ B) _FALSE_
ANSWER: A

Q19. Emergent Requirements are those that cannot be addressed by a single Component of the System, but rather they depend on how the Software Components interoperate.
A) _TRUE_ B) _FALSE_
ANSWER: A

Q20. Quantifiable Requirements refers to the Non-Functional Requirements that should be stated such that they are Precise, Verified, Clear, Unambiguous, Qualitative, and Quantitative.
A) _TRUE_ B) _FALSE_
ANSWER: A

Q21. Requirements Analysis involves the following:
A) Requirements Classification, Conceptual Modeling, and Negotiation.
B) Requirements Architectural Design and Allocation.
C) Requirements Formal Analysis.
D) All Options above are Correct.
ANSWER: D

Q22. Requirements Specification Documentation includes the following:
A) System Definition Document, System Requirements Specifications, and Software Requirements Specifications.
B) Software Definition Document, Software Requirements Specifications, and Software Requirements Specifications.
C) Requirements Architectural Design and Allocation Document.
D) All Options above are Correct.
ANSWER: A

Q23. Requirements Validation involves the following:
A) Requirements Reviews, Prototyping, Model Validation, and Acceptance Tests.
B) Requirements Reviews and Audits, Prototyping, Model Validation, and Acceptance Tests.
C) Both Options above are Correct.
D) All Options above are Incorrect.
ANSWER: B

Q24. There are several International ISO/IEC standards that support the Software Requirements Knowledge Area.
A) _TRUE_ B) _FALSE_
ANSWER: A

Q25. Software Design is the Process of defining the Architecture, Components, Interfaces, and other Characteristics of a System.
A) _TRUE_ B) _FALSE_
ANSWER: A

Q26. Software Design involves certain key design issues such as Concurrency; and Handling Events, Errors, and Exceptions.
A) _TRUE_ B) _FALSE_
ANSWER: A

Q27. Software Structure and Architecture involves Architectural Viewpoints, Styles, and Design Decisions; and Design Patterns.
A) _TRUE_ B) _FALSE_
ANSWER: A

Q28. Software Interface Design involves User Interface Principles, User Interface Design Process, and Localization/Internationalization.
A) _TRUE_ B) _FALSE_
ANSWER: A

Q29. Software Construction is the creation of working Software solutions through a combination of Coding, Verification, Validation, Unit Testing, Integration Testing, and Debugging.
A) _TRUE_ B) _FALSE_
ANSWER: A

Q30. The Software Construction Knowledge Area is connected to most of the other Software Engineering Knowledge Areas, particularly the Software Design and the Software Testing ones.
A) _TRUE_ B) _FALSE_
ANSWER: A

Q31. The Practical Considerations while constructing Software Solutions include the following:
A) Software Construction Design, Language, Coding, and Testing.
B) Software Construction Quality, Integration, Coding, and Testing.
C) Construction for Reuse and Construction with Reuse.
D) All Options above are Correct.
ANSWER: D

Q32. The Software Construction Techniques include the following:
A) State-Based and Table-Driven Construction Techniques.
B) Assumptions, Design by Contract, and Defensive Programming.
C) Both Options above are Correct.
D) All Options above are Incorrect.
ANSWER: C

Q33. Software Testing is a Dynamic Verification of a Software Solution to ensure that it behaves as expected on a finite set of TEST-CASES.
A) _TRUE_ B) _FALSE_
ANSWER: A

Q34. Dynamic Testing implies that the Software Solution is tested using a pre-selected set of inputs.
A) _TRUE_ B) _FALSE_
ANSWER: A

Q35. Finite Testing implies that the Software Solution is tested using a subset of all possible TEST-CASES selected from a large set of TEST-CASES.
A) _TRUE_ B) _FALSE_
ANSWER: A

Q36. The subset SELECTION-CRITERIA depends on the Requirements that the Software Solution must be tested for, the Validation and Verification Requirements, and any other Implicit Requirements.
A) _TRUE_ B) _FALSE_
ANSWER: A

Q37. Software Maintenance is an Integral part of a Software Solution Life-Cycle. However, it is not given enough attention in comparison with the Software Development. However, things started to change recently as more attention is being paid to Software Maintenance.
A) _TRUE_ B) _FALSE_
ANSWER: A

Q38. Reengineering and Reverse Engineering are used for the same purpose and give similar outcomes. However, the Reverse Engineering is used when there is no adequate details about the source code and the various documentation of the system that needs to be revised, restructured, and/or reengineered.
A) _TRUE_ B) _FALSE_
ANSWER: A

Q39. Software retirement indicates that the Software is no longer needed, no longer can help the users, no longer can cope with the business requirements, and so forth.
A) _TRUE_ B) _FALSE_
ANSWER: A

Q40. Software Configuration Management is a supporting the Software Life-Cycle Process that benefits the Project Management, Development and Maintenance Activities, and Quality Assurance Activities.
A) _TRUE_ B) _FALSE_
ANSWER: A

Q41. Software Configuration Auditing involves the Auditing of the Software Functional Configuration and the Software Physical Configuration.
A) _TRUE_ B) _FALSE_
ANSWER: A

Q42. Software Engineering Management is used to present a Software Project Management Approach. It is associated with the other 14 Software Engineering Knowledge Areas.
A) _TRUE_ B) _FALSE_
ANSWER: A

Q43. The IEEE-CS Software Project Management Approach composes six phases, which are the Initiation and Scope Definition Phase, Software Project Planning Phase, Software Project Enactment Phase, Review and Evaluation Phase, Closure Phase, and Software Engineering Measurement Phase.
A) _TRUE_ B) _FALSE_
ANSWER: A

Q44. The Initiation and Scope Definition Phase of the IEEE-CS Software Project Management Approach is to quickly review a RFP and decide either to go or not to go with the Software Project Opportunity associated with that RFP.
A) _TRUE_ B) _FALSE_
ANSWER: A

Q45. The Initiation and Scope Definition Phase of the IEEE-CS Software Project Management Approach includes the following Processes:
A) Determination and Negotiation of Requirements.
B) Feasibility Study and Analysis.
C) Process for the Review and Revision of Requirements.
D) All Options above are Correct.
ANSWER: D

Q46. The Software Project Planning Phase of the IEEE-CS Software Project Management Approach is to carefully review an RPF and create a detailed Plan that defines the Work needed to SUCCESSFULLY carry on and manage the Software Project associated with that RFP.

A) _TRUE_ B) _FALSE_

ANSWER: A

Q47. The Software Project Planning Phase of the IEEE-CS Software Project Management Approach includes the following Processes:

A) The Processes for Process Planning and Determining the Deliverables.

B) The Processes for estimating the Effort, Schedule, and Cost of the Project.

C) The Processes for Resource Allocations, Risk Management, and Quality Management.

D) All Options above are Correct.

ANSWER: D

Q48. The Software Project Enactment Phase of the IEEE-CS Software Project Management Approach is to execute, monitor, and control the Plan created during the earlier Software Project Planning Phase.

A) _TRUE_ B) _FALSE_

ANSWER: A

Q49. The Software Project Enactment Phase of the IEEE-CS Software Project Management Approach includes the following Processes:

A) The Processes for Plans Implementation, Software Acquisition and Supplier Contract Management, and Reporting.

B) The Processes for Implementing the Measurement Process, Monitoring, and Controlling the Project.

C) Both Options above are Correct.

D) All Options above are Incorrect.

ANSWER: C

Q50. The Review and Evaluation Phase of the IEEE-CS Software Project Management Approach is to review and evaluate the Work throughout the Project Life-Cycle and ensure that things are going SATISFACTORILY. The same thing applies to the Project Work related to its Schedule, Cost, and Quality Assurance and Control.

A) _TRUE_ B) _FALSE_

ANSWER: A

Q51. The Review and Evaluation Phase of the IEEE-CS Software Project Management Approach includes the following Processes:
A) The Processes for Determining the Requirements Satisfaction.
B) The Processes for Reviewing and Evaluating the Project and the Product Performance.
C) Both Options above are Correct.
D) All Options above are Incorrect.
ANSWER: C

Q52. The Closure Phase of the IEEE-CS Software Project Management Approach is to determine the Work that must be SUCCESSFULLY completed in order to declare the Project completion.
A) _TRUE_ B) _FALSE_
ANSWER: A

Q53. The Software Engineering Measurement Phase of the IEEE-CS Software Project Management Approach is to measure and assess the Software Solution after it is operational.
A) _TRUE_ B) _FALSE_
ANSWER: A

Q54. Software Engineering Processes are concerned with the Project Activities carried on by Software Engineers to develop, maintain, and operate Software solutions.
A) _TRUE_ B) _FALSE_
ANSWER: A

Q55. Software Engineering Processes include the following:
A) Software Requirements-specific Processes, Software Design-specific Processes, and Software Construction-specific Processes.
B) Software Testing-specific Processes, Software Maintenance-specific Processes, and Software Configuration-specific Processes.
C) Software Management-specific Processes, and Software Quality Assurance and Control-specific Processes.
D) All Options above are Correct.
ANSWER: D

Q56. Software Engineering Models and Methods are concerned with Engineering and Structuring Software Solutions. Structuring Software Solutions make them Systematic, Repeatable, and SUCCESSFUL.
A) _TRUE_ B) _FALSE_
ANSWER: A

Q57. Software Engineering Methods provide an Approach for Specifying, Designing, Constructing, Testing, Verifying, and Validating the Final Software Solutions.
A) _TRUE_ B) _FALSE_
ANSWER: A

Q58. Software Quality is concerned with the Software Engineering Perspective of:
A) Culture and Ethics, Value and Cost of Quality, Quality Models and Quality Characteristics, and Quality Improvement.
B) Software Safety, Software Quality Assurance, Software Quality Control, Software and Systems Verification and Validation, and Software Reviews and Audits.
C) Software Quality Requirements, Software Defects Characterization, Software Quality Management Techniques, Software Quality Measurements, and Software Quality Tools.
D) All Options above are Correct.
ANSWER: D

Q59. The Software Quality Management Processes include:
A) Software Quality Assurance Process.
B) Software Verification and Validation Process.
C) Software Reviews and Audits.
D) All Options above are Correct.
ANSWER: D

Q60. Software Engineering Professional Practice is concerned with the Knowledge, Skills, and Attitudes that Software Engineers must have to practice Software Engineering in a Professional, Responsible, and Ethical Manner.
A) _TRUE_ B) _FALSE_
ANSWER: A

Q61. Professionalism implies Accreditation, Certification, Licensing, Codes of Ethics and Professional Practice, and Documentation.
A) _TRUE_ B) _FALSE_
ANSWER: A

Q62. Groups of Dynamics and Psychology involve the Individual Cognitions; Handling with Problems Complexities; Interacting with Stakeholders; Dealing with Uncertainties, Ambiguities, and Multicultural Work Environments.
A) _TRUE_ B) _FALSE_
ANSWER: A

Q63. Communication Skills involve Reading, Understanding, Writing, Summarizing, Presentation, and Team Communication Skills.
A) _TRUE_ B) _FALSE_
ANSWER: A

Q64. Software Engineering Economics (SEE) is concerned with making Business Decisions related to Software Engineering. Economics is the study of Values, Costs, Resources, and their Relationship under certain Circumstances.
A) _TRUE_ B) _FALSE_
ANSWER: A

Q65. Software Engineering Economics involves the following topics:
A) Finance, Accounting, Controlling, Decision Making Process, Valuation, and Inflation.
B) Depreciation, Taxation, Cash Flow, Time-Value of Money, Efficiency, Effectiveness, and Productivity.
C) Both Options above are Correct.
D) All Options above are Incorrect.
ANSWER: C

Q66. The main Software Engineering Economics Analysis Methods include the following:
A) For-Profit Decision Analysis, Minimum Acceptable Rate of Return, Return on Investment, and Return on Capital Employed.
B) Cost-Benefit Analysis, Cost-Effectiveness Analysis, Break-Even Analysis, Multiple Attribute Evaluation, and Optimization Analysis.
C) Both Options above are Correct.
D) All Options above are Incorrect.
ANSWER: C

Q67. Among the Practical Considerations that must be taken into account when dealing with the Software Engineering Economics are the Good-Enough Principle, the Friction-Free Economies, the Ecosystems, and the Offshoring and Outsourcing.
A) _TRUE_ B) _FALSE_
ANSWER: A

Q68. Computing Foundation includes Knowledge about Computers and their underlying Principles of Hardware and Software that serve as a Framework for Software Engineering.
A) _TRUE_ B) _FALSE_
ANSWER: A

Q69. The Computing Foundations Knowledge Area includes many topics such as:

A) Problem-Solving Techniques, Abstraction, Programming Language Basics, Debugging Tools and Techniques, Data Structure and Representation, Algorithms and Complexity, and System Basic Concepts.

B) Computer Organization, Operating Systems Basics, Compiler Basics, Database Basics, Data Management, Network Communication Basics, Parallel and Distributed Computing and Basic Developer Human Factors.

C) Basic developer human factors and secure software development and maintenance.

D) All Options above are Correct.

ANSWER: D

Q70. The Mathematical Foundations Knowledge Area includes various topics such as the Sets, Relations, Functions, Basic Logic, Proof Techniques, Basic Counting, Graphs and Trees, Discrete Probability, Finite State Machines, Grammars, Numerical Precision, Accuracy, Errors, Number Theory, and Algebraic Structures.

A) _TRUE_ B) _FALSE_

ANSWER: A

Q71. The Engineering Foundations includes various topics such as the Empirical Methods and Experimental Techniques, Statistical Analysis, Measurement, Engineering Design, Modeling, Simulation, Prototyping, Standards, and Root-Cause Analysis.

A) _TRUE_ B) _FALSE_

ANSWER: A

Q72. Software Requirements are Technically Oriented and their Specifications are Application Oriented.

A) _TRUE_ B) _FALSE_

ANSWER: B

Q73. Software Configuration Management is a "Supporting-Software Life Cycle Process" that benefits Project Management, Development and Maintenance Activities except Quality Assurance Activities.

A) _TRUE_ B) _FALSE_

ANSWER: B

Q74. __________ is a property that must be exhibited by something in order to solve some problem in the real world.

A) A Software Requirement

B) Software Design

C) Software Construction

D) Software Testing

ANSWER: A

Q75. __________ is a Dynamic Verification to ensure that a Program behaves as expected on a Finite Set of Test Cases that are selected from an Infinite Execution Domain.
A) Software Requirement
B) Software Design
C) Software Construction
D) Software Testing
ANSWER: D

Q76. Which of the following Organization is associated with Software Engineering Discipline?
A) AMC
B) IEEX
C) IEEE
D) Both of A and C above are Correct
ANSWER: C

Q77. Software Quality is concerned with the __________
A) Software Systems Verification and Validation
B) Software Safety
C) Both of A and B above are Correct
D) None of the above is Correct
ANSWER: C

Q78. "SWE Methods" provides an approach to the __________
A) Problem-Solving and Procedures for Construction and Analysis of the End-Item Software
B) Systematic Specification, Design, Construction, Test, and Verification of the End-Item Software
C) Both of A and B above are Correct
D) All of the above are Correct
ANSWER: B

Q79. The following topics: Statistical Analysis; (Modeling, Simulation, and Prototyping); and Empirical Methods and Experimental Techniques belong to __________
A) Mathematical Foundations
B) Computing Foundations
C) Engineering Foundations
D) All of the above are Correct
ANSWER: C

Q80. SWE Professional Practice is concerned with Professionalism, which involves __________

A) Interacting with Stakeholders

B) Writing Skills

C) Dealing with Multicultural Environments

D) None of the above is Correct

ANSWER: D

Q81. The most Fundamental concept in Software Engineering is __________

A) The Software Life Cycle

B) The Software Requirements and Design Phases

C) The Software Testing and Maintenance Phases

D) None of the above is Correct

ANSWER: A

Q82. Which one of the following is not part of the SWE Mathematical Foundation?

A) Finite State Machines

B) Modeling, Simulation, and Prototyping

C) Basic Logic

D) Discrete Probability

E) Number Theory

ANSWER: B

Q83. Which one of the following is not part of communication Skills that SWE should practice?

A) Reading, Understanding, and Summarizing Skills

B) Writing Skills

C) Dealing with Problem Complexity

D) Presentation Skills

ANSWER: C

Q84. The process of defining the Architecture, Components, Interfaces, and other Characteristics of a System or Component is __________

A) Software Requirements

B) Software Design

C) Software Testing

D) Software Maintenance

ANSWER: B

Q85. SWE Professional Practice is concerned with Professionalism, which involves __________

A) Interacting with Stakeholders

B) Writing Skills

C) Dealing with Multicultural Environments

D) None of the above is Correct

ANSWER: D

Q86. Time, Cost, Quality, and Scope are the only possible constraints that may affect projects.

A) _TRUE_ B) _FALSE_

ANSWER: B

Appendix B.4: Chapter 5 Bank of Questions

Q1. Business Justification is one of the Scrum Areas.
A) _TRUE_ B) _FALSE_
ANSWER: B

Q2. Business Justification is one of the Scrum Aspects Area.
A) _TRUE_ B) _FALSE_
ANSWER: A

Q3. Organization is one of the Scrum Areas.
A) _TRUE_ B) _FALSE_
ANSWER: B

Q4. Organization is one of the Scrum Aspects Area.
A) _TRUE_ B) _FALSE_
ANSWER: A

Q5. Quality is one of the Scrum Areas.
A) _TRUE_ B) _FALSE_
ANSWER: B

Q6. Quality is one of the Scrum Aspects Area.
A) _TRUE_ B) _FALSE_
ANSWER: A

Q7. Change is one of the Scrum Areas.
A) _TRUE_ B) _FALSE_
ANSWER: B

Q8. Change is one of the Scrum Aspects Area.
A) _TRUE_ B) _FALSE_
ANSWER: A

Q9. Risk is one of the Scrum Areas.
A) _TRUE_ B) _FALSE_
ANSWER: B

Q10. Risk is one of the Scrum Aspects Area.
A) _TRUE_ B) _FALSE_
ANSWER: A

Q11. The . . . . for Agile Software Development was introduced in 2001 for finding better ways for Software Development.
A) Manifesto
B) Manifist
C) Manifistication
D) All Options above are Incorrect
ANSWER: A

Q12. Agile Software Development is meant to:
A) Have efficient and effective interactions among the Team Members.
B) Have an ongoing delivery of working Software Components.
C) Pay less attention to Processes, Tools, and Comprehensive Documentation.
D) All Options above are Correct
ANSWER: D

Q13. Agile Software Development welcomes and endorses CHANGE-REQUESTS received from the Customer throughout the Project Life Cycle.
A) _TRUE_ B) _FALSE_
ANSWER: A

Q14. Agile Software Development focuses on the Individuals and their interactions rather than on the tools and processes.
A) _TRUE_ B) _FALSE_
ANSWER: A

Q15. Agile Software Development focuses on the continuous delivery of working Software rather than on developing comprehensive Documentation.
A) _TRUE_ B) _FALSE_
ANSWER: A

Q16. Agile Software Development focuses on collaborating with the Customer rather than on negotiating the Project Contract.
A) _TRUE_ B) _FALSE_
ANSWER: A

Q17. Agile Software Development focuses on welcoming changes and responding to them rather than on complying with the Project Plans.
A) _TRUE_ B) _FALSE_
ANSWER: A

Q18. The main principles of Agile include the following:
A) The highest priority should be satisfying the Customer through continuous delivery of valuable Software.
B) CHANGE-REQUESTS are welcomed during the Project.
C) Delivery of working Software Components is incremental and ongoing.
D) All Options above are Correct.
ANSWER: D

Q19. The main principles of Agile include the following:
A) Projects must be carried on around motivated Individuals to get the Projects completed.
B) The Primary Measure of Progress should be the Outcome working Software Components.
C) Agility can be enhanced through following up on Technical and Design Excellence and Developments.
D) All Options above are Correct.
ANSWER: D

Q20. The main principles of Agile include the following:
A) Increasing the amount of completed work is Essential and that can be achieved through using simple Development and Management Methods.
B) Team Self-Organization helps the Emergence of the best Architectures and Designs.
C) Regularly, the Team becomes more and more Effective and Efficient.
D) All Options above are Correct.
ANSWER: D

Q21. Which one of the following statements is valid about the KANBAN Method?
A) The KANBAN Method indicates the use of visual Tools to track and improve Production.
B) The KANBAN Method was introduced by Taiichi Ohno, the founder of the Toyota Production Systems (TPS).
C) Examples of the KANBAN visual Tools are the Task Cards, ScrumBOARDS, and BURNDOWN-CHARTs.
D) All Options above are Correct.
ANSWER: D

Q22. Which one of the following statements is valid about the Lean KANBAN Method?

A) The Lean KANBAN Method merges the use of the KANBAN Visualization Methods and the Lean Principles to create an Incremental Visual Management System.

B) The Lean KANBAN optimizes an Organizational System to produce valuable Results using its Resources and Needs, while reducing waste.

C) The Lean KANBAN Method involves increasing the Productivity by reducing delays, early detection of errors, and reducing the total amount of effort required to finish a Task.

D) All Options above are Correct.

ANSWER: D

Q23. Which one of the following statements is valid about the Crystal Methods?

A) The Crystal Methods of Software Development were introduced during the early 1990s by Alistair Cockburn. They are People-centric, lightweight, and easy to adapt. Their Development Processes and Tools can be tuned to cope with the Project Requirements.

B) The Crystal Methods include the Crystal Clear, Crystal Orange Web, Crystal Diamond, Crystal Yellow, Crystal Red, Crystal Sapphire, Crystal Orange, and Crystal Maroon.

C) The Crystal Methods have four Roles, which are the Executive Sponsor, Lead Designer, Developers, and Experienced Users.

D) All Options above are Correct.

ANSWER: D

Q24. Which one of the following statements is valid about the Dynamic Systems Development Methods (DSDM)?

A) The Dynamic Systems Development Methods)DSDM(Framework was first published in 1995. The DSDM represents the Project Work Quality and effort in terms of Cost and time.

B) The DSDM adjusts the Project Deliverables to prioritize the Project Deliverables into four categories, which are the MUST-HAVE, SHOULD-HAVE, COULD-HAVE, and WILL-NOT-HAVE.

C) The DSDM has six phases, which are the Pre-Project Phase, Feasibility Phase, Foundations Phase, Exploration-and-Engineering Phase, Deployment Phase, and Benefit-Assessment Phase.

D) All Options above are Correct.

ANSWER: D

Q25. Which one of the following statements is valid about the Extreme Programming Method (XP)?

A) The Extreme Programming (XP) Method was introduced in Chrysler Corporation and in the 1990s it became familiar. It makes it possible to control the Cost of changing Software.

B) The XP Method facilitates the Incremental Development, Flexible Scheduling, Automated Code Testing, Verbal Communications, Evolving Design, and Close Collaboration. It involves Communications, Simplicity, Feedback, and Courage.

C) The XP Method has four Roles, which are Customer, Developer, Tracker, and Coach.

D) All Options above are Correct.

ANSWER: D

Q26. Which one of the following statements is valid about the Feature Driven Development Method (FDD)?

A) The Feature-Driven Development (FDD) Method was introduced in 1997 by Jeff DeLuca. The FDD operates on the principle of completing a Project by breaking it down into small functions that can be delivered in less than two weeks.

B) The FDD Method has two principles, which are the Software Development is a Human Activity, and Software Development is a client-valued Functionality.

C) The FDD Method has six Roles, which are the Project Manager, Development Manager, Class Owners, Chief Architect, Chief Programmers, and Domain Experts.

D) The FDD Method is an iterative Approach that includes Developing an overall model, Building a Feature list, and Planning, Designing, and building Feature by Feature.

E) All Options above are Correct.

ANSWER: E

Q27. Which one of the following statements is valid about the Test Driven Development Method (TDD)?

A) The Test Driven Development (TDD) Method is known also as Test-First Development (TFD). It was introduced by Kent Beck as a Software Development Method that involves writing automated test code first and developing the least amount of code necessary to pass that test later.

B) The entire Project is broken down into small and client-valued Features to be developed in a very short Development cycle.

C) The TDD Method has two levels, which are the Acceptance TDD (ATDD), and Developer TDD (DTDD).

D) All Options above are Correct.

ANSWER: D

Q28. Which one of the following statements is valid about the Adaptive Software Development Method (ASD)?

A) The Adaptive Software Development (ASD) Method was based on the rapid application Development work of Jim Highsmith and Sam Bayer.

B) The ASD Method involves constant adaptation of Processes to the work in hands, provision of solutions to Problems surfacing in large Projects, and iterative and incremental Development with continuous prototyping.

C) The ASD Method has three Phases, which are the Speculate Phase, Collaborate Phase, and Learn Phase.

D) All Options above are Correct.

ANSWER: D

Q29. Which one of the following statements is valid about the Agile Unified Process Method (AUP)?

A) The Agile Unified Process AUP Method evolved from IBM's Rational Unified Process.

B) The AUP combines Industry-Endorsed Agile Methods such as Test Driven Development TDD, Agile Modeling, Agile Change Management, and Database Refactoring to deliver a Quality working Product.

C) The AUP Method has four Phases, which are the Inception, Elaboration, Construction, and Transition.

D) All Options above are Correct.

ANSWER: D

Q30. Which one of the following statements is valid about the Domain Driven Design Method (DDD)?

A) The Domain Driven Design DDD Method is an Agile Development Approach for handling complex Designs with Implementation linked to an evolving Model.

B) The DDD Method was conceptualized in 2004 by Eric Evans. Domain is defined as an area of Activity to which the User applies a Program or Functionality.

C) Both Options above are Correct.

D) All Options above are Incorrect.

ANSWER: C

Q31. A Scrum Project is a collaborative effort to create a new Product, Solution, or Service in accordance with the Project Vision Statement.

A) _TRUE_ B) _FALSE_

ANSWER: A

Q32. Scrum is a well-known Agile Method that is adaptive, iterative, fast, flexible, and effective Method that continuously delivers value throughout a Project.

A) _TRUE_ B) _FALSE_

ANSWER: A

Q33. The Scrum Framework supports the development of Products, Solutions, and/or Services regardless the Projects complexities.

A) _TRUE_ B) _FALSE_

ANSWER: A

Q34. In the case of a Scrum Project has six Sprints such that each Sprint has two User Stories, and each User Story has a time-box of two weeks and generates at least one deliverable; what is the Project overall duration?

A) 6 Sprints * 2 User Stories per Sprint * 2 Weeks per User Story = 24 Weeks

B) 6 Sprints * 4 Weeks per Sprint = 24 Weeks

C) Both Options above are Valid

D) All Options above are Incorrect

ANSWER: C

Q35. In the case of a Scrum Project has six Sprints such that each Sprint has two User Stories, and each User Story has a time-box of 2 weeks and generates at least one deliverable; what is the total number of User Stories included in the Prioritized Product Backlog?

A) 6 Sprints * 2 User Stories per Sprint = 12 User Stories

B) 6 Sprints * 2 User Stories per Sprint * 2 Deliverables per User Story = 24 Deliverables

C) Both Options above are Correct

D) All Options above are Incorrect

ANSWER: A

Q36. In the case of a Scrum Project has six Sprints such that each Sprint has two User Stories, and each User Story has a time-box of 2 weeks and generates at least one deliverable; what is the number of Product Increments produced and delivered throughout the Project life cycle?

A) 6 Sprints * 2 User Stories per Sprint * at least 1 Deliverable per User Story = at least 12 Product Increments

B) 6 Sprints * at least 2 Deliverables per Sprint = at least 12 Product Increments

C) Both Options above are Correct

D) All Options above are Incorrect

ANSWER: C

Q37. In the case of a Scrum Project has six Sprints such that each Sprint has two User Stories, and each User Story has a time-box of 2 weeks and generates at least one deliverable; what is the number of Stakeholders Meetings that take place throughout the Project life cycle?
A) 1 Stakeholders Meeting
B) 1 Stakeholders Meeting per Sprint
C) 1 Stakeholders Meeting per User Story
D) All Options above are Incorrect
ANSWER: A

Q38. In the case of a Scrum Project has six Sprints such that each Sprint has two User Stories, and each User Story has a time-box of 2 weeks and generates at least one deliverable; what is the number of Sprint Planning Meetings that take place throughout the Project life cycle?
A) At least 6 Sprint Planning Meetings (e.g., at least one per Sprint).
B) At most 6 Sprint Planning Meetings (e.g., at most one per Sprint).
C) Exactly 6 Sprint Planning Meetings (e.g., exactly one per Sprint).
D) All Options above are Incorrect.
ANSWER: A

Q39. In the case of a Scrum Project has six Sprints such that each Sprint has User Stories, and each User Story has a time-box of 2 weeks and generates at least one deliverable; what is the number of Sprint Review Meetings that take place throughout the Project life cycle?
A) At least 6 Sprint Review Meetings (e.g., at least one per Sprint).
B) At most 6 Sprint Review Meetings (e.g., at most one per Sprint).
C) Exactly 6 Sprint Review Meetings (e.g., exactly one per Sprint).
D) All Options above are Incorrect.
ANSWER: A

Q40. In the case of a Scrum Project has six Sprints such that each Sprint has two User Stories, and each User Story has a time-box of 2 weeks and generates at least one deliverable; what is the number of Sprint Retrospect Meetings that take place throughout the Project life cycle?
A) At least 6 Sprint Retrospect Meetings (e.g., at least one per Sprint).
B) At most 6 Sprint Retrospect Meetings (e.g., at most one per Sprint).
C) Exactly 6 Sprint Retrospect Meetings (e.g., exactly one per Sprint).
D) All Options above are Incorrect.
ANSWER: C

Q41. What is the number of Project Retrospect Meetings that take place throughout the Project life cycle?
A) At least 6 Project Retrospect Meetings (e.g., at least one per Sprint).
B) At most 6 Project Retrospect Meetings (e.g., at most one per Sprint).
C) Exactly 6 Project Retrospect Meetings (e.g., exactly one per Sprint).
D) Exactly 1 Project Retrospect Meeting.
ANSWER: D

Q42. The main Benefits of using the Scrum Method in Software Development and in Software Project Management include Adaptability, Transparency, Continuous Feedback, Continuous Improvement, Continuous Delivery of Effective Deliverables and Value, Efficient Development Process, Motivation among Employees, and Fast Problem Solving.
A) _TRUE_ B) _FALSE_
ANSWER: A

Q43. The Empirical Process Control Scrum principle describes the Scrum philosophy based on the concepts of:
A) Transparency, Inspection, and Adaptation.
B) Transparency, Inspection, and Portation.
C) Transparency, Inspection, and Adaptation.
D) All Options above are Incorrect.
ANSWER: C

Q44. Transparency indicates that all Aspects of any of the Scrum Processes can be observed by anyone.
A) _TRUE_ B) _FALSE_
ANSWER: A

Q45. Transparency in Scrum can be accomplished using the following:
A) The Project Vision Statement that is viewable by all Stakeholders and the Team, the Prioritized Product Backlog's User-Stories that are viewable by everyone within and outside the Team, and the Release Planning Schedule that can be coordinated across multiple Scrum Teams.
B) The Team Progress Visibility through the use of INFORMATION-RADIATORS such as ScrumBOARDS and BURNDOWN-CHARTS, and the Daily Standup Meetings during which, all Team Members can present what they have done yesterday, what they will do today, and any Issues they are facing with.
C) The Sprint Review Meetings, during which, the Team can demonstrate their Deliverables to the Product Owner and Stakeholders seeking their Approval before the Team can proceed with deploying these Deliverables at the Customer site.
D) All Options above are Correct.
ANSWER: D

Q46. Inspection in SCRUM can be accomplished by:
A) Using a common ScrumBOARD and other INFORMATION RADIATORS that show the progress of the Scrum Team with the current Sprint.
B) Collecting the feedback from the customer and from other Stakeholders during the Processes of "Develop Epics," "Create Prioritized Product Backlog," and "Conduct Release Planning."
C) Inspecting and approving the Deliverables by the Product Owner during the "Demonstrate and Validate Sprint" Process.
D) All Options above are Correct.
ANSWER: D

Q47. The Leadership Style that is appropriate for Scrum Projects is the Servant-Leadership-Style by which results are reached by focusing on the Scrum Team's needs.
A) _TRUE_ B) _FALSE_
ANSWER: A

Q48. In Scrum, the dimension of Collaboration includes Awareness, Articulation, and Appropriation.
A) _TRUE_ B) _FALSE_
ANSWER: A

Q49. Awareness indicates that each member of a Scrum Team should be aware of the Duties, Tasks, and Activities that are carried on by each Member in his/her Team.
A) _TRUE_ B) _FALSE_
ANSWER: A

Q50. Articulation indicates that during a Sprint, the work must be divided carefully among the Team Members in a way that their individual Outcomes can be easily integrated and presented.
A) _TRUE_ B) _FALSE_
ANSWER: A

Q51. Appropriation indicates that Resources must be used appropriately such that no or minimal efforts and time are wasted.
A) _TRUE_ B) _FALSE_
ANSWER: A

Q52. In Scrum, the Product Owner has to interpret the Project Stakeholders" needs while prioritizing the User-Stories in the Prioritized Product Backlog based on the Risks associated with each User-Story, the Value and Impact of each User-Story, and the Dependencies among the User-Stories.
A) _TRUE_ B) _FALSE_
ANSWER: A

Q53. Sprints Time-Boxing indicates that:
A) A Sprint is a time-boxed iteration that lasts for one to six weeks.
B) A Sprint is a time-boxed iteration that lasts for 4 weeks.
C) A Sprint is a time-boxed iteration that lasts for 6 weeks.
D) All Options above are Incorrect.
ANSWER: A

Q54. Daily Standup Meetings Time-Boxing indicates that:
A) A Daily Standup Meeting is a 15 minutes daily meeting, during which the Team Members meet and report the Progress of their portions of the Project.
B) A Daily Standup Meeting is a 4 hours daily meeting, during which the Team Members meet and report the Progress of their portions of the Project.
C) A Daily Standup Meeting is a multi-hours daily meeting, during which the Team Members meet and report the Progress of their portions of the Project.
D) All Options above are Incorrect.
ANSWER: A

Q55. Sprint Planning Meetings Time-Boxing indicates that:
A) A Sprint Planning Meeting is a 6-h single meeting conducted prior to each Sprint.
B) A Sprint Planning Meeting is a 4-h single meeting conducted prior to each Sprint.
C) A Sprint Planning Meeting is divided into two sub-meetings: an Objective Definition Sub-Meeting, during which the Product Owner explains the highest priority User-Stories in the Prioritized Product Backlog to the Scrum Team; and a Task Estimation Sub-Meeting, during which the Scrum Team decides on “how” to complete the selected Prioritized Product Backlog Items to fulfill the Sprint Objective(s).
D) All Options above are Correct.
ANSWER: C

Q56. Sprint Review Meetings Time-Boxing indicates that:
A) A Sprint Review Meeting is a 4-hour meeting.
B) A Sprint Review Meeting that is conducted to allow the Scrum Team to demonstrate the Deliverables of the current Sprint to the Product Owner, before he/she accepts or rejects these Deliverables in accordance with the Acceptance-Criteria.
C) A 4 hours meeting, during which the Scrum Team demonstrate the Deliverables of the current Sprint to the Product Owner, before he/she accepts or rejects these Deliverables in accordance with the Acceptance-Criteria.
D) All Options above are Correct.
ANSWER: D

Q57. A Sprint Retrospect Meetings Time-Boxing refers to a 4-h Meeting prior to the end of each Sprint after shipping the deliverables of that Sprint to the Customer. During a Sprint Retrospect Meeting, the Scrum Team Members review and reflect on the current-and-about-to-complete Sprint in terms of the Processes-followed, Tools-used, Collaborative-efforts, and Communication-Mechanisms-used.
A) _TRUE_ B) _FALSE_
ANSWER: A

Q58. Scrum Organization refers to the Scrum Organizational Structure in terms of the Scrum's Portfolios, Programs, and Projects.
A) _TRUE_ B) _FALSE_
ANSWER: A

Q59. Scrum Organization refers to the various Scrum Core and Optional Roles.
A) _TRUE_ B) _FALSE_
ANSWER: A

Q60. The Scrum Core Roles are the Product Owner, Scrum Master, and Scrum Team.
A) _TRUE_ B) _FALSE_
ANSWER: A

Q61. The Scrum Optional Roles are the Stakeholders, Customer, Users, Sponsor, Vendors, Scrum Guidance Body (SGB), Chief Product Owner, and Chief Scrum Master.
A) _TRUE_ B) _FALSE_
ANSWER: A

Q62. The Product Owner must communicate the Product Functional Requirements (e.g., User-Stories) to the Scrum Team clearly and effectively.
A) _TRUE_ B) _FALSE_
ANSWER: A

Q63. The Product Owner must precisely define the Product's Acceptance-Criteria.
A) _TRUE_ B) _FALSE_
ANSWER: A

Q64. The Product Owner must ensure that the Product's Deliverables (e.g., at the Sprint-Level and at the Project-Level) satisfy the Product's Acceptance-Criteria clearly and certainly.
A) _TRUE_ B) _FALSE_
ANSWER: A

Q65. During the Create Project Vision Process, the Product Owner must define the Project Vision and facilitate the creation of the Project Charter and its Budget.
A) _TRUE_ B) _FALSE_
ANSWER: A

Q66. During the Identify Scrum Master and Stakeholders Process, the Product Owner must help select a Scrum Master for the Project and identify the Project Stakeholders.
A) _TRUE_ B) _FALSE_
ANSWER: A

Q67. During the Form Scrum Team Process, the Product Owner must help select the Scrum Team Members, help develop a Collaboration Plan, and help develop the Team Building Plan jointly with the Scrum Master.
A) _TRUE_ B) _FALSE_
ANSWER: A

Q68. During the Develop Epics Process, the Product Owner must create a set of Epics used to develop and promote the targeted Product.
A) _TRUE_ B) _FALSE_
ANSWER: A

Q69. During the Create Prioritized Product Backlog Process, the Product Owner must prioritize the User-Stories and place them into the Prioritized Product Backlog.
A) _TRUE_ B) _FALSE_
ANSWER: A

Q70. During the Conduct Release Planning Process, the Product Owner must create the Release Planning Schedule and help determine the Sprint's length.
A) _TRUE_ B) _FALSE_
ANSWER: A

Q71. During the Create User-Stories Process, the Product Owner must help create the User-Stories and define the ACCEPTANCE-CRITERIA for User-Stories.
A) _TRUE_ B) _FALSE_
ANSWER: A

Q72. During the "Approve, Estimate, and Commit User-Stories" Process, the Product Owner must approve the User-Stories and encourage the Scrum Team to commit to them.
A) _TRUE_ B) _FALSE_
ANSWER: A

Q73. During the Create Tasks Process, the Product Owner must explain the User-Stories to the Scrum Team while the Team is deciding on the Tasks and Activities needed to complete each of them.
A) _TRUE_ B) _FALSE_
ANSWER: A

Q74. During the Estimate Tasks Process, the Product Owner must guide the Scrum Team and help them estimate and decide on the efforts required to carry-on the Tasks and Activities needed to complete each USER-STORY.
A) _TRUE_ B) _FALSE_
ANSWER: A

Q75. During the Create Sprint Backlog Process, the Product Owner must clarify the User-Stories to the Scrum Team while the Team is creating the Sprint Backlog.
A) _TRUE_ B) _FALSE_
ANSWER: A

Q76. During the Create Deliverables Process, the Product Owner must clarify the Business Requirements to the Scrum Team and provide them with the details of the required Deliverables out of each Sprint.
A) _TRUE_ B) _FALSE_
ANSWER: A

Q77. During the Groom Prioritized Product Backlog Process, the Product Owner must groom the Prioritized Product Backlog in accordance with the feedback and recommendations made during the Retrospect Meetings at the end of each Sprint.
A) _TRUE_ B) _FALSE_
ANSWER: A

Q78. During the Demonstrate and Validate Sprints Process, the Product Owner must accept or reject the proposed Deliverables of each Sprint, provide feedback to the Scrum Master and Scrum Team, and accordingly update the Release Plan and the Prioritized Product Backlog.
A) _TRUE_ B) _FALSE_
ANSWER: A

Q79. During the Ship Deliverables Process, the Product Owner must facilitate the deployment of the Product Releases in coordination with the Customer.
A) _TRUE_ B) _FALSE_
ANSWER: A

Q80. During the Retrospect Project Process, the Product Owner must participate in all Sprint Retrospect Meetings as well as in the overall Project Retrospect Meeting.
A) _TRUE_ B) _FALSE_
ANSWER: A

Q81. The Product Owner must be:

A) A Scrum Expert, Knowledgeable in the Project Business Domain, Possessing Excellent Communication Skills, Capable of Handling Uncertainties, and Possessing Strong Negotiation Skills.

B) Approachable, Proactive, Decisive, Pragmatic, and Goal-Oriented.

C) Both Options above are Correct.

D) All Options above are Incorrect.

ANSWER: C

Q82. The Scrum Master must be a Servant-Leader coaching and motivating his/her Team.

A) _TRUE_ B) _FALSE_

ANSWER: A

Q83. The Scrum Master must facilitate a Protective and Productive Work Environment for his/her Team.

A) _TRUE_ B) _FALSE_

ANSWER: A

Q84. The Scrum Master must resolve and eliminate Issues and external influencing Factors.

A) _TRUE_ B) _FALSE_

ANSWER: A

Q85. The Scrum Master must guard the Scrum Principles, Aspects, and Processes.

A) _TRUE_ B) _FALSE_

ANSWER: A

Q86. During the Identify Scrum Master and Stakeholders Process, the Scrum Master must help the Product Owner identify the Project Stakeholders.

A) _TRUE_ B) _FALSE_

ANSWER: A

Q87. During the Form Scrum Team Process, the Scrum Master must facilitate the selection of the Scrum Team Members and the creation of the Scrum Team Collaboration and Building Plan. Also, he/she must ensure that the Project has backup Resources.

A) _TRUE_ B) _FALSE_

ANSWER: A

Q88. During the Develop Epics Process, the Scrum Master must help the Product Owner create the Project Epics.

A) _TRUE_ B) _FALSE_

ANSWER: A

Q89. During the Create Prioritized Product Backlog Process, the Scrum Master must help the Product Owner create the Prioritized Product Backlog, define the Deliverables Acceptance-Criteria, and define the Project Done-Criteria.
A) _TRUE_ B) _FALSE_
ANSWER: A

Q90. During the Conduct Release Planning Process, the Scrum Master must coordinate the creation of the "Release Planning Schedule" and determine the Sprint's length.
A) _TRUE_ B) _FALSE_
ANSWER: A

Q91. During the Create User-Stories Process, the Scrum Master must help the Product Owner create and implement the User-Stories.
A) _TRUE_ B) _FALSE_
ANSWER: A

Q92. During the "Approve, Estimate and Commit User-Stories" Process, the Scrum Master must facilitate the Scrum Team Meetings to estimate and plan for the implementation of the User-Stories.
A) _TRUE_ B) _FALSE_
ANSWER: A

Q93. During the Create Tasks Process, the Scrum Master must help the Scrum Team identify and create the Tasks-List for the current Sprint.
A) _TRUE_ B) _FALSE_
ANSWER: A

Q94. During the Estimate Tasks Process, the Scrum Master must help the Scrum Team estimate the needed effort to complete the current Sprint.
A) _TRUE_ B) _FALSE_
ANSWER: A

Q95. During the Create Sprint Backlog Process, the Scrum Master must help the Scrum Team select the right User-Stories from the Prioritized Product Backlog for inclusion in the current Sprint's Backlog.
A) _TRUE_ B) _FALSE_
ANSWER: A

Q96. During the Create Deliverables Process, the Scrum Master must support the Scrum Team and help them create the current Sprint's Deliverables, and update the ScrumBOARD and the Impediment Log Document.
A) _TRUE_ B) _FALSE_
ANSWER: A

Q97. During the Conduct Daily Standup Meetings Process, the Scrum Master must ensure updating both of the ScrumBOARD and the Impediment Log Document.
A) _TRUE_ B) _FALSE_
ANSWER: A

Q98. During the Groom Prioritized Product Backlog Process, the Scrum Master must facilitate the Review Meetings to review the Prioritized Product Backlog.
A) _TRUE_ B) _FALSE_
ANSWER: A

Q99. During the Convene Scrum of Scrums Process, the Scrum Master must participate in the Scrum of Scrums Meetings, and discuss and resolve any Issues that affect his/her Scrum Team.
A) _TRUE_ B) _FALSE_
ANSWER: A

Q100. During the Demonstrate and Validate Sprints Process, the Scrum Master must enable the Scrum Team to demonstrate the Deliverables of the current Sprint to the Product Owner and to solicit the endorsement of these Deliverables before shipping them to the Customer.
A) _TRUE_ B) _FALSE_
ANSWER: A

Q101. During the Sprint Retrospect Process, the Scrum Master must improve the Work Environment of the incoming Sprints based on the Lessons Learned from the current Sprints.
A) _TRUE_ B) _FALSE_
ANSWER: A

Q102. During the Project Retrospect Process, the Scrum Master must improve the Work Environment of the future Projects based on the Lessons Learned from the current Project.
A) _TRUE_ B) _FALSE_
ANSWER: A

Q103. The Scrum Master must be a Scrum Expert, Servant-Leader, Moderator, Problem Solver, Approachable, Motivator, Perceptive, Mentor, and Introspective.
A) _TRUE_ B) _FALSE_
ANSWER: A

Q104. The Scrum Team must develop the Product, Solution, and/or Service in accordance with the User-Stories in the current Sprint's Backlog.
A) _TRUE_ B) _FALSE_
ANSWER: A

Q105. During the Form Scrum Team Process, the Scrum Team must help create the Collaboration Plan and create the Team Building Plan.
A) _TRUE_ B) _FALSE_
ANSWER: A

Q106. During the Develop Epics Process, the Scrum Team must clearly understand the Epics.
A) _TRUE_ B) _FALSE_
ANSWER: A

Q107. During the Prioritized Product Backlog Process, the Scrum Team must clearly understand the User-Stories in the Prioritized Product Backlog.
A) _TRUE_ B) _FALSE_
ANSWER: A

Q108. During the Conduct Release Planning Process, the Scrum Team must consider and agree on the Sprint's length as proposed by the Scrum Master.
A) _TRUE_ B) _FALSE_
ANSWER: A

Q109. During the Create User-Stories Process, the Scrum Team must help the Product Owner create the Project User-Stories.
A) _TRUE_ B) _FALSE_
ANSWER: A

Q110. During the "Approve, Estimate, and Commit User-Stories" Process, the Scrum Team must estimate the effort and time required for each Task.
A) _TRUE_ B) _FALSE_
ANSWER: A

Q111. During the Create Tasks Process, the Scrum Team must decide on and develop the Task list for each of the User-Stories, while taking into consideration any dependencies among them.
A) _TRUE_ B) _FALSE_
ANSWER: A

Q112. During the Estimate Tasks Process, the Scrum Team must estimate the effort and time for each Task.
A) _TRUE_ B) _FALSE_
ANSWER: A

Q113. During the Create Sprint Backlog Process, the Scrum Team must select the right User-Stories from the Prioritized Product Backlog and include them in the incoming Sprint's Backlog.
A) _TRUE_ B) _FALSE_
ANSWER: A

Q114. During the Create Deliverables Process, the Scrum Team must create the Deliverables, identify Risks and implement necessary Risks Mitigation Plans, and keep the Impediment Log Document updated.
A) _TRUE_ B) _FALSE_
ANSWER: A

Q115. During the Conduct Daily Standup Process, the Scrum Team must keep the BURNDOWN-CHART, ScrumBOARD, and Impediment Log Document updated, discuss any Issues raised by the participating Team Members, identify Risks, and submit CHANGE-REQUESTS whenever that is necessary.
A) _TRUE_ B) _FALSE_
ANSWER: A

Q116. During the Groom Prioritized Backlog Process, the Scrum Team must participate in the Review Meetings of the Prioritized Product Backlog.
A) _TRUE_ B) _FALSE_
ANSWER: A

Q117. During the Convene Scrum of Scrums Process, the Scrum Team must provide the Scrum Master with the necessary Inputs on Issues to be discussed and resolved at the Scrum of Scrums Meetings.
A) _TRUE_ B) _FALSE_
ANSWER: A

Q118. During the "Demonstrate and Validate Sprints" Process, the Scrum Team must demonstrate the completed Deliverables to the Product Owner, and seek his/her Approval and Endorsement.
A) _TRUE_ B) _FALSE_
ANSWER: A

Q119. During the Sprint Retrospect Process, the Scrum Team must provide the Product Owner and the Scrum Master with their suggestions for Improvements that would benefit the incoming Sprints.
A) _TRUE_ B) _FALSE_
ANSWER: A

Q120. During the Project Retrospect Process, the Scrum Team must participate in the Project Retrospect Meeting.
A) _TRUE_ B) _FALSE_
ANSWER: A

Q121. The Scrum Team Members must be Knowledgeable in the Scrum areas (e.g., Principles, Aspects, and Processes), Collaborative, Self-Organized, Motivated, Proactive, Technical Expert, Team Player, Responsive, Goal-Oriented, and Introspective.
A) _TRUE_ B) _FALSE_
ANSWER: A

Q122. Consider a two-layered Scrum of Scrums Project Structure with nine Scrum Teams equally distributed across three Scrum of Scrums groups aggregated into a single Scrum of Scrums of Scrums group. Then, the Scrum Project population would be:
A) In terms of the Scrum Teams' Members, there are nine Scrum Teams. Each Team has up to nine members. Thus, the count of the overall Scrum Teams' members cannot exceed 81.
B) In terms of the Product Owners/Scrum Masters, there are nine Scrum Teams. Thus, there should be nine Product Owners and nine Scrum Masters.
C) In terms of the Chief Product Owners/Chief Scrum Masters, there are three Scrum of Scrums and one Scrum of Scrums of Scrums. Thus, there should be four Chief Product Owners and four Chief Scrum Master.
D) All Options above are Correct.
ANSWER: D

Q123. The Chief Product Owner must coordinate and supervise multiple Product Owners when the Project is large and has multiple Scrums.
A) _TRUE_ B) _FALSE_
ANSWER: A

Q124. The Chief Product Owner must interface with the Program Product Owner to ensure that the Project is in full alignment with the Program Goals and Objectives.
A) _TRUE_ B) _FALSE_
ANSWER: A

Q125. The Chief Scrum Master must gather and communicate Information across the Scrum Teams within the same Project.
A) _TRUE_ B) _FALSE_
ANSWER: A

Q126. In Scrum, the Roles that are directly associated with the Portfolios include the:
A) Portfolio Product Owner and Portfolio Scrum Master.
B) Program Product Owner and Program Scrum Master.
C) Portfolio and Program Product Owner, and Portfolio and Program Scrum Master.
D) All Options above are Incorrect.
ANSWER: A

Q127. In Scrum, the Roles that are directly associated with the Programs include the:
A) Portfolio Product Owner and Portfolio Scrum Master.
B) Program Product Owner and Program Scrum Master.
C) Portfolio and Program Product Owner and Portfolio and Program Scrum Master.
D) All Options above are Incorrect.
ANSWER: B

Q128. In Scrum, the Roles that are directly associated with the Projects include the:
A) Program/Project Product Owner and Program/Project Scrum Master.
B) Portfolio/Program/Project Product Owner and Portfolio/Program/Project Scrum Master.
C) Chief Product Owner, Chief Scrum Master, Product Owner, Scrum Master, and Scrum Team.
D) All Options above are Incorrect.
ANSWER: C

Q129. With Tukman's Theory of Team Development, the Forming Stage is characterized with the following statement:
A) The "Roles and Responsibilities" of the individual Team Members are not clear nor yet defined.
B) There exist much confusion, and "Less Agreements" among the Team Members.
C) The Team would be in a good shape and can work together on the Project.
D) The Team would be working together consistently and coherently with high Productivity.
ANSWER: A

Q130. With Tukman's Theory of Team Development, the Storming Stage is characterized with the following statement:
A) The "Roles and Responsibilities" of the individual Team Members are not clear nor yet defined.
B) There exist much confusion and "Less Agreements" among the Team Members.
C) The Team would be in a good shape and can work together on the Project.
D) The Team would be working together consistently and coherently with high Productivity.
ANSWER: B

Q131. With Tukman's Theory of Team Development, the Norming Stage is characterized with the following statement:

A) The "Roles and Responsibilities" of the individual Team Members are not clear nor yet defined.

B) There exist much confusion, and "Less Agreements" among the Team Members.

C) The Team would be in a good shape and can work together on the Project.

D) The Team would be working together consistently and coherently with high Productivity.

ANSWER: C

Q132. With Tukman's Theory of Team Development, the Performing Stage is characterized with the following statement:

A) The "Roles and Responsibilities" of the individual Team Members are not clear nor yet defined.

B) There exist much confusion, and "Less Agreements" among the Team Members.

C) The Team would be in a good shape and can work together on the Project.

D) The Team would be working together consistently and coherently with high Productivity.

ANSWER: D

Q133. With Maslow's Theory of Needs Hierarchy, the Physiological Layer is characterized with the following statement:

A) Physiological needs include food, water, cloths, shelter, sleeping, and so forth,

B) Safety needs include economic security; and protection from harm, disease, and violence.

C) Feeling needs include feeling loved, belonging to someone, and so forth.

D) Having the ability to feel self-esteemed, having good reputation, respected by others, and so forth.

E) Being able to do what one can do best and fully realize self-potential.

F) All Options above are Incorrect.

ANSWER: A

Q134. With Maslow's Theory of Needs Hierarchy, the Safety Layer is characterized with the following statement:

A) Physiological needs include food, water, cloths, shelter, sleeping, and so forth,

B) Safety needs include economic security; and protection from harm, disease, and violence.

C) Feeling needs include feeling loved, belonging to someone, and so forth.

D) Having the ability to feel self-esteemed, having good reputation, respected by others, and so forth.

E) Being able to do what one can do best and fully realize self-potential.

F) All Options above are Incorrect.

ANSWER: B

Q135. With Maslow's Theory of Needs Hierarchy, the "Feeling Loved/Belonging" Layer is characterized with the following statement:

A) Physiological needs include food, water, cloths, shelter, sleeping, and so forth,

B) Safety needs include economic security; and protection from harm, disease, and violence.

C) Feeling needs include feeling loved, belonging to someone, and so forth.

D) Having the ability to feel self-esteemed, having good reputation, respected by others, and so forth.

E) Being able to do what one can do best and fully realize self-potential.

F) All Options above are Incorrect.

ANSWER: C

Q136. With Maslow's Theory of Needs Hierarchy, the "Ability to Feel Esteemed" Layer is characterized with the following statement:

A) Physiological needs include food, water, cloths, shelter, sleeping, and so forth,

B) Safety needs include economic security; and protection from harm, disease, and violence.

C) Feeling needs include feeling loved, belonging to someone, and so forth.

D) Having the ability to feel self-esteemed, having good reputation, respected by others, and so forth.

E) Being able to do what one can do best and fully realize self-potential.

F) All Options above are Incorrect.

ANSWER: D

Q137. With Maslow's Theory of Needs Hierarchy, the Self-Actualization Layer is characterized with the following statement:
A) Physiological needs include food, water, cloths, shelter, sleeping, and so forth,
B) Safety needs include economic security; and protection from harm, disease, and violence.
C) Feeling needs include feeling loved, belonging to someone, and so forth.
D) Having the ability to feel self-esteemed, having good reputation, respected by others, and so forth.
E) Being able to do what one can do best and fully realize self-potential.
F) All Options above are Incorrect.
ANSWER: E

Q138. Theory X Leadership Style indicates that leaders would assume that their Team Members are unmotivated and would avoid work whenever that is possible. This Style of Leadership is not suitable for Scrum Projects.
A) _TRUE_ B) _FALSE_
ANSWER: A

Q139. Theory Y Leadership Style indicates that leaders would assume that their Team Members are self-motivated and seek work and would stand for challenge and accept more Responsibilities. This Style of Leadership is indeed suitable for Scrum Projects.
A) _TRUE_ B) _FALSE_
ANSWER: A

Q140. The Win-Win Conflict Management Method indicates the following:
A) Team Members can cooperatively deal directly with problems to resolve any disagreements. Scrum Projects need Work Environments that encourage the Scrum Team Members to openly discuss problems and work through them to reach Win-Win Outcomes.
B) In case a Team Member feels that his/her contributions are not recognized or that he/she is not well-treated, then he/she may lose focus and stop his/her Effective contribution to the Project and agree (e.g., even if he/she is in disagreement) to whatever he/she is told to do. This Method is not appropriate for Scrum Projects.
C) In some situations, Team Members attempt to search for Solutions that bring only some degree of Satisfaction to them and to the other disputing Parties. This Approach involves "give and take" to satisfy every Team Member. Scrum Meetings are there to ensure solving problems through mutual Discussions.
D) A Scrum Master may think that he/she is a de facto Manager and impose his/her viewpoint on others. This Method results in a Win-Lose Outcome. This Method is not recommended for Scrum Projects.
ANSWER: A

Q141. The Lose-Win Conflict Management Method indicates the following:

A) Team Members can cooperatively deal directly with problems to resolve any disagreements. Scrum Projects need Work Environments that encourage the Scrum Team Members to openly discuss problems and work through them to reach Win-Win Outcomes.

B) In case a Team Member feels that his/her contributions are not recognized or that he/she is not well-treated, then he/she may lose focus and stop his/her Effective contribution to the Project and agree (e.g., even if he/she is in disagreement) to whatever he/she is told to do. This Method is not appropriate for Scrum Projects.

C) In some situations, Team Members attempt to search for Solutions that bring only some degree of Satisfaction to them and to the other disputing Parties. This Approach involves "give and take" to satisfy every Team Member. Scrum Meetings are there to ensure solving problems through mutual Discussions.

D) A Scrum Master may think that he/she is a de facto Manager and impose his/her viewpoint on others. This Method results in a Win-Lose Outcome. This Method is not recommended for Scrum Projects.

ANSWER: B

Q142. The Lose-Lose Conflict Management Method indicates the following:

A) Team Members can cooperatively deal directly with problems to resolve any disagreements. Scrum Projects need Work Environments that encourage the Scrum Team Members to openly discuss problems and work through them to reach Win-Win Outcomes.

B) In case a Team Member feels that his/her contributions are not recognized or that he/she is not well-treated, then he/she may lose focus and stop his/her Effective contribution to the Project and agree (e.g., even if he/she is in disagreement) to whatever he/she is told to do. This Method is not appropriate for Scrum Projects.

C) In some situations, Team Members attempt to search for Solutions that bring only some degree of Satisfaction to them and to the other disputing Parties. This Approach involves "give and take" to satisfy every Team Member. Scrum Meetings are there to ensure solving problems through mutual Discussions.

D) A Scrum Master may think that he/she is a de facto Manager and impose his/her viewpoint on others. This Method results in a Win-Lose Outcome. This Method is not recommended for Scrum Projects.

ANSWER: C

Q143. The Win-Lose Conflict Management Method indicates the following:
A) Team Members can cooperatively deal directly with problems to resolve any disagreements. Scrum Projects need Work Environments that encourage the Scrum Team Members to openly discuss problems and work through them to reach Win-Win Outcomes.
B) In case a Team Member feels that his/her contributions are not recognized or that he/she is not well-treated, then he/she may lose focus and stop his/her Effective contribution to the Project and agree (e.g., even if he/she is in disagreement) to whatever he/she is told to do. This Method is not appropriate for Scrum Projects.
C) In some situations, Team Members attempt to search for Solutions that bring only some degree of Satisfaction to them and to the other disputing Parties. This Approach involves "give and take" to satisfy every Team Member. Scrum Meetings are there to ensure solving problems through mutual Discussions.
D) A Scrum Master may think that he/she is a de facto Manager and impose his/her viewpoint on others. This Method results in a Win-Lose Outcome. This Method is not recommended for Scrum Projects.
ANSWER: D

Q144. The Servant Leadership Style is best characterized by the following statement:
A) Leaders listen and heal their Team Members; and they must be persuasive, aware of what is going to happen and of what is going on.
B) When time is limited, leaders would delegate some of their Planning and Decision-Making Responsibilities to some of their competent Team Members.
C) Leaders of this Style would make Decisions on their own with little without involving their Team Members.
D) Leaders of this Style would instruct their Team Members to what Tasks are required and when and how to perform them.
ANSWER: A

Q145. The Delegating Leadership Style is best characterized by the following statement:
A) Leaders listen and heal their Team Members; and they must be persuasive, aware of what is going to happen and of what is going on.
B) When time is limited, leaders would delegate some of their Planning and Decision-Making Responsibilities to some of their competent Team Members.
C) Leaders of this Style would make Decisions on their own with little without involving their Team Members.
D) Leaders of this Style would instruct their Team Members to what Tasks are required and when and how to perform them.
ANSWER: B

Q146. The Autocratic Leadership Style is best characterized by the following statement:

A) Leaders listen and heal their Team Members; and they must be persuasive, aware of what is going to happen and of what is going on.

B) When time is limited, leaders would delegate some of their Planning and Decision-Making Responsibilities to some of their competent Team Members.

C) Leaders of this Style would make Decisions on their own with little without involving their Team Members.

D) Leaders of this Style would instruct their Team Members to what Tasks are required and when and how to perform them.

ANSWER: C

Q147. The Directing Leadership Style is best characterized by the following statement:

A) Leaders listen and heal their Team Members; and they must be persuasive, aware of what is going to happen and of what is going on.

B) When time is limited, leaders would delegate some of their Planning and Decision-Making Responsibilities to some of their competent Team Members.

C) Leaders of this Style would make Decisions on their own with little without involving their Team Members.

D) Leaders of this Style would instruct their Team Members to what Tasks are required and when and how to perform them.

ANSWER: D

Q148. The Laissez-Faire Leadership Style is best characterized by the following statement:

A) Leaders would leave their Team Members unsupervised with little or no interfere in their work.

B) Leaders support and monitor their Team Members by listening to them, assisting and encouraging them, and by maintaining positive outlooks when encountering Uncertainties.

C) Leaders ensure the completion of all Tasks before the deadlines.

D) Leaders establish Authority with respect.

ANSWER: A

Q149. The "Coaching and Supportive" Leadership Style is best characterized by the following statement:

A) Leaders would leave their Team Members unsupervised with little or no interfere in their work.

B) Leaders support and monitor their Team Members by listening to them, assisting and encouraging them, and by maintaining positive outlooks when encountering Uncertainties.

C) Leaders ensure the completion of all Tasks before the deadlines.

D) Leaders establish Authority with respect.

ANSWER: B

Q150. The Task-Oriented Leadership Style is best characterized by the following statement:
A) Leaders would leave their Team Members unsupervised with little or no interfere in their work.
B) Leaders support and monitor their Team Members by listening to them, assisting and encouraging them, and by maintaining positive outlooks when encountering Uncertainties.
C) Leaders ensure the completion of all Tasks before the deadlines.
D) Leaders establish Authority with respect.
ANSWER: C

Q151. The Assertive Leadership Style is best characterized by the following statement:
A) Leaders would leave their Team Members unsupervised with little or no interfere in their work.
B) Leaders support and monitor their Team Members by listening to them, assisting and encouraging them, and by maintaining positive outlooks when encountering Uncertainties.
C) Leaders ensure the completion of all Tasks before the deadlines.
D) Leaders establish Authority with respect.
ANSWER: D

Q152. The Servant-Leader Leadership Style is the most suitable Leadership Style for the Scrum Master.
A) _TRUE_ B) _FALSE_
ANSWER: A

Q153. The responsibilities of the Portfolio Product Owner with respect to SCRUM Business Justification include:
A) Delivering value and creating Business Justification for his/her Portfolio.
B) Providing value Guidance and approving Business Justification for Programs.
C) Both Options above are Correct.
D) All Options above are Incorrect.
ANSWER: C

Q154. The responsibilities of the Program Product Owner with respect to SCRUM Business Justification include:
A) Delivering value and creating Business Justification for his/her Program.
B) Providing value Guidance and approving Business Justification for Projects under his/her Program.
C) Both Options above are Correct.
D) All Options above are Incorrect.
ANSWER: C

Q155. The responsibilities of the SCRUM Master with respect to SCRUM Business Justification include:

A) Facilitating the creation of the Project Deliverables.

B) Managing Risks and CHANGE-REQUESTS.

C) Both Options above are Correct.

D) All Options above are Incorrect.

ANSWER: C

Q156. The responsibilities of the SCRUM Team with respect to SCRUM Business Justification include:

A) Creating the Project Deliverables.

B) Demonstrating them to the Product Owner.

C) Implementing them for the Customer.

D) All Options above are Correct.

ANSWER: D

Q157. The responsibilities of the Stakeholder with respect to Scrum Business Justification include:

A) Stakeholders include Customers, Users, and Sponsors.

B) Sponsors are responsible for funding their Scrum Projects and monitoring them to confirm that they realized their Benefits.

C) Customers and Users are involved in defining the Prioritized list of User-Stories in the Prioritized Product Backlog.

D) All Options above are Correct.

ANSWER: D

Q158. Justifying the Business Needs for a Scrum Project possesses three steps: assessing and presenting the Project's Business Case, justifying the Project's Value, and confirming the Project's Benefits Realization.

A) _TRUE_ B) _FALSE_

ANSWER: A

Q159. Justifying the Business need for a Scrum Project is based on several sets of factors including the following:

A) The Project Reasoning (e.g., factors that make a Project justifiable and meaningful); and the Business Needs Factors (e.g., the Project's needed Business Outcomes in accordance with that Project's Vision Statement).

B) The Project Benefits (e.g., the expected Benefits from a Project such as a measurable improvement in the Product); and the Major Risks (e.g., any uncertainties and/or events that were not planned for in a Project, but impact the SUCCESS of that Project).

C) The Project Timescales (e.g., the Duration of a Project) and the Project Costs (e.g., the Costs of Investments and Development of a Project).

D) All Options above are Correct.

ANSWER: D

Q160. Scrum Quality refers to:

A) Ensuring that the Deliverables and Product of a Scrum Project satisfy the predefined Acceptance-Criteria of that Project.

B) Ensuring that the Business Value of that Project is achieved in accordance with the Customer Expectations.

C) Involving Quality Planning, Quality Control, and Quality Assurance.

D) All Options above are Correct.

ANSWER: D

Q161. The Factors used to determine the Scope Quality Requirements for a Scrum Project include the following:

A) The "Business Needs" that the Project must satisfy, and the ability of the Customer to satisfy these Business Needs.

B) The Stakeholders' current and potential future Business Needs.

C) Both Options above are Correct.

D) All Options above are Incorrect.

ANSWER: C

Q162. In respect to SCRUM Quality, A . . . . is responsible for setting the Acceptance-Criteria for the Portfolio, and for reviewing the Portfolio Deliverables.

A) Portfolio Product Owner

B) Portfolio Scrum Master

C) Program Product Owner

D) Program Scrum Master

ANSWER: A

Q163. In respect to SCRUM Quality, A . . . . is responsible for ensuring that a sustainable focus is maintained on the Quality of Features of the overall Portfolio.
A) Portfolio Product Owner
B) Portfolio Scrum Master
C) Program Product Owner
D) Program Scrum Master
ANSWER: B

Q164. In respect to SCRUM Quality, a . . . . is responsible for setting the Acceptance-Criteria for the Program and for reviewing the Program Deliverables.
A) Portfolio Product Owner
B) Portfolio Scrum Master
C) Program Product Owner
D) Program Scrum Master
ANSWER: C

Q165. In respect to SCRUM Quality, A . . . . is responsible for ensuring that a sustainable focus is maintained on the Quality of Features of the overall Program.
A) Portfolio Product Owner
B) Portfolio Scrum Master
C) Program Product Owner
D) Program Scrum Master
ANSWER: D

Q166. In respect to SCRUM Quality, a . . . . is responsible for stating and clearly defining the Business Needs and Requirements in the Prioritized Product Backlog; for ensuring that the Deliverables satisfy the Quality Requirements; for setting the Acceptance-Criteria for the entire Project; for facilitating the creation of the User-Stories' Acceptance-Criteria; and for reviewing and validating the Deliverables.
A) Product Owner
B) Scrum Master
C) Scrum Team
D) Scrum Guidance Body (SGB)
ANSWER: A

Q167. In respect to SCRUM Quality, a . . . . is responsible for estimating the obstacles that may impact the Quality of Deliverables and Processes; for ensuring that a sustainable focus is maintained on the Quality of the Project Features; and for ensuring that the Scrum Processes are followed by all of the Scrum Team Members.
A) Product Owner
B) Scrum Master
C) Scrum Team
D) Scrum Guidance Body (SGB)
ANSWER: B

Q168. In respect to SCRUM Quality, a . . . . is responsible for developing and maintaining all Sprints Deliverables; for encouraging Effective Communications such that the Business Needs and Requirements are made clear and understandable; for sharing knowledge and familiarizing the Team Members with these Features; for benefiting from the experience of others; and for making appropriate and smooth changes to Deliverables.
A) Product Owner
B) Scrum Master
C) Scrum Team
D) Scrum Guidance Body (SGB)
ANSWER: C

Q169. In respect to SCRUM Quality, a . . . . provides a Framework for developing the Acceptance-Criteria.
A) Product Owner
B) Scrum Master
C) Scrum Team
D) Scrum Guidance Body (SGB)
ANSWER: D

Q170. In respect to SCRUM Quality, The . . . . review and accept the Deliverables and the Outcome Product.
A) Product Owner
B) Scrum Master
C) Scrum Team
D) Stakeholders
ANSWER: D

Q171. Which of the following statements is valid in respect to the Scrum Change?

A) Scrum Change involves welcoming and endorsing the CHANGE-REQUESTS received from the Stakeholders.

B) Stakeholders can request changes at any time throughout the Project Life-Cycle.

C) It is almost impossible that Stakeholders can precisely define their Business Requirements upfront.

D) All Statements above are valid.

ANSWER: D

Q172. In respect to SCRUM Change, a .... is responsible for providing the Portfolio CHANGE-REQUESTS; and for approving the amended, removed, and/or added Products in accordance with the Portfolio Requirements.

A) Portfolio Product Owner

B) Portfolio Scrum Master

C) Program Product Owner

D) Program Scrum Master

ANSWER: A

Q173. In respect to SCRUM Change, a .... is responsible for facilitating the identification, assessment, and management of the Portfolio CHANGE-REQUESTS.

A) Portfolio Product Owner

B) Portfolio Scrum Master

C) Program Product Owner

D) Program Scrum Master

ANSWER: B

Q174. In respect to SCRUM Change, a .... is responsible for providing the Program CHANGE-REQUESTS; and for approving the amended, removed, and added Products in accordance with the Program Requirements.

A) Portfolio Product Owner

B) Portfolio Scrum Master

C) Program Product Owner

D) Program Scrum Master

ANSWER: C

Q175. In respect to SCRUM Change, a .... is responsible for facilitating the identification, assessment, and management of the Program CHANGE-REQUESTS.

A) Portfolio Product Owner

B) Portfolio Scrum Master

C) Program Product Owner

D) Program Scrum Master

ANSWER: D

Q176. In respect to SCRUM Change, a .... is responsible for providing the Project CHANGE-REQUESTS; for assessing the impact of the Portfolio, Program, and/or Project CHANGE-REQUESTS; and for prioritizing the User-Stories in the Prioritized Product Backlog.
A) Product Owner
B) Scrum Master
C) Scrum Team
D) Stakeholder
ANSWER: A

Q177. In respect to SCRUM Change, a .... is responsible for facilitating the identification, assessment, and escalation of the Scrum Team CHANGE-REQUESTS.
A) Product Owner
B) Scrum Master
C) Scrum Team
D) Stakeholder
ANSWER: B

Q178. In respect to SCRUM Change, the .... is responsible for suggesting changes during the two Processes of Create Deliverables and Conduct Daily Standup.
A) Product Owner
B) Scrum Master
C) Scrum Team
D) Stakeholder
ANSWER: C

Q179. In respect to SCRUM Change, a .... is responsible for providing CHANGE-REQUESTS and for helping prioritize these CHANGE-REQUESTS.
A) Product Owner
B) Scrum Master
C) Scrum Team
D) Stakeholder
ANSWER: D

Q180. In respect to SCRUM Change, the .... provides an overall Guidance for the Change Management Procedures throughout the Project Life-Cycle.
A) Product Owner
B) Scrum Master
C) Scrum Team
D) Scrum Guidance Body (SGB)
ANSWER: D

Q181. Which of the following statements is valid in respect to Risks?

A) A Risk is an expected event that may impact the Project Objectives.

B) If the Risk's Impact is Negative, then that Risk is a Threat. While, if the Risk's Impact is Positive, then that Risk is an Opportunity.

C) If the Risk's Impact is Negative, then that Risk is an Opportunity. While, if the Risk's Impact is Positive, then that Risk is a Threat.

D) All statements above are valid.

ANSWER: B

Q182. Which of the following statements is valid in respect to Scrum Risks?

A) A Risk is a future unexpected and/or uncertain event with no current Impact, but it needs to be planned for encountering it, just in case it happens in the future. While an Issue is a present event that has current Impact and needs immediate action to encounter it.

B) A Risk is a future unexpected and/or uncertain event with no current Impact, but it needs to be planned for encountering it, just in case it happens in the future. While a Threat is a present event that has current Impact and needs immediate action to encounter it.

C) A Threat is a future unexpected and/or uncertain event with current Impact, but it needs to be planned for encountering it, just in case it happens in the future. While an Issue is a present event that has current Impact and needs immediate action to encounter it.

D) All statements above are valid.

ANSWER: A

Q183. Risk Management Process possesses five steps: Risk identification, Risk assessment, Risk prioritization, Risk mitigation, and Risk Communication.

A) _TRUE_ B) _FALSE_

ANSWER: A

Q184. Risk Identification can be achieved through the following:

A) Reviewing the Lessons Learned from the Outcomes of the Sprint/Project Retrospect Meetings Outcomes, from the Risk Checklists, and from the Risk Prompt Lists.

B) Brainstorming Sessions.

C) Deriving a Risk Breakdown Structure.

D) All Options above are Correct.

ANSWER: A

Q185. Risk Assessment can be achieved through and by using the following:
A) Risk Meetings, Probability Trees, and Pareto Analysis.
B) Probability Impact Grid and Expected Monitory Value (EMV).
C) Both Options above are Correct.
D) All Options above are Incorrect.
ANSWER: C

Q186. Risk Meetings indicates that the Product Owner can easily prioritize Risks by Meeting with the Scrum Team and the concerned Stakeholders.
A) _TRUE_ B) _FALSE_
ANSWER: A

Q187. A Probability Tree composes all Risks and their Outcomes are represented with branches in that Tree. Then the values of these Outcomes are summed together to calculate the overall anticipated Impact of any Risk to the Project.
A) _TRUE_ B) _FALSE_
ANSWER: A

Q188. Pareto Analysis is used to rank Risks by their Impact Magnitudes.
A) _TRUE_ B) _FALSE_
ANSWER: A

Q189. Probability Impact Grid is used to rank Risks by their Severity. A Risk Severity is calculated by multiplying its Probability by its Impact Rating Value.
A) _TRUE_ B) _FALSE_
ANSWER: A

Q190. Expected Monitory Value (EMV) is used to rank Risks by their expected Monitory Values. A Risk expected Monitory Value is calculated by multiplying the Risk Impact in Dollars with its Probability in Percentage.
A) _TRUE_ B) _FALSE_
ANSWER: A

Q191. Risks are taken into consideration while creating or updating a Prioritized Product Backlog during the "Create Prioritized Product Backlog" Process or during the "Groom Prioritized Product Backlog" Process.
A) _TRUE_ B) _FALSE_
ANSWER: A

Q192. Prioritizing Risks consideration is associated with the User-Stories they may be impacted by these Risks. Thus, the "Prioritized Product Backlog" is called Risk Adjusted Prioritized Product Backlog.
A) _TRUE_ B) _FALSE_
ANSWER: A

Q193. Which of the following statements is valid in respect to Risk Mitigation?

A) Scrum allows for early detection of Failures. Therefore, it has a Natural Mitigation Feature built in.

B) Risks are mitigated by implementing a number of Proactive and/or Reactive Responses.

C) When a Risk Event occurs, a Reactive Response (e.g., as a Plan B) may be formulated and used.

D) Risks are accepted when their Probabilities and/or Impacts are too low.

E) All statements above are valid.

ANSWER: E

Q194. Which of the following statements is valid in respect to Risk Communication?

A) It is extremely important to communicate Information related to the Risks with the Project Stakeholders. This includes the potential Impact of each Risk and the Plans for Responding to them.

B) Risks Communication should take place in parallel with the four sequential steps of Risk Identification, Risk Assessment, Risk Prioritization, and Risk Mitigation.

C) The Scrum Team may need to discuss during the Daily Standup Meetings the identified Risks with the Scrum Master.

D) The Product Owner needs to prioritize these Risks and communicate them to the Scrum Team.

E) All statements above are valid.

ANSWER: E

Q195. Scrum Flexibility can reduce the Probability, Impact, and Severity of Risks that are relevant to the Business Environment.

A) _TRUE_ B) _FALSE_

ANSWER: A

Q196. Scrum Regular Feedback can reduce the Probability, Impact, and Severity of Risks that are relevant to the Stakeholders' Expectations.

A) _TRUE_ B) _FALSE_

ANSWER: A

Q197. Scrum Team Ownership can reduce the Probability, Impact, and Severity of Risks that are relevant to the Sprints' Tasks, Resources, and Schedule Estimations.

A) _TRUE_ B) _FALSE_

ANSWER: A

Q198. Scrum Transparency can reduce the Probability, Impact, and Severity of Risks that are hard to detect.

A) _TRUE_ B) _FALSE_

ANSWER: A

Q199. Scrum Iterative Delivery can reduce the Probability, Impact, and Severity of Risks relevant to Investments.
A) _TRUE_ B) _FALSE_
ANSWER: A

Q200. In respect to SCRUM Risk, When the .... Risks are identified, the .... Product Owner must assess the Probability and Impact of each Risk; prioritize these Risks; and determine the right Responses. The .... Product Owner must communicate these Risks to the relevant Stakeholders, Program's Teams, and Projects' Teams.
A) Portfolio/Portfolio/Portfolio
B) Program/Program/Program
C) Project/Project/Project
D) All Options above are Incorrect.
ANSWER: A

Q201. In respect to SCRUM Risk, when the .... Risks are identified, the .... Product Owner must enter them into the .... Risk Adjusted Prioritized Product Backlog; assess their Probabilities and Impacts in order to prioritize them; and determine the Programs' appropriate Responses. The .... Product Owner must communicate the Risks to relevant Stakeholders and the Project Teams.
A) Portfolio/Portfolio/Portfolio/Portfolio
B) Program/Program/Program/Program
C) Project/Project/Project/Project
D) All Options above are Incorrect.
ANSWER: B

Q202. In respect to SCRUM Risk, a .... is responsible for capturing and assessing the Portfolio's Risks and for prioritizing and communicating them to the Stakeholders and the Program and Project Teams.
A) Portfolio Product Owner
B) Portfolio Scrum Master
C) Program Product Owner
D) Program Scrum Master
ANSWER: A

Q203. In respect to SCRUM Risk, a .... is responsible for helping in identifying, assessing, and communicating the Portfolio's potential Risks.
A) Portfolio Product Owner
B) Portfolio Scrum Master
C) Program Product Owner
D) Program Scrum Master
ANSWER: B

Q204. In respect to SCRUM Risk, a .... is responsible for capturing and assessing the Program's potential Risks, and for prioritizing and communicating them to the Stakeholders and to the Project Teams.
A) Portfolio Product Owner
B) Portfolio Scrum Master
C) Program Product Owner
D) Program Scrum Master
ANSWER: C

Q205. In respect to SCRUM Risk, a .... is responsible for helping in identifying, assessing, and escalating the Program's potential Risks.
A) Portfolio Product Owner
B) Portfolio Scrum Master
C) Program Product Owner
D) Program Scrum Master
ANSWER: D

Q206. In respect to SCRUM Risk, a .... is responsible for capturing and assessing the Project potential Risks; for prioritizing and communicating them to the Stakeholders, Program and Portfolio Teams; and for ensuring that the Project Risk Levels are acceptable.
A) Product Owner
B) Scrum Master
C) Scrum Team
D) Stakeholders
ANSWER: A

Q207. In respect to SCRUM Risk, a .... is responsible for helping in identifying and escalating the Project potential Risks to the Scrum Team.
A) Product Owner
B) Scrum Master
C) Scrum Team
D) Stakeholders
ANSWER: B

Q208. In respect to SCRUM Risk, a .... is responsible for identifying the Project potential Risks during the Product Development and for Implementing their Management Activities as advised by the Product Owner.
A) Product Owner
B) Scrum Master
C) Scrum Team
D) Stakeholders
ANSWER: C

Q209. In respect to SCRUM Risk, ... are responsible for interacting with the Scrum Team to provide them with their Inputs in respect to Managing the Project potential Risks.
A) Product Owner
B) Scrum Master
C) Scrum Team
D) Stakeholders
ANSWER: D

Q210. In respect to SCRUM Risk, the .... provides Guidance in respect to the Procedures of managing the potential Risks.
A) Product Owner
B) Scrum Master
C) Scrum Team
D) Scrum Guidance Body (SGB)
ANSWER: D

Q211. The Scrum Initiate Phase has six processes: Create Project Vision, Identify Scrum Master and Stakeholders, Form Scrum Team, Develop Epics, Create Prioritized Product Backlog, and Conduct Release Planning.
A) _TRUE_ B) _FALSE_
ANSWER: A

Q212. The "Create Project Vision" Process is used to create the Project Business Vision Statement.
A) _TRUE_ B) _FALSE_
ANSWER: A

Q213. The main Inputs to the “Create Project Vision” Process include the following:

A) The Project Business Case.

B) Inputs from the Program Product Owner, the Program Stakeholders, the Program Scrum Master, the Chief Product Owner, and from the Scrum Guidance Body (SGB).

C) The Program Product Backlog, a Trial Project, and a Proof of Concept, the Company Vision and Mission, and a Market Study.

D) All Options above are Correct.

ANSWER: D

Q214. Which of the following statements is valid in respect to the SWOT Analysis Method?

A) SWOT Analysis is one of the main methods used during the “Create Project Vision” Process.

B) SWOT Analysis is a structured Project Planning Method used to assess the Strengths, Weaknesses, Opportunities, and Threats of a Project.

C) SWOT Analysis helps the Project Stakeholders and Decision makers to finalize the Project lists of Processes, Methods and Tools, Objectives, Priorities, Changes, and Risks.

D) All Options above are Correct.

ANSWER: D

Q215. The main Outputs of the “Create Project Vision” Process include the following:

A) The identified Product Owner.

B) The Project Vision Statement, the Project Charter, and the Project Budget.

C) Both Options above are Correct.

D) All Options above are Incorrect.

ANSWER: D

Q216. The “Identify Scrum Master and Stakeholders” Process is used to identify the Project Scrum Master and Stakeholders, mainly by the Project Product Owner.

A) _TRUE_ B) _FALSE_

ANSWER: A

Q217. The main Inputs to the "Identify Scrum Master and Stakeholder" Process include the following:

A) Inputs from the Product Owner, the Program Product Owner, the Scrum Master, the Chief Product Owner, the Chief Scrum Master, the Program Stakeholders, and the Scrum Guidance Body (SGB).

B) The Project Vision Statement, the Organizational Resource Matrix, the Skills Requirement Matrix, the Human Resources Availability and Commitments, and the needed Human Resources and their Requirements.

C) Both Options above are Correct.

D) All Options above are Incorrect.

ANSWER: D

Q218. The SELECTION-CRITERIA used to select a Scrum Master is based on His/her Problem-Solving Skills, Availability and Commitment; and his/her Servant Leadership Style.

A) _TRUE_ B) _FALSE_

ANSWER: A

Q219. The main Outputs of the "Identify Scrum Master and Stakeholder" Process include the following:

A) A list of the identified Scrum Master.

B) A list of the identified Stakeholders.

C) Both Options above are Correct.

D) All Options above are Incorrect.

ANSWER: C

Q220. The "Form Scrum Team" Process is used to identify the Project Scrum Team Members, mainly by the Project Product Owner and the Scrum Master.

A) _TRUE_ B) _FALSE_

ANSWER: A

Q221. The main Inputs to the "Form Scrum Team" include the following:

A) Inputs from the Product Owner, the Chief Product Owner, the Scrum Master, and the Scrum Guidance Body (SGB).

B) The Project Vision Statement, the needed Human Resources and their Requirements, and the Human Resources Availability and Commitments.

C) The Organizational Resource Matrix, the Skills Requirement Matrix, and the Resource Requirements.

D) All Options above are Correct.

ANSWER: D

Q222. During the "Form Scrum Team" Process, the SRUM-TEAM-SELECTION-CRITERIA Factors include the following:
A) The Knowledge and Expertise in various fields.
B) The ability to determine the SUCCESS of Self-Organizing Teams.
C) Being Independent, Self-Motivated, Customer-Focused, Responsible, Collaborative Independent Thinking, and Team players.
D) All Options above are Correct.
ANSWER: D

Q223. The main Outputs of the "Form Scrum Team" include the following:
A) A list of the identified Scrum Team Members and a list of their identified backup individuals.
B) A Collaboration Plan and a Team Building Plan.
C) Both Options above are Correct.
D) All Options above are Incorrect.
ANSWER: C

Q224. Which of the following statements is valid in respect to the "Develop Epics" Process?
A) The "Develop Epics" Process is used to develop and articulate promotional SUCCESS stories out of the upcoming Product as well as large-size User-Stories.
B) The Project Product Owner is the sole responsible role for the "Develop Epics" Process. However, the Scrum Master and the Scrum Team provide their Inputs in this respect.
C) Both Options above are Correct.
D) All Options above are Incorrect.
ANSWER: C

Q225. The main Inputs to the "Develop Epics" Process include the following:
A) Inputs from the Scrum Master, the Stakeholders, the Scrum Guidance Body (SGB), and from previous similar Projects.
B) All involved Laws and Regulations and applicable Contracts.
C) The Project Vision Statement, Program Product Backlog, Program and Portfolio Risks, and the lists of approved and unapproved CHANGE-REQUESTS.
D) All Options above are Correct.
ANSWER: D

Q226. The applicable Contracts involved during the “Develop Epics” Process include the following:

A) Incremental Delivery Contract, Joint-Venture Contract, and Development in Phases Contract.

B) Incremental Contract, Joint-Venture Contract, and Incentive and Penalty Contract.

C) Both Options above are Correct.

D) All Options above are Incorrect.

ANSWER: C

Q227. Which of the following statements is valid in respect to the “Create Prioritized Product Backlog” Process?

A) The “Create Prioritized Product Backlog” Process is used to create the Project Prioritized Product Backlog.

B) The Project Product Owner is the sole responsible Role for the “Create Prioritized Product Backlog” Process. However, the Scrum Master and the Scrum Team provide their Inputs in this respect.

C) Both Options above are Correct.

D) All Options above are Incorrect.

ANSWER: C

Q228. The main Inputs to the “Create Prioritized Product Backlog” Process include the following:

A) Inputs from the Scrum Master, the Scrum Team, and the Stakeholders.

B) Epics, but not Persona.

C) The Project Vision Statement, the Program Product Backlog, the Business Requirements, the list of approved CHANGE-Requests, the list of identified Risks.

D) All Options above are Correct.

ANSWER: C

Q229. Which of the following statements is valid in respect to the “MoSCoW” Prioritization Scheme?

A) The “MoSCoW Prioritization Scheme” derives its name from the first letters of the phrases “MUST HAVE,” “SHOULD HAVE,” “COULD HAVE,” and “WILL NOT HAVE.”

B) The “MoSCoW Prioritization Scheme” is used to prioritize the User-Stories in Scrum Projects.

C) Both statements above are valid.

D) All statements above are invalid.

ANSWER: C

Q230. The main Outputs of the "Create Prioritized Product Backlog" Process include a Prioritized Product Backlog and a Done-Criteria.
A) _TRUE_ B) _FALSE_
ANSWER: A

Q231. The "Conduct Release Planning" Process is used to devise the Project Release Plan that indicates when the Product Increments/Deliverables will be delivered to the Customer.
A) _TRUE_ B) _FALSE_
ANSWER: A

Q232. The main Inputs to the "Conduct Release Planning" Process include the following:
A) Inputs from the Scrum Master, the Stakeholders, the Program Product Owner, the Program Scrum Master, the Chief Product Owner, and the Scrum Guidance Body (SBG).
B) The Business Requirements, the Holiday Calendar, the Program Product Backlog, and the Done-Criteria.
C) The Prioritized Product Backlog and the Project Vision Statement.
D) All Options above are Correct.
ANSWER: D

Q233. Developing a Release Plan needs the use of some predefined Release Prioritization Methods that are Industry and Organization specific and determined by the Organization's Senior Management.
A) _TRUE_ B) _FALSE_
ANSWER: A

Q234. The main Outputs of the "Conduct Release Planning" Process include a Release Planning Schedule, the Sprint Length, the Release Targeted Customers, and a refined version of the Prioritized Product Backlog.
A) _TRUE_ B) _FALSE_
ANSWER: A

Q235. The Scrum Plan and Estimate Phase involves five Processes: Create User-Stories, Approve, Estimate, and Commit User-Stories, Create Tasks, Estimate Tasks, and Create Sprint Backlog.
A) _TRUE_ B) _FALSE_
ANSWER: A

Q236. The "Create User-Stories" Process is used to define how the Project Product Owner can create the Project User-Stories before he/she prioritize them into the Project Prioritized Product Backlog.
A) _TRUE_ B) _FALSE_
ANSWER: A

Q237. The main Inputs to the "Create User-Stories" Process include the following:
A) Inputs from the Scrum Team, the Stakeholders, and the Scrum Guidance Body (SGB).
B) The Prioritized Product Backlog, the Done-Criteria, the Business Requirements, the involved Laws and Regulations, and the applicable Contracts.
C) Both Options above are Correct.
D) All Options above are Incorrect.
ANSWER: C

Q238. The "Approve, Estimate, and Commit User-Stories" Process is used to select User-Stories from the Prioritized Product Backlog and approve them into the current Sprint Backlog; estimate their Tasks, needed Resources, and time; and having the Scrum Team to commit to work towards working out these User-Stories to create their anticipated Deliverables.
A) _TRUE_ B) _FALSE_
ANSWER: A

Q239. The main Inputs to the "Approve, Estimate, and Commit User-Stories" include the following:
A) Inputs from the Scrum Team and the Scrum Guidance Body (SGB).
B) The User-Stories and their Acceptance-Criteria.
C) Both Options above are Correct.
D) All Options above are Incorrect.
ANSWER: C

Q240. The main Output of the "Approve, Estimate, and Commit User-Stories" is the Approved/Estimated/Committed User-Stories.
A) _TRUE_ B) _FALSE_
ANSWER: A

Q241. The "Create Tasks" Process is used to create the Tasks required for accomplishing the User-Stories of the current Sprint.
A) _TRUE_ B) _FALSE_
ANSWER: A

Q242. The main Inputs to the "Create Tasks" Process are the following:
A) Inputs from the Scrum Team.
B) A set of Approved/Estimated/Committed User-Stories.
C) Both Options above are Correct.
D) All Options above are Incorrect.
ANSWER: C

Q243. During the "Create Task" Process, the Decomposition Method is used to break down High-Level Tasks into Lower-Level Tasks.
A) _TRUE_ B) _FALSE_
ANSWER: A

Q244. The main Outputs of the "Create Tasks" Process include a list of the created Tasks, a list of the Updated Approved/Estimated/Committed User-Stories, and a list of Dependencies among the created Tasks.
A) _TRUE_ B) _FALSE_
ANSWER: A

Q245. The "Estimate Tasks" Process is used to estimate the Tasks required for accomplishing the User-Stories of the current Sprint.
A) _TRUE_ B) _FALSE_
ANSWER: A

Q246. The main Inputs to the "Estimate Tasks" Process include the following:
A) Inputs from the Scrum Team and the Scrum Guidance Body (SGB).
B) The list of the created Tasks, the list of identified Risks, and the list of identified Dependencies among the created Tasks.
C) The User-Stories Acceptance-Criteria.
D) All Options above are Correct.
ANSWER: D

Q247. The main Outputs of the "Estimate Tasks" Process include the "Effort Estimated Task List" and the "Updated Tasks List."
A) _TRUE_ B) _FALSE_
ANSWER: A

Q248. The "Create Sprint Backlog" Process is used to select the right USER-STORIES from the Prioritized Backlog and place them into the current Sprint's Backlog.
A) _TRUE_ B) _FALSE_
ANSWER: A

Q249. The main Inputs to the “Create Sprint Backlog” Process include the following:
A) Inputs from the SCRUM Team.
B) The Dependencies among the created Tasks.
C) The Sprint Length and the Team Calendar.
D) All Options above are Correct.
E) All Options above are Incorrect.
ANSWER: D

Q250. The main Outputs of the “Create Sprint Backlog” Process include a Sprint Backlog and BURNDOWN-CHART.
A) _TRUE_ B) _FALSE_
ANSWER: A

Q251. The SCRUM Implement Phase involves three Processes, which are the Create Deliverables, Conduct Daily Standup, and Groom Prioritized Product Backlog.
A) _TRUE_ B) _FALSE_
ANSWER: A

Q252. The “Create Deliverables” Process is used to carry out the work in the current Sprint and to create its associated Product Increments.
A) _TRUE_ B) _FALSE_
ANSWER: A

Q253. The Inputs to the “Create Deliverables” Process include the following:
A) Inputs from the SCRUM Team and the SCRUM Guidance Body (SGB)
B) Scrumboard, Sprint Backlog, and Impediment Log.
C) Release Planning Schedule and Inter-Tasks Dependencies.
D) All Options above are Correct.
ANSWER: D

Q254. A Scrumboard depicts the progress of the Sprint estimated tasks as follows:

A) A column for the Tasks that are not yet started (e.g., TO-DO Tasks); a second column for the Tasks that are being carried out, but not completed (e.g., IN-PROGRESS Tasks); a third column for the Tasks that are completed and still being tested (e.g., TESTING Tasks); and a fourth column for the Tasks that have completed and successfully tested (e.g., DONE Tasks).

B) A column for the Tasks that are not yet started (e.g., NOT-STARTED Tasks); a second column for the Tasks that are being carried out, but not completed (e.g., BEING-EXECUTED Tasks); a third column for the Tasks that are completed and still being tested (e.g., COMPLETED-BUT-BEING-TESTED Tasks); and a fourth column for the Tasks that have completed and successfully tested (e.g., SUCCESSFULLY-COMPLETED Tasks).

C) Both Options above are Correct in terms of their meanings.

D) All Options above are Incorrect.

ANSWER: C

Q255. In respect to the Impediment Log Document, which of the following statements is valid?

A) Impediments impact the Productivity of the SCRUM Team.

B) In order for the SCRUM Team to continue working effectively, all Impediments must be identified/resolved/removed.

C) Impediments can be Internal such as having Lack of Communications or External such as having Issues with Software Licenses.

D) All Statements above are valid.

ANSWER: D

Q256. At the end of each Sprint, at least one Product Increment or Deliverable is completed (e.g., they must possess all Features and Functionality defined in the USER-STORIES of that Sprint), then these Deliverables must be tested SUCCESSFULLY.

A) _TRUE_ B) _FALSE_

ANSWER: A

Q257. The "Conduct Daily Standup" Process is used for the holding Sprint Daily Standup Meetings.

A) _TRUE_ B) _FALSE_

ANSWER: A

Q258. The main Inputs to the “Conduct Daily Standup” Process include the following:
A) Inputs from the SCRUM Team, the SCRUM Master, and the Product Owner.
B) The Sprint Burndown Chart, the Impediment Log Document, the Scrumboard, and the Inter-Tasks Dependencies.
C) Both Options above are Correct.
D) All Options above are Incorrect.
ANSWER: C

Q259. A Daily Standup Meeting is a 15-min short daily Meeting, during which each of the SCRUM Team Members answers the following daily questions: “What did I complete yesterday?”; “What will I complete today?”; and “What impediments or obstacles (if any) am I currently facing?”
A) _TRUE_ B) _FALSE_
ANSWER: A

Q260. When a SCRUM Team Members are co-located in the same working location, that location is commonly called the . . . .
A) War Room.
B) Operations Room.
C) Development Room.
D) The Retrospect Room.
ANSWER: A

Q261. The “Groom Prioritized Product Backlog” Process is used by the Project Product Owner to update the Project Prioritized Product Backlog in accordance with the Feedbacks and Outcomes of the Sprint Retrospect Meetings.
A) _TRUE_ B) _FALSE_
ANSWER: A

Q262. The main Inputs to the “Groom Prioritized Product Backlog” Process include the following:
A) Inputs from the SCRUM Team and the SCRUM Guidance Body (SGB) and the Release Planning Schedule.
B) The Prioritized Product Backlog, the Rejected Deliverables, the Approved and the Un-Approved Change-Requests, and the identified Risks.
C) Both Options above are Correct.
D) All Options above are Incorrect.
ANSWER: C

Q263. The main Outputs of the "Groom Prioritized Product Backlog" Process include the following:
A) An updated version of the Prioritized Product Backlog.
B) An updated version of the Release Planning Schedule.
C) Both Options above are Correct.
D) All Options above are Incorrect.
ANSWER: C

Q264. The SCRUM Review and Retrospect Phase involves three processes, which are the Convene SCRUM of SCRUMs, Demonstrate and Validate Sprint, and Sprint Retrospect.
A) _TRUE_ B) _FALSE_
ANSWER: A

Q265. The "Convene SCRUM of SCRUMs" Process is used by the Project Chief Product Owner to meet with all Product Owners and SCRUM Masters of the Project multiple SCRUMs.
A) _TRUE_ B) _FALSE_
ANSWER: A

Q266. The main Inputs to the "Convene SCRUM of SCRUMs" Process include the following:
A) Representatives of the SCRUM Masters and the SCRUM Teams, and Inputs from the Chief SCRUM Master and the Chief Product Owners.
B) The Outputs from the Sprint Retrospect Process.
C) The Meeting Agenda, the Impediment Log Document, and the Inter-Tasks Dependencies.
D) All Options above are Correct.
ANSWER: D

Q267. Each SCRUM Team representative should update the Meeting attendees in respect to his/her SCRUM Team's Activities and Challenges. This involves the following: (i) The work that has been going on since the last SoS Meeting; (ii) The work that has been planned for through the next SoS Meeting; (iii) The work that was supposed to be completed in response to Needs from other SCRUM Teams, but remained uncompleted; (iv) The work that has been planned for, but may impact some or all of the other SCRUM Teams.
A) _TRUE_ B) _FALSE_
ANSWER: A

Q268. The main Outputs of the "Convene SCRUM of SCRUMs" Process include the following:
A) A list of all resolved issues.
B) An updated Impediment Log Document.
C) An updated list of the Inter-Tasks Dependencies.
D) All Options above are Correct.
ANSWER: D

Q269. The "Demonstrate and Validate Sprint" Process is used to have the SCRUM Team Members to present and demonstrate to the Product Owner the Deliverables of the current Sprint to solicit his/her Approval before shipping them to the Customer.
A) _TRUE_ B) _FALSE_
ANSWER: A

Q270. The main Inputs to the "Demonstrate and Validate Sprint" Process include the following:
A) Inputs from the SCRUM Team, the Stakeholders, and the SCRUM Guidance Body (SGB).
B) The Sprint Backlog, the Sprint Deliverables, and the Release Planning Schedule.
C) The DONE Criteria and User-Stories Acceptance-Criteria.
D) All Options above are Correct.
ANSWER: D

Q271. The main Outputs of the "Demonstrate and Validate Sprint" Process include the following:
A) A list of all accepted Deliverables along with another list of all rejected Deliverables.
B) An updated list of the Inter-Tasks Dependencies, an updated list of the Risks, and an updated version of the Release Planning Schedule.
C) Both Options above are Correct.
D) All Options above are Incorrect.
ANSWER: C

Q272. The "Sprint Retrospect" Process is used to have the Project Product Owner to meet with the SCRUM Master, the SCRUM Team, and the concerned Stakeholders to discuss what happened during the current Sprint and to assess the Customer's feedback after the current Sprint's Deliverables have been shipped to the Customer and deployed for them. The learned lessons here are used with the upcoming Sprints. In addition, they can be used to make grooming recommendations to the Product Owner to update the Project Prioritized Product Backlog.
A) _TRUE_ B) _FALSE_
ANSWER: A

Q273. The main Inputs to the "Sprint Retrospect" Process include the following:
A) Inputs from the SCRUM Master, the SCRUM Team, the Product Owner, and the SCRUM Guidance Body (SGB).
B) The Outputs from the "Demonstrate and Validate Sprint" Process.
C) Both Options above are Correct.
D) All Options above are Incorrect.
ANSWER: C

Q274. At a Sprint Retrospect Meeting, a (n) .... would want to participate in the Sprint Retrospect Meeting and learn about everything discussed.
A) Explorer B) Shopper C) Vacationer D) Prisoner
ANSWER: A

Q275. At a Sprint Retrospect Meeting, a (n) .... would want to listen to everything, but choose what he/she would want to take away from the Sprint Retrospect Meeting.
A) Explorer B) Shopper C) Vacationer D) Prisoner
ANSWER: B

Q276. At a Sprint Retrospect Meeting, a (n) .... would want to relax and take it easy in the Sprint Retrospect Meeting.
A) Explorer B) Shopper C) Vacationer D) Prisoner
ANSWER: C

Q277. At a Sprint Retrospect Meeting, a (n) .... is attending the Sprint Retrospect Meeting because it is a must to attend, but he would want to be somewhere else.
A) Explorer B) Shopper C) Vacationer D) Prisoner
ANSWER: D

Q278. Examples of the Metrics and Measuring Methods used during a Sprint Retrospect Meeting include Team Velocity, Done Success Rate, Estimation Effectiveness, Review Feedback Ratings, and Team Morale Ratings.
A) _TRUE_ B) _FALSE_
ANSWER: A

Q279. The .... refers to the number of USER-STORIES completed in a single Sprint.
A) Team Velocity
B) Done SUCCESS Rate
C) Estimation Effectiveness
D) Review Feedback Ratings
E) Team Morale Ratings
ANSWER: A

Q280. The .... refers to the percentage of USER-STORIES that have been completed vs. the ones that were committed to by the SCRUM Team.
A) Team Velocity
B) Done SUCCESS Rate
C) Estimation Effectiveness
D) Review Feedback Ratings
E) Team Morale Ratings
ANSWER: B

Q281. The .... refers to the percentage of deviations between the estimated and actual times spent on the Tasks and USER-STORIES.
A) Team Velocity
B) Done SUCCESS Rate
C) Estimation Effectiveness
D) Review Feedback Ratings
E) Team Morale Ratings
ANSWER: C

Q282. The .... refers to the Feedback from the Stakeholders using quantitative and/or qualitative ratings.
A) Team Velocity
B) Done SUCCESS Rate
C) Estimation Effectiveness
D) Review Feedback Ratings
E) Team Morale Ratings
ANSWER: D

Q283. The .... refers to the Results from the SCRUM Team Morale Self-Assessments.
A) Team Velocity
B) Done SUCCESS Rate
C) Estimation Effectiveness
D) Review Feedback Ratings
E) Team Morale Ratings
ANSWER: E

Q284. The main Outputs of the "Sprint Retrospect" Process include the following:
A) An agreed-upon list of actionable improvements and their assignments and due dates.
B) A list of proposed non-functional items for the Prioritized Product Backlog and a set of Sprint Retrospect Log Documents.
C) A set of Lessons Learned and an updated version of the SCRUM Guidance Body (SGB) recommendations.
D) All Options above are Correct.
ANSWER: D

Q285. The SCRUM Release Phase involves two Processes, which are the Ship Deliverables and Retrospect Project.
A) _TRUE_ B) _FALSE_
ANSWER: A

Q286. The "Ship Deliverables" Process enables the Project Product Owner to make the necessary Arrangements for the SCRUM Team to ship the current Sprint's approved Deliverables to the Customer after approved at the Sprint Review Meeting.
A) _TRUE_ B) _FALSE_
ANSWER: A

Q287. The main Inputs to the "Ship Deliverables" Process include the following:
A) Inputs from the Product Owner, the Stakeholders, the SCRUM Master, the SCRUM Team, and from the SCEUM Guidance Body (SGB).
B) The Accepted Deliverables and the User-Stories Acceptance-Criteria.
C) Both Options above are Correct.
D) All Options above are Incorrect.
ANSWER: C

Q288. The main Outputs of the "Ship Deliverables" Process include the following:
A) A working Deliverables Agreement and a set of working Deliverables.
B) A set of Product Releases.
C) Both Options above are Correct.
D) All Options above are Incorrect.
ANSWER: C

Q289. The "Retrospect Project" Process enables the Project Product Owner to discuss and learn from what happened throughout the Project from the start to the end. All Lessons Learned here are used for future upcoming Projects.
A) _TRUE_ B) _FALSE_
ANSWER: A

Q290. The main Inputs to the “Retrospect Project” Process include Inputs from the SCRUM Team, the SCRUM Master, the Chief Product Owner, the Chief SCRUM Master, the Stakeholders, and from the SCRUM Guidance Body (SGB).

A) _TRUE_ B) _FALSE_

ANSWER: A

Q291. The main Outputs of the “Retrospect Project” Process include the following:

A) A list of the agreed-upon actionable improvements and their assignments and due dates.

B) A set of proposed non-functional items for the Program Product Backlog and for the Prioritized Product Backlog.

C) An updated version of the SCRUM Guidance Recommendations.

D) All Options above are Correct.

ANSWER: D

Appendix C: Reading materials and references

[1] Project Management Institute, Inc. *A Guide to the Project Management Body of Knowledge*. 14 Campus Boulevard, Newton Square, Pennsylvania 19073-3299, USA. Link: https://www.pmi.org/pmbok-guide -Standards/foundational/pmbok [Last seen on March 5, 2023]

[2] Project Management Institute, Inc. *Code of Ethics and Professional Conduct*. 14 Campus Boulevard, Newton Square, Pennsylvania 19073-3299, USA. Link: http://www.pmi.org/codeofEthics.

[3] ISO/IEC, Software Engineering – Guide to the software engineering body of knowledge (SWEBOK), ISO/IEC TR 19759:2015(E).

[4] SCRUMstudy, A Guide to SCRUM BODY OF KNOWLEDGE (SBOK GUIDE), 2016 Edition. ISBN: 978-0-9899252-0-4.

[5] Radaideh, M., "Benchmarking the software engineering undergraduate program curriculum at Jordan university of science and technology with the IEEE software engineering body of knowledge: (software engineering knowledge areas #11–15)". The 2021 International Conference on Computational Science and Computational Intelligence (CSCI'21: December 15–17, 2021; Las Vegas, USA) https://www.american-cse.org/csci2021/. Published by the IEEE CPS – Benchmarking the software engineering undergraduate program curriculum at Jordan university of science and technology with the IEEE software engineering body of knowledge: (software engineering knowledge areas #11–15) | IEEE Conference Publication | IEEE Xplore.

[6] Radaideh, M., "Shifting the Paradigms from Teaching Project Management to Teaching Software Project Management at the Jordan University of Science and Technology According to the IEEE Software Engineering Management Knowledge Area". The 2021 International Conference on Computational Science and Computational Intelligence (CSCI'21: December 15–17, 2021; Las Vegas, USA) https://www.american-cse.org/csci2021/. Published by the IEEE CPS – Shifting the paradigms from teaching project management to teaching software project management at Jordan university of science and technology based on the IEEE software engineering management knowledge area | IEEE Conference Publication | IEEE Xplore.

[7] Radaideh, M., "Benchmarking the Software Engineering Undergraduate Program Curriculum at Jordan University of Science and Technology with the IEEE Software Engineering Body of Knowledge (Software Engineering Knowledge Areas #1–5)". Published in the Springer Nature Research Book Series: Advances in Software Engineering, Education, and e-Learning, pp. 747–768, September 2021 (https://link.springer.com/chapter/10.1007/978-3-030-70873-3_53).

[8] Radaideh, M., "Benchmarking the Software Engineering Undergraduate Program Curriculum at Jordan University of Science and Technology with the IEEE Software Engineering Body of Knowledge (SWE Knowledge Areas #6–10)". Published in the Springer Nature Research Book Series: Advances in Software Engineering, Education, and e-Learning, pp. 85–100, September 2021 (https://link.springer.com/chapter/10.1007/978-3-030-70873-3_7).

Definitions

- ISO/IEC JTC 1/SC7 is the major source of international standards on software and systems engineering.
- ISO: The International Organization for Standardization
- IEC: The International Electrotechnical Commission
- JTC: Joint Technical Committee, a child of the ISO and IEC.
- International Standards: Documents containing requirements that must be satisfied in order to claim conformance.

https://doi.org/10.1515/9783111206868-008

Index

https://doi.org/10.1515/9783111206868-009

www.ingramcontent.com/pod-product-compliance
Lightning Source LLC
Chambersburg PA
CBHW061150240926
51846CB00006B/975